Frontiers in Medicinal Chemistry

Volume 8

Edited by

Atta-ur-Rahman, *FRS*

Honorary Life Fellow, Kings College, University of Cambridge, UK

Mohammad Iqbal Choudhary

H.E.J. Research Institute of Chemistry, International Center for Chemical and Biological Sciences, University of Karachi, Pakistan

and

Allen B. Reitz

Chemical Biology, Institute of Hepatitis and Virus Research, CEO, Fox Chase Chemical Diversity Center, Incorporation Doylestown, PA 18902, USA

CONTENTS

PREFACE

Volume 8 of *Frontiers in Medicinal Chemistry* comprises 6 chapters on topics of high importance in the fields of medicinal chemistry and early-stage drug discovery research. The topics and authors were selected from highly cited publications in the Bentham journals *Curr. Med. Chem., Curr. Pharm. Des.,* and *Curr. Top. Med. Chem.* The original authors were given the opportunity to rewrite their contributions, particularly updating them with more modern insights and references that have emerged in the intervening period of time.

The first chapter by Dr. Gary Caldwell is a tour de force review on the use of ADME optimization and toxicity assessment in drug discovery research. Dr. Caldwell has written multiple chapters on this ever evolving topic and has served as a book editor for Bentham in the past. His chapter in this volume of *Frontiers* is a valued and updated reference guide in the field. Chapters 2 and 3 by Briasoulis *et al.* and Escribano-Lopez *et al.*, respectively, cover oxidative stress-mediated approaches to treat cardiovascular diseases such as atherosclerosis. The fourth chapter by Potenza *et al.* is on diabetes, and in particular endothelial dysfunction and the associated implications as to mechanism and therapeutic targets. In Chapter 5, Jokanović and Petrović review pyridinium oximes as cholinesterase reactivators for the treatment of organophosphorus poisoning. Finally, Rodik and colleagues discuss the current and potential uses of the cyclic macrocyclic oligomer calixarenes in biomedical research.

I would like to express my gratitude to all the authors for their excellent contributions. I would also like to thank the entire team of Bentham Science Publishers, particularly Mr. Omer Shafi (Assistant Manager Publications), Mr. Shehzad Naqvi (Senior Manager Publications) and team leader Mr. Mahmood Alam (Director Publications) for their excellent efforts. We are confident that this volume will receive wide appreciation from students and researchers.

Atta-ur-Rahman, *FRS*
Honorary Life Fellow, Kings College,
University of Cambridge,
UK

M. Iqbal Choudhary
H.E.J. Research Institute of Chemistry,
International Center for Chemical and Biological Sciences,
University of Karachi,
Pakistan

Dr. Allen B. Reitz
Chemical Biology, Institute of Hepatitis and Virus Research
CEO, Fox Chase Chemical Diversity Center, Incorporation
Doylestown, PA 18902, USA

CONTRIBUTORS

Alexandros Briasoulis	1st Cardiology Department, University of Athens Medical School, "Hippokration" Hospital, Athens, Greece
Antonio Hernández-Mijares	Service of Endocrinology, University Hospital Dr. Peset, Foundation for the Promotion of Health and Biomedical Research in the Valencian Region (FISABIO), Valencia, Spain
Carmela Nacci	Department of Biomedical Sciences and Human Oncology – Pharmacology Section, Medical School – University of Bari "Aldo Moro", Bari, Italy
Celia Bañuls	Service of Endocrinology, University Hospital Dr. Peset, Foundation for the Promotion of Health and Biomedical Research in the Valencian Region (FISABIO), Valencia, Spain
Charalampos Antoniades	1st Cardiology Department, University of Athens Medical School, "Hippokration" Hospital, Athens, Greece
Dimitris Tousoulis	1st Cardiology Department, University of Athens Medical School, "Hippokration" Hospital, Athens, Greece
Eva Sola	Service of Endocrinology, University Hospital Dr. Peset, Foundation for the Promotion of Health and Biomedical Research in the Valencian Region (FISABIO), Valencia, Spain
Evangelos Oikonomou	1st Cardiology Department, University of Athens Medical School, "Hippokration" Hospital, Athens, Greece
Gary W. Caldwell	Janssen Research & Development, L.L.C. Discovery Sciences, Lead Discovery Department, Welsh and McKean Roads, P.O. Box 776, Spring House, PA 19477, USA
George-Angelo Papamikroulis	1st Cardiology Department, University of Athens Medical School, "Hippokration" Hospital, Athens, Greece
Irene Escribano-Lopez	Service of Endocrinology, University Hospital Dr. Peset, Foundation for the Promotion of Health and Biomedical Research in the Valencian Region (FISABIO), Valencia, Spain
Luca Sgarra	Department of Biomedical Sciences and Human Oncology – Pharmacology Section, Medical School – University of Bari "Aldo Moro", Bari, Italy

Maria Antonietta De Salvia	Department of Biomedical Sciences and Human Oncology – Pharmacology Section, Medical School – University of Bari "Aldo Moro", Bari, Italy
Maria Assunta Potenza	Department of Biomedical Sciences and Human Oncology – Pharmacology Section, Medical School – University of Bari "Aldo Moro", Bari, Italy
Milagros Rocha	Service of Endocrinology, University Hospital Dr. Peset, Foundation for the Promotion of Health and Biomedical Research in the Valencian Region (FISABIO), Valencia, Spain
Milan Jokanović	Faculty of Medicine, University of Nish, Nish, Serbia
Monica Montagnani	Department of Biomedical Sciences and Human Oncology – Pharmacology Section, Medical School – University of Bari "Aldo Moro", Bari, Italy
Noelia Diaz-Morales	Service of Endocrinology, University Hospital Dr. Peset, Foundation for the Promotion of Health and Biomedical Research in the Valencian Region (FISABIO), Valencia, Spain
Radica Stepanović Petrović	Faculty of Pharmacy, University of Belgrade, Belgrade, Serbia
Raquel Castelló	Service of Endocrinology, University Hospital Dr. Peset, Foundation for the Promotion of Health and Biomedical Research in the Valencian Region (FISABIO), Valencia, Spain
Roman V. Rodik	Institute of Organic Chemistry, National Academy of Sciences of Ukraine, Ukrain
Rosa Falcón	Service of Endocrinology, University Hospital Dr. Peset, Foundation for the Promotion of Health and Biomedical Research in the Valencian Region (FISABIO), Valencia, Spain
Sandra Lopez-Domenech	Service of Endocrinology, University Hospital Dr. Peset, Foundation for the Promotion of Health and Biomedical Research in the Valencian Region (FISABIO), Valencia, Spain
Susana Rovira-Llopis	Service of Endocrinology, University Hospital Dr. Peset, Foundation for the Promotion of Health and Biomedical Research in the Valencian Region (FISABIO), Valencia, Spain

Valentina Leo	Department of Biomedical Sciences and Human Oncology – Pharmacology Section, Medical School – University of Bari "Aldo Moro", Bari, Italy
Victor M. Víctor	Service of Endocrinology, University Hospital Dr. Peset, Foundation for the Promotion of Health and Biomedical Research in the Valencian Region (FISABIO), Valencia, Spain
Vitaly I. Kalchenko	Institute of Organic Chemistry, National Academy of Sciences of Ukraine, Ukrain
Vyacheslav I. Boyko	Institute of Organic Chemistry, National Academy of Sciences of Ukraine, Ukrain

Frontiers in Medicinal Chemistry

CHAPTER 1

ADME Optimization and Toxicity Assessment in Drug Discovery

Gary W. Caldwell*

Janssen Research & Development, L.L.C. Discovery Sciences, Lead Discovery Department, Welsh and McKean Roads, P.O. Box 776, Spring House, PA 19477, USA

Abstract: Integrating physicochemical, absorption, distribution, metabolism, excretion, pharmacokinetics, and toxicity data into the drug discovery/preclinical development decision process in order to reduce the attrition rates of new chemical entities (NCEs) in clinical development is reviewed. The review is organized around the three main stage gates in a small molecule target-based approach including hit-to-lead (H2L), lead optimization (LO) and the final stage gate for selecting NCEs for entry into Phase I clinical trials. The preclinical *in silico* computational methods and *in vitro* cellular assays utilized at each stage gate are discussed from a drug discovery perspective. Preclinical assays utilized at the H2L and LO stage gates must have turn-around-times within a timeframe that is consistent with the iterative cycle of the research projects and consume small quantities of compounds while at the final NCE stage gate more traditional assays are used. Unfortunately, many preclinical assays are ambiguous in predicting human preclinical data since they contain a significant amount of false-positive and false-negative information and, therefore, are not easily translatable from cellular/animals to humans. Thus, understanding the limitations of these preclinical assays is a must for all medicinal chemists for developing go/no-go selection criteria and drug-design optimization strategies to advance small molecule drug candidates through the various stage gates of a target-based screening approach.

Keywords: ADME, blood brain barrier, cell toxicity, CYP450, DNA binding, efflux, formulation, hERG, hit-to-lead, *in silico* computational methods, *in vitro* assays, *in vivo* animal studies, induction, inhibition, Irwin test, lead optimization, metabolic stability, mutagenicity, new chemical entity, oral bioavailability, oxidative stress, permeability, pharmaceutical industry, physicochemical, protein binding, reactive intermediates, screening, solubility, target selection, target-based screening.

***Corresponding author Gray W. Caldwell:** Janssen Research & Development, L.L.C. Discovery Sciences, Lead Discovery Department, Welsh and McKean Roads, P.O. Box 776, Spring House, PA 19477, USA; Tel: 215-628-5537; Fax: 215-628-7064; E-mail: gcaldwel@ITS.jnj.com

Atta-ur-Rahman, M. Iqbal Choudhary and Allen B. Reitz (Eds.)

INTRODUCTION

Moving forward in the 21st century, the goal of the pharmaceutical industry continues to be the discovery and delivery of life-saving medicines (i.e., drugs) that improves global public health care. The pharmaceutical process that produces drugs for patients can be broadly categorized into a discovery/preclinical development step, a clinical drug development step, and a commercialization step (Figs. **1** and **2**). Each step in the process works somewhat independently from each other; however, the entire end-to-end pharmaceutical process has overarching common goals for each of the three steps to minimize timelines and financial investments [1]. The drug discovery/preclinical development step in a small molecule target-based screening approach is managed by subdividing the discovery/preclinical development step into stage gate research processes such as target selection, high-throughput screening (HTS), hit-to-lead (H2L), lead optimization (LO), and new chemical entity (NCE) selection. The stage gates utilize compound selection criteria based primarily on *in silico* computational methods, *in vitro* assays, and *in vivo* animal models to advance drug candidates from H2L to the LO and finally to the selection of an NCE. Thus, the main goal of the drug discovery/preclinical development step is to select NCEs that have a high probability of becoming a safe drug after receiving more extensive and time-consuming clinical testing. The clinical development step is subdivided into human safety clinical trials (Phase I), and drug efficacy clinical trials (Phase II, and III). The commercialization step involves the development of marketing strategies that ensures physicians can quickly and safely prescribe new drugs to patients. In some cases, post-marketing surveillance trials (i.e., Phase IV) are conducted to monitor long-term effectiveness and impact of the drug on the quality of life of patients. In the US, the movement of NCEs from the drug discovery/preclinical development step to clinical trial steps requires the filing of an Investigational New Drug application (IND) with the US Food and Drug Administration (FDA). The movement of drugs from clinical trials to commercialization requires the filing of a New Drug Application (NDA) followed by final FDA approval of the drug to a new molecular entity (NME) status [2]. The time to complete the entire pharmaceutical process for small molecule drug candidates to become NMEs - on the average - ranges from 10 to 15 years with the discovery/preclinical development step accounting for approximately 3 to 5 years and the clinical development and commercialization steps accounting for 7 to 10 years [3-5]. The cost to produce a single NME has increased steadily over the years with financial investments today ranging from 1 to 2 billion dollars [6]. About one-third of total expenditures for the pharmaceutical process is spent on the discovery/preclinical development step and two-thirds on the clinical development and commercialization steps. The average success rate for the pharmaceutical industry to discover and market a small molecule drug - for all therapeutic areas - has been estimated to range from 7% to 11% [7]. Therefore, while the pharmaceutical process to discover and deliver drugs to patients has an underlying scientific industrialization structure, the success rate is very low primarily due to incomplete knowledge of disease pathology and pharmacology with serendipity continuing to play a significant role in the process.

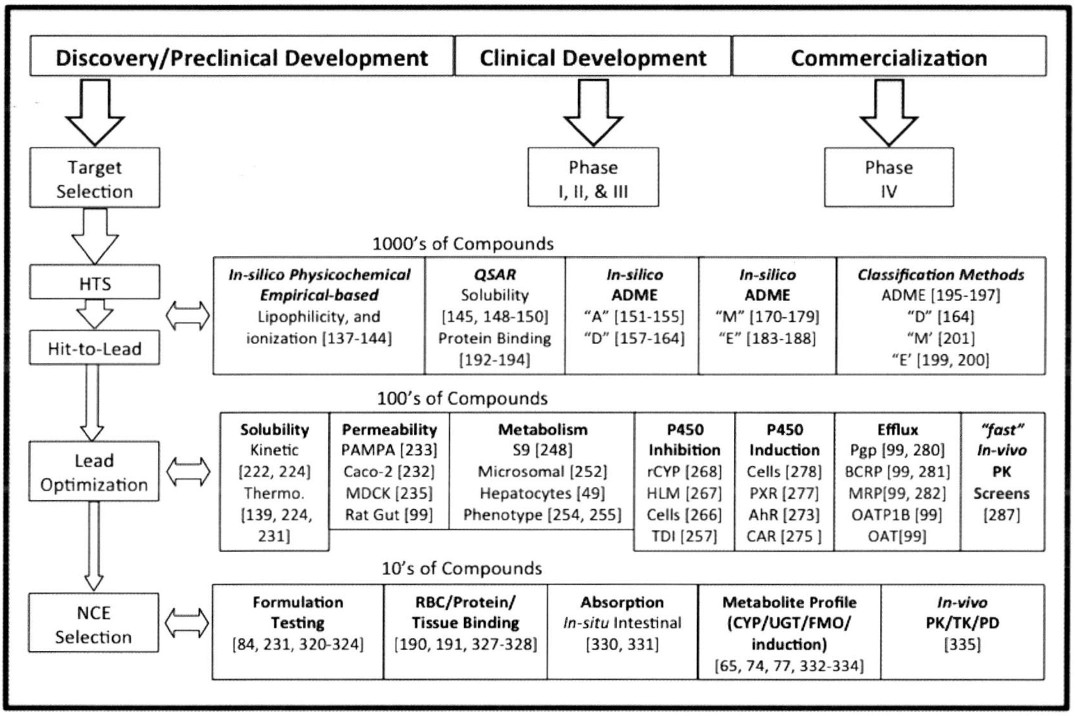

Figure 1: Preclinical drug discovery assays for physicochemical, ADME, and PK endpoints. See text for an explanation of the figure.

The FDA has approved approximately 1,408 small molecule NMEs between the years 1950 to 2014 [8]. On the average, the pharmaceutical process has produced 20 to 25 small molecule NMEs annually for the past 60 years and this average value has not trended upward as expected based on advancements in the understanding of disease pharmacology, chemistry, biology, informatics, computational and analytical technologies [4-6]. This absence of sustained productivity, significant attrition rates, government regulatory regulations, and rising costs for NMEs - particularly within the last 25 years - has generated great concern within the pharmaceutical industry as evident by the many publications that have analyzed the metrics and trends that drive research and development (R&D) productivity [9-28]. The R&D productivity metric and trends suggest that the high attrition rates of NCEs in clinical development tend to be governed by unacceptable safety (Phase I and II) - possibly due to off-target effects or failure of animal models to translate to humans, lack of efficacy (Phase II) - possibly due to limiting target engagement issues or selecting targets that do not significantly influence the disease pharmacology, and economic reasons (Phase I and II) – possibly due to budget/resource constraints or a change in the company's portfolio [20-26, 29]. It should be kept in mind that it is an oversimplification to assume that a single scientific factor governs the attrition rates of NCEs in clinical trials. It is more likely that attrition rates are governed by multiple factors since drug efficacy and safety deficiencies are related in part to pharmacokinetics (PK), toxicokinetics (TK), and drug-drug interactions (DDI), which are related in part to absorption, distribution, metabolism and excretion (ADME) properties of the drug [30].

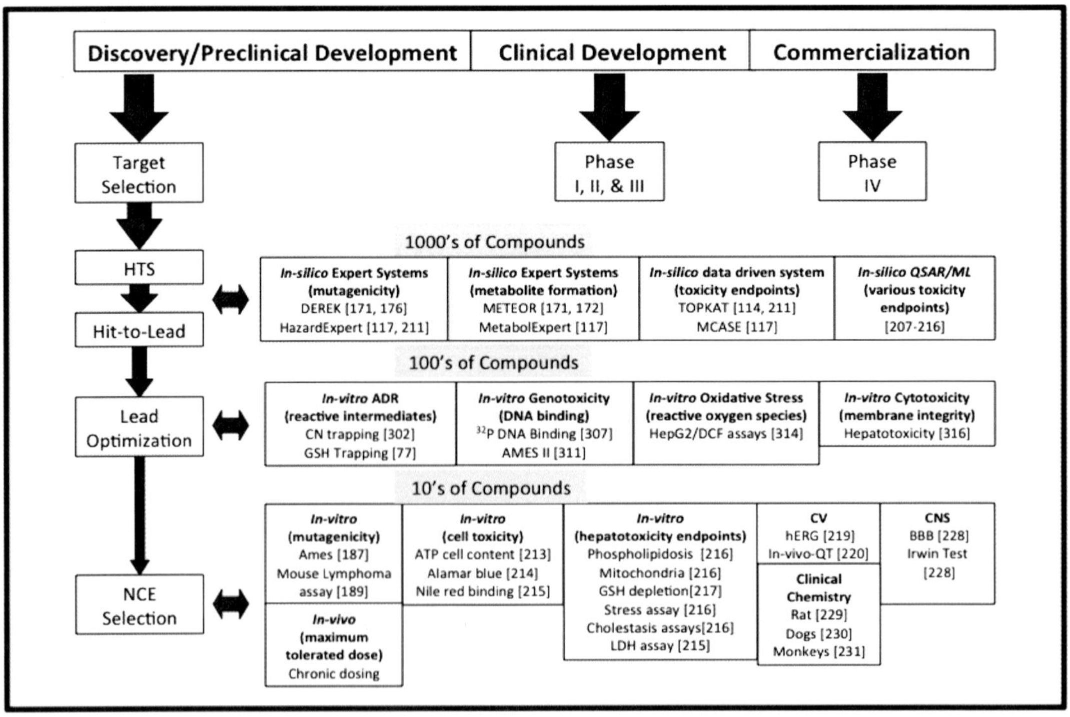

Figure 2: Preclinical drug discovery assays for toxicity endpoints. See text for an explanation of the figure.

For the past 25 years, thousands of scientists have conducted research to design surrogate *in silico* computational methods, *in vitro* ADME, and *in vitro* toxicity assays to predict human *in vivo* PK/efficacy and TK/safety at the discovery/preclinical development step [31-100]. Therefore, most pharmaceutical companies today use panels of well-characterized *in silico* computational methods, *in vitro* ADME, and *in vitro* toxicity screens in parallel with *in vivo* animal PK, TK, efficacy and safety assays to identify drug candidates that have the potential of becoming NMEs at the discovery/preclinical development step (Figs. **1** and **2**) [52, 56, 61, 64, 67, 72, 83, 99]. It is interesting to note that during the early 1990s, approximately 40% of NCE failures in Phase I clinical trials were attributed to ADME, and/or PK defects [9-11, 21]. After pharmaceutical companies started using panels of well-characterized ADME/PK screens to make earlier and smarter go/no-go decisions for drug candidates, the attrition rates of NCEs due to ADME/PK failures in Phase I clinical trials declined to approximately 11% in early 2000s [14, 21, 29]. While an 11% ADME/PK attrition rate is still significant, it highlights the fact that using *in silico*/*in vitro*/*in vivo* screens to uncover and correct ADME/PK defects in drug candidates early in the drug discovery process can be accomplished; however, correcting only ADME/PK defects is not enough to significantly increase the overall R&D productivity of NMEs. It suggests that the reason for the lack of increase in R&D productivity seen today stems from target validation issues and/or the failure of efficacy/safety *in silico* computational methods, *in vitro* cellular assays, and *in vivo* animal models to translate accurately to *in vivo* human predictions [3-5, 100].

The aim of this review is to provide a selected overview - based on the author's experience - of *in silico* computational methods, *in vitro* ADME assays, and *in vitro* toxicity strategies that have been successful or unsuccessful from a drug discovery/preclinical development perspective [75, 99, 100]. Pharmaceutical scientists have recognized for many years that unacceptable physicochemical (LogP, pKa, solubility, etc.), ADME, PK, and toxicity properties are serious problems in advancing drug candidates to NCE status. It has been recognized that many of these pharmaceutical defects can be addressed at different stage gates during the drug discovery process [56, 64, 65, 81, 93, 95]. Since different types of preclinical activities take place at each of these difference stage gates, it is important to distinguish these preclinical activities and to recognize their advantages and disadvantages in designing selection criteria and drug-design optimization strategies. Therefore, we will address the following questions:

- What *in silico/in vitro/in vivo* ADME and toxicity assays do we use in early- and late-phase drug discovery?
- What are the advantages or disadvantages of these assays?
- What types of decisions can be made using these assays?

Using the drug discovery concept outlined in Figs. (**1** and **2**) to select drug candidates for clinical development, we will focus on these questions for the H2L, LO and finally the NCE selection stage gates. Understanding the limitations of preclinical assays is important for all medicinal chemists in establishing selection criteria and drug-design optimization strategies to advance small molecule drug candidates through the drug discovery/preclinical development step.

DRUG DISCOVERY/PRECLINICAL DEVELOPMENT

Since the drug discovery/preclinical development step of the pharmaceutical process is relatively inexpensive compared to the clinical development and commercialization steps, eliminating flawed drug candidates at this step is important since it will reduce overall R&D costs. The drug discovery/preclinical development step for a small molecule target–based screening approach starts with the selection of appropriate therapeutic targets (Figs. **1** and **2**) [22]. A therapeutic target is defined as a particular biological target (i.e., enzyme, receptor, ion channel, etc.) that is hypothesized to be linked to the disease pharmacology. The target selection process is the starting point of all drug discovery programs and is highly influenced or even mandated by company franchises since it has to occur a decade or longer before the launch of an NME. In a target-based screening approach, target selection is typically a two-step process involving target identification and validation of its link to the disease pharmacology [101- 103]. During the last decade, genomics, proteomics, and metabolomics approaches have drastically improved the target identification step [104]; however, the validation step may require many years of research to understand fully the disease pharmacology and thus, target validation remains the bottleneck in the process. Typically, the first stage gate, after target selection, is high throughput screening (HTS) of a primary binding-assay [105-108] using corporate chemical libraries to identify promising compounds that bind to the target [109-111]. The goal here is to screen large compound libraries in a relatively short amount of time attempting to find diverse compounds or families of structurally related compounds with significant potency against the selected target (i.e., hits). These hit compounds are clustered together based on similar structural motifs and ranked based on their potency to identify subsets of chemotypes. Hit compounds are filtered with substructure database searching algorithms (i.e., CAS SciFinder or PubChem) using similarity metrics to remove known drugs, compounds that contain reactive moieties (i.e., aldehydes, alkyl halides, Michael acceptors and so on), and promiscuous compounds that are known to inhibit many types of protein-targets [112]. There is a variety of *in silico* computational methods that can be used to filter hit compounds to identify and possibly to eliminate hits with potentially poor physicochemical, ADME and toxicity properties [100]. Once the hit or clustered chemotype have been validated, using various secondary target-related screens to eliminate false-positives and to confirm concentration-dependent potency activity, H2L compound series are declared. The H2L stage gate in combination with the LO stage gate drives the NCE drug candidate selection process. There are a variety of *in silico* and *in vitro* physicochemical, ADME and toxicity assays that can be used to guide the selection criteria and the drug-design optimization strategies to advance small molecule drug candidates through the H2L and LO stage gates. From these data, several drug candidates are selected as potential NCEs and will be screened to improve pharmacodynamic (PD) and PK/TK properties against the selected target, to establish efficacy and to eliminate any toxicity events using a range of *in vitro* assays and *in vivo* animal models. The overall goal of the final stage gate involves the selection of an NCE and collecting all necessary preclinical data for an IND submission of the NCE.

DRUG DISCOVERY PRECLINICAL SCREENS

Drug discovery preclinical screens must have turn-around-times within a timeframe that is consistent with the iterative cycle of drug discovery research projects. In addition, the data generated from these screens must have high enough quality to make the necessary go/no-go decisions for the advancement of drug candidates through the various stage gates within a target-based screening approach. Since no single factor can account for the NCE failures observed in clinical development, it is necessary to select a range of full and abbreviated ADME/PK and toxicity *in silico/in vitro/in vivo* assays that can provide maximum information in the discovery phase. Pharmaceutical companies use panels of well-characterized ADME/PK and toxicity screens to make better and earlier preclinical go/no-go decisions for drug candidates [52, 56, 61, 64, 67, 72, 75, 83, 99]. A scheme based on this drug discovery concept is shown in Figs. (**1** and **2**). Different types of assays are applied to each stage gate of the drug discovery/preclinical development step; that is, in our example, at the screening and H2L stage gate, the drug LO stage gate, and finally the NCE selection stage gate. At the screening and H2L stage gate, one needs to use assays that have sample throughputs that are relatively high since thousands of compounds can be selected at this stage of the process. Typically, *in silico* computational methods or "ultra-fast" *in vitro* assays - using compound library dimethylsulfoxide (DMSO) stock solutions - are used to sort compound properties into a high/low classification scheme. At the drug LO stage gate, one needs to use "fast" assays that have sample throughputs in the hundreds per day and consume small amounts of compounds in the range of 1 to 2 milligrams. "Fast" assays are defined as those that have turn-around-times within a timeframe that is consistent with the iterative cycle of drug discovery research projects. Using these types of "fast" *in vitro* assays, typically one would like to prioritize compounds from high-to-low to establish structure-activity relationships. The drug LO stage gate assays include physicochemical assays, abbreviated formulation screens, abbreviated *in vivo* PK assays, *in vitro* ADME, efflux, and P450 interactions (inhibition and induction) assays, to name a few. In the final selection stage gate, a few drug candidates are tested in assays requiring gram amounts of compounds for full *in vivo* PK and TK studies, acute and chronic dosing studies, drug-drug interaction, drug-serum protein binding, receptor (or enzyme) selectivity screening, and secondary genetic toxicity testing, safety pharmacology assays such as cardiovascular effects (CV) and central nervous system effects (CNS), developmental and reproductive toxicity testing, and clinical pathology.

Primary assays such as "ultra-fast" and "fast" *in vitro* or abbreviated *in vivo* assays should have reasonable sample throughput, require minimal amounts of compound and are validated; that is, the strengths and the weaknesses of each assay are fully evaluated to understand the false positives and false negative rates. For a drug candidate to pass a primary assay, the selection criteria must be clearly understood and established for each assay. It should be understood if the results from the assays could be prioritized from high-to-low or simply classified into a high/low classification scheme. The accuracy and precision of each assay need to be understood to establish selection criteria. Due to biological variability, the acceptable degree of accuracy for some assays might be as high as 10-fold with the precision, as measured by the percent coefficient of variation (% CV), ranging from 30% to 40%. Also, the precision may be more variable due to lab-to-lab biological sample variation. While there are many "fast" *in vitro* or abbreviated *in vivo* assays that are currently used in drug discovery, there are very few "ultra-fast" *in vitro* assays [75, 99]. The traditional *in silico* and *in vitro* physicochemical/ADME/toxicity assays utilized in H2L, LO, and NCE selection stage gates will be reviewed in the following sections.

H2L STAGE GATE (EARLY DRUG DISCOVERY)

For the past 25 years, due to the development of HTS in discovery, medicinal chemists have been faced with the daunting challenge of selecting good LO chemical starting points from thousands of possible starting points (i.e., hits). A significant amount of effort in the pharmaceutical industry and academic labs has been devoted to developing computer-based *in silico* approaches to calculate various physicochemical, ADME, PK and toxicity properties of compounds directly from their structures [113-130]. Having this type of data helps to prioritize hits not only based on their potency against the target but also based on their "druggability" [40, 113]. In addition, *in silico* approaches have the capability to detect and identify problematic structural motifs before they are synthesized, and also have the potential to rapidly screen virtual libraries of drug candidates to discover new hits [131-136]. There are many *in silico* tools in different combinations that can be used to establish selection criteria at the H2L stage gate [100]. In fact, there are so many different *in silico* techniques that have been developed in the last 15 years, many medicinal and computational chemists struggle to select those techniques that are beneficial for designing new compounds. Here, we will review only a few selected *in silico* approaches and examine their strengths and weaknesses.

In Silico Physicochemical Properties: LogP, LogD, pK_a, and S_0

Properties of virtual compounds are, in many cases, calculated based on the two-dimensional (2D) structure of the compound and simple empirical-based *in silico* models. Empirical-based models involve summing the contributions of structural units (i.e., atoms or fragments) within a compound to predict quantitatively the properties of the compound as a whole. The values of the structural units are derived from experimental data of known compounds for each property of interest. Once the structural units are known for a particular property, they can be combined in different ways to create real or virtual compounds and thus, by summing the value of the structural units in the compound its property is calculated. Structural properties of compounds, using an empirical-based approach, includes molecular weight (MW), hydrogen bonding characteristics, flexibility index, and molecular polar surface area, to name a few. These types of properties are easily calculated using the 2D structure of compounds and structural unit values of the property of interest [40, 113].

Physicochemical properties including lipophilicity (LogP or LogD), acid ionization ((pK_a = -Log [H^+]), and intrinsic solubility (S_0) which are the most important properties of hits are also calculated from empirical-based 2D structural units [137-142]. Many reasonable *in silico* empirical-based models for calculating LogP, LogD, and pKa values have been published in the literature and are commercially available in software packages [143]. Unfortunately, the typical accuracy errors between these *in silico* LogP, LogD, and pKa models and experimental data are somewhat large ranging between 0.1 and 1 log unit. While this type of error does not interfere with prioritizing hits into a high/low classification scheme, caution should be taken by medicinal chemists when ranking hits in a narrow range (i.e., high to low) using calculated LogP, LogD, and pKa values. The prediction of aqueous solubility of a compound, from its structure, is a much more challenging problem than the prediction of lipophilicity since aqueous solubility involves the partitioning of the solid form of the compound into water [139, 144, 145]. The solubility of a compound depends on the charge state of the molecule, the pH of the solvent, its solid-state form (i.e., amorphous or crystalline) and the temperature of the surrounding environment. The prediction of the solid-state form of a compound from its structure has not been solved [139]. In addition, typical experimental errors for aqueous solubility measurements can be approximately ±1.0 LogS units or larger. Thus, calculating the contributions of structural units within a compound to predict quantitatively solubility has a great deal of error embedded in it. Calculated intrinsic solubility values of hits using empirical-based models should not be used at key decision points in the drug discovery process. The quantitative structure–property relationship (QSPR) and quantitative structure–activity relationship (QSAR) approaches are regression models that relate a set of compound predictors such as, physicochemical properties (i.e., LogP and MW) and theoretical molecular descriptors (i.e., molecular orbital energies) to the compound's property of interest [146]. The QSAR approach has been used many times to predict intrinsic aqueous solubility (neutral form of the molecule) with models ranging from simple linear models using small sets of molecular descriptors to non-linear models using large sets of molecular descriptors [146-150]. It is generally concluded that these types of QSAR or QSPR methods cannot be used to globally predict aqueous solubility for pharmaceutical compounds with sufficient accuracy due to deficiencies in the algorithms and/or descriptor sets of the individual models [147, 150]. Calculated intrinsic solubility values of hits using QSAR or QSPR models should be considered only a very rough estimation of their true value. An interesting approach for improving the prediction of aqueous solubility of virtual compounds is based on the idea of using several QSAR models in combination with a random forest classification approach [151]. In this machine learning application, ten commercially available QSAR models were used to generate aqueous solubility data for all compounds. The best QSAR model was selected for a virtual compound based on the predictions for structurally similar compounds using decision trees generated by the random forest method. In general, *in silico* solubility predictions cannot currently be used as a substitute for experimental measurements at key decision points in the drug discovery process.

In Silico ADME: "A" in ADME

In silico ADME has been reviewed many times by others and readers are encouraged to examine these publications [114-130]. The "A" in ADME represents the absorption process which involves the extent and the transfer rate of a compound from the gastrointestinal (GI) fluid across primarily the jejunum and the ileum segments of the small intestine into the portal blood system. The driving forces for this process are the compound concentration gradients between the GI and the portal blood supply, electrical differences, the hydrostatic pressure gradients introduced by the presence of the compound and external factors such as dissolution rates of the compounds, stomach emptying rates and food consumption. For a compound to reach the general systemic circulation, it must penetrate cell barriers either by passing between cells (paracellular route) or through them (transcellular route). When compounds utilize the transcellular route, they cross the membrane barrier either by mechanisms involving the active participation of components of the membranes (i.e., energy-consuming carrier-mediated processes) or by passive processes (i.e., non-energy-consuming diffusion processes). While carrier-mediated transmembrane processes play an important role for some compounds, passive diffusion through the bilayer membranes is the dominant process in the disposition of most small molecules.

The extent of absorption of a compound is measured as the percentage of orally dosed compound absorbed (%Fa) across the small intestine into the portal blood system. In a typical experiment, the concentration of the compound being orally administered is known. By removing a blood sample *via* a catheter inserted into the hepatic portal vein, the concentration of the compound in the portal blood system can be determined. The %Fa value is calculated as the ratio of the orally dosed concentration of compound divided by the portal vein concentration of compounds. %Fa can be considered a function of two key components: the permeability and the solubility of the compound. The permeability of a molecule is a measure of the ability of a molecule to cross a cell membrane barrier and is expressed in velocity units. Permeability is primarily a function of the compound's LogD, MW, and the aqueous diffusion constant of the compound. Various *in vitro* permeability assays have been developed that mimic the relevant characteristics of *in vivo* absorption including the human colon adenocarcinoma (Caco-2) cell assay and the Madin-Darby canine kidney (MDCK) cell assay. These will be discussed along with other methods to determine permeability in a later section.

There is a great deal of interest in *in silico* methods for predicting the %Fa or the permeability of hits from their 2D chemical structures [152-156]. Many of these methods use QSAR models with molecular descriptors (i.e., polar surface area, molecular size, flexibility, hydrogen bonding capacity and so on) along with a variety of calculated physiochemical properties such as LogP and pKa [148]. In addition, these models use building methodologies such as linear or non-linear regression techniques, classification trees, partial least squares, and neural networks. Generally it is found that these QSAR %Fa or QSAR permeability models predict the training data set reasonably well, but typically do not predict new data sets with acceptable accuracy [157]. This result is not unexpected for %Fa since it is not a discrete fundamental property of the compound and, therefore, cannot be predicted solely from a structural representation of the compound. %Fa is a complex biological process highly dependent upon many internal and external factors. However, the permeability of a compound is a discrete property of the compound and accurate predictions of it by *in silico* methods may be developed in the future. In general, *in silico* absorption predictions cannot currently be used as a substitute for experimental measurements at key decision points in the drug discovery process.

In Silico ADME: "D" in ADME

The "D" in ADME represents the distribution process of a compound in the body. When a dose of a compound enters the systemic circulation (i.e., blood), the compound is distributed to all parts of the body. A theoretical volume of distribution at steady state (Vd_{ss}) is assumed as the volume into which a drug is distributed in the body. Thus, a high value of Vd_{ss} (> 42 L) for humans indicates that a compound is highly likely to be distributed throughout body tissues whereas a low value (< 3 L) suggests it is predominantly located in the systemic circulation. The Vd_{ss} in conjunction with the clearance of the compound determines the half-life of a compound in the body. In addition, the distribution of a compound is an important parameter to understand since it is generally a prerequisite for a compound to pass from the blood into other fluids and tissues before it can induce/exert its pharmacological action. The Vd_{ss} is a function of the tissue to plasma binding capacity of the compound as well as the ease with which the compound crosses the membrane and to the organ/tissue blood perfusion rates. The binding capacity of a compound is related to its physicochemical properties (pKa, Log P, aqueous solubility, etc.) and the affinity property of the compound for organs/tissues (blood protein binding and cell binding). In general, there is a large variability in the Vd_{ss} parameter in humans and animals. The clinical error range in Vd_{ss} human data is typically 2- to 3-fold from the mean values. There is a great deal of interest in predicting Vd_{ss} from chemical structures [158-165]. For example, Lombardo and co-workers [157, 161] have used a variety of QSAR linear and non-linear statistical techniques to predict Vd_{ss} from chemical structures. They have shown that Vd_{ss} can be predicted within a geometric mean 2-fold error. This magnitude of the error is on the same order as experimental data and can be used as a substitute for experimental measurements at key decision points in the drug discovery process.

In Silico ADME: "M" in ADME

The "M" in ADME represents the metabolism of a compound by enzymatic reactions. Hydrophobic compounds usually have low aqueous solubility under physiological conditions and must be converted into more hydrophilic molecules before they are excreted from the body. This biotransformation is typically denoted as Phase 1 and Phase 2 enzymatic reactions. The Phase 1 enzymatic reactions convert the hydrophobic parent compound to a more polar metabolite by oxidation, reduction, or hydrolysis chemical reactions. These reactions expose or introduce a polar functional group (-OH, $-NH_2$, -SH, or $-CO_2H$), and usually result in only a small increase in the hydrophilicity of the compound. During the last four decades, extensive studies of Phase 1 reaction types have shown that there are many cytochrome P450 enzymes (CYP450s) [166] and non-P450 enzymes (i.e., flavin monooxygenases (FMOs) and monoamine oxidases) [167-169] that have very broad substrate selectivity, catalytic versatility, and play important roles in the xenobiotic metabolism. The CYP450s are the predominant pathways for drug metabolism of most small molecule drugs, and the seven most common CYP450s are 1A2, 2B6, 2C8, 2C9, 2C19, 2D6, and 3A4. Phase 2 drug-metabolizing enzymes are generally transferases and are responsible for conjugating a compound or its Phase 1 metabolites with a highly polar molecule such as, glucuronidation, sulphation, and acetylation where they are mediated by the enzymes glucuronyltransferases (UGT), sulfotransferases (SULT), and N-acetyl-transferases (NAT), respectively [170]. The ability to predict the metabolic fate of a compound using *in silico* methods would allow medicinal chemists the ability to establish drug-design strategies that increased the half-life of a compound in the body or to eliminate reactive metabolites that might exhibit toxicity.

Several computational approaches have been created for predicting the metabolic fate of a compound. Expert systems are software packages that rely on a set of programmed rules, which have been distilled from available metabolism/ toxicological knowledge and human expert judgments to predict metabolites and their toxicity. This computational approach has been more successful than other approaches since it is intuitively appealing to most users. Software packages, such as TIMES, METEOR, METASITE, and METABOLEXPERT are commercially available [170-176]. These types of expert models typically predict a large ensemble of metabolites that are possible for a compound. In many cases, the ensemble of metabolites contains most of the experimentally found metabolites and metabolites that were not found experimentally. These false-positive metabolites may not have materialized *in vivo* because of numerous biological factors including low drug-enzyme reaction rates, competitive reactions, and enzyme inhibition. On the other hand, the false-negative metabolites may not have been generated due to the limitation of the structural restrictions in the program. Nonetheless, metabolite expert systems give medicinal chemists the most likely sites of metabolism on a given molecule. QSAR models for predicting the site of compound metabolism have undergone significant advances during the past few years [177-180]. However, most of the QSAR based models lack sufficient interpretability and offer poor predictability for novel compounds. In general, *in silico* models for predicting drug metabolism of virtual compounds can predict a set of potential metabolite candidates for medicinal chemists to design synthetic procedures to metabolically stabilize them but the end-user cannot be assured that these metabolites have any *in vivo* significance.

In Silico ADME: "E" in ADME

The "E" in ADME represents the excretion of a compound by the liver, kidney, and other organs from the body. After oral absorption, compounds are eliminated from the body *via* excretion and/or metabolism processes. We define the compound elimination process as the irreversible removal of compounds from the body and denote this process as the total clearance (CL_T) of the compound. The CL_T of a compound has units expressed in volume per time (e.g., L/hr). Therefore, CL_T is defined as the volume of blood cleared of compound from the body per unit of time. We can assume that the CL_T is the sum of the renal clearance (CL_R)

and non-renal clearances (CL_{NR}) [33]. Non-renal clearance includes biliary excretion (i.e., leading to excretion in feces), lung excretion, and biotransformation (i.e., metabolism leading to excretion in urine or feces). The CL_T of a compound can be measured using *in vivo* animal models. For humans, when $CL_T < 25$ L/hr, one can state that the drug has a low extraction from the body either *via* the kidneys, non-rental processes or both. When > 87 L/hr, one can state that the drug has a high extraction from the body either *via* the kidneys, non-rental processes or both. The advantage of the clearance concept is that it describes all the processes of compound elimination from the body without identifying the individual mechanisms involved.

The most important practical application of CL_T derives from the fact that this PK parameter determines the average concentration at steady state for a given dose rate. In other words, if accurate assessment of CL_T can be made before the compound is administered to patients, then a better decision on the appropriate first-dose can be made, which would lead to improved safety and efficacy of new NCEs. The clinical range in CL_T human data is typically 3- to 5-folds in the mean value due to inter-individual variations. The CL_T of a compound is extremely difficult to model using QSAR techniques since CL_T has a strong structural dependence on metabolism. Therefore, QSAR modeling of CL_T or its components CL_R and CL_{NR} have been limited. The largest component of non-renal clearance for many compounds is the biotransformation of the compound in the liver, and, therefore, non-renal clearance is often associated with hepatic clearance (CL_H). The hepatic clearance is composed of two factors: compounds must first be cleared from the liver by being metabolized and then must be cleared from the body by being excreted. The rate of metabolism of a compound is called the intrinsic clearance (CL_{int}) and is a pure measure of hepatic enzyme activity towards a compound. The CL_{int} of a compound can be determined by *in vitro* measurements based on liver microsomes (i.e., Phase 1 metabolism only) or hepatocytes (i.e., Phase 1 and 2 metabolism) [181, 182]. QSAR models for predicting CL_{int} have been reported in the literature [183-185]. However, these QSAR based models lack sufficient interpretability and offer poor predictability for novel compounds and thus, cannot be used as a substitute for experimental measurements at key decision points in the drug discovery process.

Physiologically Based Pharmacokinetic (PBPK) ADME Predictions

Physiologically based pharmacokinetic (PBPK) modeling is another way of predicting ADME properties. PBPK models are based on the anatomical and physiological structure of the body. By mathematically applying the principles of enzyme kinetics, mass transport, and fluid dynamics, PBPK is capable of simulating the ADME fate of a compound in the body. Software packages such as GastroPlus [186] and Simcyp [187] are commercially available. It has been shown for 33 drugs that the use of physicochemical parameters generated with the ACD Labs software [143] combined with Simcyp software [187] produced accurate hepatic clearance values [188]. These authors concluded that this *in silico* method could be used to predict intrinsic hepatic clearance at the early drug discovery stage. While these whole-body simulation techniques are clearly important in predicting ADME properties, there still appears to be a reluctance to use these models in a drug discovery environment.

In Silico Drug-Protein Binding

The pharmacodynamic and the toxicological action of drugs can be significantly influenced by the binding of drugs to various blood proteins and tissues [189, 191]. For example, drugs that are highly bound to blood proteins have smaller Vd_{ss} compared to drugs with low binding to blood proteins. Thus, drug-protein binding effects the volume of distribution of a drug, the half-life of a drug, and potentially, contributes to the drug's inter-individual variability. Drugs with high tissue affinities may also accumulate in lipid or adipose tissue. Premature displacement of drugs from blood proteins or tissues by other drugs or endogenous compounds may result in an increase in free drug in the systemic system that may diffuse to pharmacologic receptor sites thus causing an unexpected intense or unwanted drug response. This situation is particularly relevant for drugs with high extraction ratios and non-hepatic clearance mechanisms. Also, the formation of reactive metabolites of drugs that irreversibly bind to proteins or tissues may cause certain types of toxicity. Therefore, an understanding of the binding characteristics of drugs and metabolites to blood proteins such as, human serum albumin (HSA), α-acid glycoprotein (AGP), lipoproteins, and immunoglobulins, and to tissues such as, lipid or adipose is useful information in the design and administration of therapeutic drugs. In addition, knowledge of protein binding in both plasma and microsomal components is needed in the prediction of $CL_{int.}$ Many QSAR models have been developed to predict the amount of binding to plasma proteins [192-194] and many of these methods are available in software packages such as ACD Labs [143], ADMET Predictor [183] and Simcyp [187]. From a global point of view, these QSAR models can prove useful to help to identify compounds with lower binding on the average. However, they cannot be used as a substitute for experimental measurements at key decision points in the drug discovery process.

In Silico ADME Classification Methods

Classification methods, whereby compounds using experimentally established cutoff values are classified into high/low schemes, are very practical at the hit generation stage gate to make go/no-go decisions to advance compounds. Machine learning (ML) methods have shown great potential in predicting the classification of compounds into broad schemes by correlating ADME properties to molecular descriptors [195-197]. A few examples of ML methods include techniques such as, linear discriminant analysis (LDA), k nearest neighbor (kNN), artificial neural networks (ANN), support vector machine (SVM), random forest (RF), and naïve Bayesian (NB) with kNN, SVM, and RF being frequently used to correlate ADME properties. There are also ML regression models that combine QSAR methods with RF [151] and principal component analysis (PCA) [197]. Some ADME and PK classification methods include human intestinal absorption [198], Vd_{ss} [164], clearance [199, 200], human liver microsomal stability [201], P450 inhibition [202], drug-HSA protein binding [203], brain-to-plasma ratios [204], P-glycoprotein inhibitors [205], and oral bioavailability [206], to name a few examples. Classification methods appear to have a distinct advantage over quantitative predictions of ADME parameters since the latter has room to account for inter-individual variations of complex biological systems. In summary, the use of ML methods for creating classification models to classify ADME properties into high/low schemes can be used at the H2L stage gate. Unfortunately, these methods are currently underutilized; however, it is expected that these ML *in silico* methods will be highly utilized in the future.

In Silico Toxicity Properties

In silico prediction or simulation of various toxicity properties continues to receive a significant amount of attention in the pharmaceutical industry and academic labs (Fig. **2**) [117, 119, 125, 129, 130, 171, 175, 207-216]. The major areas of research include genotoxicity, carcinogenicity, skin sensitization, reproductive and developmental toxicity and hepatotoxicity. *In silico* toxicity predictions include models such as structure-activity relationships (SARs) (i.e., knowledge- and rule-based expert systems), QSAR and QSARs combined with ML techniques [207]. SAR methods are typically qualitative in nature and are used to demonstrate that a moiety (or fragment) of a compound is associated with a particular toxicity (i.e., structural alerts). In principle, *in silico* models to predict any toxicity endpoint can be derived using SAR and QSAR techniques if a quality experimental database of structural diverse drug candidates exist and the mechanism of action of the toxicity endpoint is known [208, 209]. Obtaining a full range of diverse structures with mechanistic information is typically difficult for most toxicity endpoints; however, there are many structural searchable toxicity databases that are available to at least attempt to construct predictive models [210, 211]. Expert systems are software packages that include SAR and, in some cases, QSAR methods. For example, software packages such as DEREK *Nexus*, HazardExpert, OncoLogic, TOPKAT, MCASE, ToxAlerts, OpenTox, and PASS are publicly or commercially available for the prediction of mutagenicity, carcinogenicity and other toxicity endpoints [117, 171, 210-213]. For example, the software package DEREK is a knowledge-based and rule-based expert system that predicts if an inputted chemical structure contains structural moieties that bind to DNA resulting in general genotoxicity, mutagenicity, or chromosome damage. An illustrative example of DEREK is provided by studies of 266 compounds with known mutagenicity. The results indicated that 81 of 114 mutagens (71%) and 117 of 152 non-mutagens (77%) were correctly identified [171]. Many QSAR models have been developed to predict particular toxicity endpoints of compounds [207, 211]. For example, the hERG receptor is a potassium channel and drug blockade may lead to QT prolongation and potentially to fatal arrhythmias (i.e., torsade de pointes). A QSAR model was established using hERG IC_{50} binding data for 90 drug compounds with a correlation of $r^2 = 0.912$ [212]. The hERG QSAR model was tested with an additional 20 drug candidates with known hERG IC_{50} (i.e., 50% inhibitory effect concentration) binding values. Comparison of the known and predicted hERG IC_{50} binding values indicated that the model performed well with a correlation of $r^2 = 0.848$. Another approach to *in silico* toxicity predictions is ML modeling or ML modeling with SAR and QSAR models to create algorithms for a given toxicity endpoint [214-216]. For example, using the RF classification model for prediction of mutagenic properties of compounds showed that the model performed well with 89.23% accuracy, 89% precision and ROC of 95.3% [216]. Using QSAR/ML methods, one takes advantage of models that have been generated with different techniques (i.e., expert and data-driven systems) and training data sets.

When using *in silico* methods to predict toxicological activity there are three main points that need careful consideration in the assessment:

- Was the *in silico* model constructed using diverse structural compounds from quality experimental data?

- Is the mechanism of action known for the toxicological activity?

- What are the *in vivo* or *in vitro* compound exposures necessary to elicit the toxicological activity?

If these three questions can be answered - for example, Ames genotoxicity or hERG inhibition - then *in silico* toxicity predictions can be used to alert medicinal chemists of potential structural issues and direct their synthetic efforts in a different direction. However, it should be kept in mind that software packages such as DEREK and MCASE were designed to be sensitive at hazard identification and not drug design. Strictly using the structural alerts from these programs would require medicinal chemists not to use heterocyclic amines in their drug design strategies. With partial information on these three questions, *in silico* predictions should not be used as a substitute for experimental measurements at key decision points in the drug discovery process.

LEAD OPTIMIZATION STAGE GATE

Many preclinical assays have been developed to support drug optimization specifically in a drug discovery environment [32-100]. Once useful structural prototypes are obtained, from the hit stage gate, analog synthesis around these templates is initiated to improve *in vitro* or *in vivo* efficacy/potency and ADME/PK/toxicity properties. Typically, hundreds of analogs are synthesized for each hit chemotype to select drug candidates that have the correct combination of pharmacodynamic (PD) and ADME/PK/toxicity properties. Since these structural analogs are synthesized typically over a period of 1 to 2 years in the low milligram range, lower sample throughput and less compound intense *in vivo* or *in vitro* assays for physicochemical [84, 92, 99, 141, 145, 217], absorption [44, 99, 218], metabolism [36, 42, 49, 77, 99], *in vivo* pharmacokinetics [50, 219-221], drug-drug interactions [65, 68, 93], drug distribution [161-164] and toxicity [34, 54, 71, 81, 90, 96-98] can be used. The common ADME/PK/toxicity assays used at the LO stage gate are shown in Figs. **1** and **2**. Since all of these assays have been fully discussed in the literature, we will review only a few selected assays [75, 99]. When interpreting the data from these ADME/PK/toxicity assays, one quickly realizes that many of them have "large gray zones"; that is, the data is ambiguous and reliable go/no-go decisions are not possible. Also, the assays are typically plagued with a significant percentage of false positive and false negative results. Keeping these problems in mind, the strategy used to select drug candidates utilizing these assays is to assume that the drug candidate with the most positive physicochemical and ADME attributes is superior to those drug candidates with fewer attributes [52, 67].

Solubility

Numerous drug discovery thermodynamic and kinetic solubility protocols have been described in the literature [40, 56, 84, 92, 106, 141, 145, 222, 223]. The "gold standard" is the thermodynamic shake-flask method [139, 224]. Briefly, a compound is added to a buffer until saturation occurs, and this thermostatic solution is shaken until equilibrium is established between the buffer and the solid phase. The buffer is separated from the solid and the concentration of compound in the buffer is determined by an appropriate analytical method (i.e., LC-MS-UV, a rapid multi-wavelength UV plate reader or a potentiometric acid/base titration). The experimentally determined solubility of a compound strongly depends on the temperature, the physical characteristics of the solid (i.e., polymorphism, melting point, particle size, etc.), the composition of the buffer (i.e., pH, co-solvents, additives, ionic strength), the technique used to separate the solid from the buffer (i.e., filtering, centrifuging), the shaking intensity, and the amount of time the solution is shaken [225]. The accuracy of thermodynamic assays is limited since typically, the solid characteristics drug candidates are not determined at this stage gate. Kinetic solubility assays do not have a stable liquid/solid phase interface and are typically measured in the present of organic solvents. Briefly, a small volume of a DMSO stock solution (i.e., compound is dissolved in DMSO at a known concentration) is added to a buffer of interest and any undissolved particles are detected using light scattering techniques (i.e., nephelometric assay) [40]. Other versions of this assay include filtering the buffer solution to remove undissolved particles and measuring the dissolved compound by UV absorption [224]. Thermodynamic solubility assays are useful for establishing *in vivo* PK/TK formulations [226-231] while kinetic solubility assays are useful for interpreting ADME and toxicity *in vitro* assays [222]. It should be emphasized that interpreting *in vitro* ADME and toxicity data without knowledge of the solubility of the drug candidate in the assay is a serious mistake.

In Vitro Permeability

Various *in vitro* permeability assays have been developed that mimic the relevant characteristics of *in vivo* absorption [47, 232-240]. The parallel artificial membrane permeability assay (PAMPA) [73, 233, 237], the human colon adenocarcinoma (Caco-2) cells assay [232, 234, 236], the Madin-Darby canine kidney (MDCK) cell assay [235, 238], and the rat everted gut assay [233] have been used for the assessment of drug permeability. Presently, the Caco-2 monolayer cell culture model is the "gold standard" for drug permeability and is widely used in drug discovery for the prediction of human intestinal permeability. Briefly, the permeability of drug candidates across differentiated monolayers of Caco-2 cells is measured on fully differentiated cells grown on permeable filter supports. Compounds are applied in separate experiments to the apical (A) or basolateral (B) side of the monolayer and incubated for several hours. The concentration of each compound on each side of the monolayer is measured by liquid chromatography (LC) or LC tandem mass spectrometry (LC/MS/MS) techniques [47]. The permeability (Papp) is calculated in the apical to basolateral (A → B) and basolateral to apical (B → A) directions. Thus, the Caco-2 assay is a surrogate assay for a compound permeability across the human intestinal epithelium. Since there are transporters on the apical (i.e., P-gp, MRP-2, BCRP) and basolateral (i.e., MRP-1, PepT1) sides, the Caco-2 model can be used for the identification of substrates, inhibitors, and inducers of intestinal transporters. The Caco-2 assay has been developed in a 3-5 day culturing 24 well format as well as a 10 to 21-day system using a 96-well plate format [232, 240]. Many laboratories have shown reasonable correlations between *in vitro* permeability coefficient values in Caco-2 monolayers and human *in vivo* absorption (%Fa). Unfortunately, large inter-laboratory variability in Caco-2 permeability measurements has plagued the model; however, the situation continues to get better [239]. For the Biopharmaceutical Classification System (BCS), classification of intestinal permeability [241] using the *in vitro* Caco-2 model [235] or the *in situ* single pass perfused rat intestinal model [75, 99] are relevant [242]. Both Caco-2 and MDCK cell models exhibited the same rank order relationship between *in vitro* permeability values and human permeability values for many test compounds [236]. An estimation of oral absorption of drug candidates can be made using Caco-2 or MDCK permeability data along with thermodynamic solubility data. Due to the inter-laboratory variability of the *in vitro* cellular permeability data, these absorption predictions should be classified into high/low schemes using experimentally established permeability cutoff values from benchmark drugs.

In Vitro Metabolism Stability

The rate of *in vivo* biotransformation of drug candidates is an important determinant of oral bioavailability as well as systemic clearance. The systemic clearance parameter, which is used to characterize drug disposition, is the sum of all organ clearances that contribute to the elimination of a drug (i.e., hepatic and renal clearances). Therefore, several *in vitro* hepatic systems utilizing liver slices [243, 244], isolated hepatocytes [49, 245, 246], hepatic membrane subcellular fractions [247, 248] and purified CYP450 enzyme preparations [55] have been developed to mimic *in vivo* clearance. It should be remembered that liver slices and hepatocytes contain both Phase 1 and 2 activated enzymes while hepatic membrane subcellular fractions typically contain only Phase 1 activated enzymes. There are methods to activate Phase 2 enzymes in hepatic membrane subcellular fractions by utilizing the pore-forming peptide alamethicin [249]. Hepatic subcellular fractions can be prepared in large batches and stored frozen for many months with little loss of CYP450 activity. Presently, the metabolic stability assay using human liver microsomes (HLM) preparation is the "gold standard" for drug stability and is widely used in drug discovery for the prediction of human hepatic clearance. Isolated hepatocytes and HLM assays have been adapted to a 96-well plate format in combination with LC/MS/MS to assay the metabolic stability of drug candidates [49, 250, 251]. In many cases, the metabolic stability of a compound is simply rank-ordered based on the rate of disappearance of the parent drug in the present of hepatic membrane subcellular fractions or hepatocytes. A common strategy is to sort *in vitro* metabolic stability data into a high and low category based on a benchmark drug where it's *in vivo* and *in vitro* metabolic stability is well known. If the metabolic stability of a drug candidate is substantially less than the benchmark drug, then this drug candidate could be given a lower priority in favor of another analog compound.

It should be remembered that, in some cases, there will be no correlation between *in vitro* metabolic stability in HLM or cells and the oral bioavailability or the clearance. The *in vivo* biotransformation rate of disappearance of parent drug is dependent on many factors. For example, the *in vitro* microsomal and hepatic cell systems have limited ranges of metabolizing activities as compared to *in vivo* metabolizing activities. Other factors such as physicochemical properties and absorption (to name a few) need to be considered before drug candidates are eliminated. It is recommended that ranking methods be used to sort metabolic stability data into high/low clearance schemes. For example, Kerns and co-workers [252, 253] have examined the correlation between high throughput single time point *in vitro* rat microsomal stability data and abbreviated *in vivo* rat clearance data for a set of 306 drug discovery compounds. The single time point *in vitro* rat microsomal stability data was converted into half-life data ($t_{1/2}$). A drug candidate with a low $t_{1/2} < 15$ minutes was interpreted as being metabolized rapidly and thus, should have a high *in vivo* clearance CL > 25 mL/min/kg. A drug candidate with a high $t_{1/2} > 15$ minutes was interpreted as being metabolized slowly and thus, should have a low *in vivo* clearance CL < 25 mL/min/kg. Their results showed that a significant number of drug candidates with microsomal stability $t_{1/2} < 15$ minutes also produced *in vivo* CL > 25 mL/min/kg. For drug candidates with $t_{1/2}$ > 15 minutes, there was no *in vivo* CL correlation. Therefore, the microsomal stability assay can identify drug candidates with significant *in vivo* clearance liabilities but has a significant false-positive rate. Care should be taken not to over-interpret microsomal stability data, particularly when it is used to predict the hepatic clearance (i.e., CL_H) of a compound.

In Vitro Metabolism Reaction Phenotyping

Reaction phenotyping is the process of identifying *in vitro* drug-metabolizing enzymes involved in the clearance of a drug and is routinely used in drug discovery [254, 255]. Phase 1 monooxygenase enzyme reactions convert hydrophobic compounds to more polar metabolites by the introduction of a polar functional group (-OH, $-NH_2$, -SH, or $-CO_2H$) by oxidation, reduction, or hydrolysis chemical reactions [166]. Phase 2 reactions are primarily transferase reactions that conjugate a highly polar molecule to a compound. While compounds containing conjugate accepting groups such as hydroxyl, carboxyl, carbonyl, sulfuryl, amino, and imino tend to be a substrate for Phase 2 reactions, it is more common that *in vivo* Phase 2 reactions occur after compounds are metabolized by Phase 1 enzymes. In humans and other mammalian species, CYP450s and FMOs are the predominant enzyme families for Phase 1 drug metabolism while glucuronidation catalyzed by UGTs is the major Phase 2 reaction pathway [166, 168]. The combinations of these Phase 1 and 2 reaction pathways determine the clearance mechanism of most xenobiotics. Briefly, CYPP450s are a superfamily of diverse heme-containing enzymes that are membrane-bound primarily in the endoplasmic reticulum region. These enzymes require β-nicotinamide adenine dinucleotide 2′-phosphate (NADPH) as a cofactor. CYP450s are involved in the metabolism and potential bioactivation of a wide range of compounds. Although many thousands of CYP450s have been identified (i.e. approximately 21,000) across mammalian species, only 18 isoforms are known to be involved in human metabolism (i.e., CYP1-8, CYP11, CYP17, CYP19-21, CYP24, CYP26, CYP27, CYP39, CYP46, and CYP51). Three of these CYP families CYP1 (i.e., 1A2), CYP2 (i.e., 2A6, 2B6, 2E1, 2C8, 2C9, and 2D6) and CYP3 (i.e., 3A4 and 3A5) account for about 70% of human hepatic microsomes CYPs with CYP3 accounting for approximately 30%. These CYPs are the major enzymes responsible for the metabolism of most marketed drugs. Note also that the four CYP families CYP1, CYP2, CYP3 and CYP4 account for about 99% of rat hepatic microsomes with CYP2 accounting for approximately 60%. The FMOs are a family of five enzymes, FMO1-5, located in the hepatic endoplasmic reticulum region, requiring the cofactor NADPH and are expressed to varying levels in other extrahepatic tissue. The UGT superfamily of enzymes that catalyze the conjugation of D-glucuronic acid to various polar compounds can be divided into several families and subfamilies. Human UGTs belong to 2 subfamilies UGT1and 2, which are predominantly involved in glucuronidation. The subfamily of enzymes UGT1A1, 1A4, 1A9, 2B7, and 2B15 have the greatest significance in Phase 2 elimination. The UGTs catalyzes the glucuronidation of compounds by transferring a glucuronic acid moiety from the cofactor uridine 5'-diphosphoglucuronic acid (UDPGA) to substrates and thus, forming glucuronides which are water-soluble and readily excreted *via* urine or bile.

In general, CYP, FMO and UGT reaction phenotyping assays are useful in predicting drug-drug interaction as well as identifying polymorphism enzymes that are associated with variability in drug disposition. The strategy used for reaction phenotyping involves the use of recombinant enzymes, correlation analyzes with well-characterized microsomal samples, and selective isoform inhibition. These methods can be performed separately or in combination to provide complementary and supportive information [256]. At the LO stage gate, recombinant CYP enzymes 3A4, 2B6, 2C6, 2C9, and 2D6 can be used as a rapid screen to identify the major enzymes involved in biotransformation. It is better to have compounds metabolized by several CYP450s instead of one enzyme to avoid potential polymorphism issues. At the NCE selection stage gate, all three methods are typically used.

In Vitro CYP450 Inhibition

Metabolism of xenobiotics by CYP450s plays a vital role in the clearance of compounds from the body. A general feature of CYP450 enzymes is that a single enzyme can metabolize multiple drugs. For example, when two drugs are taken *in vivo* at the same time a competition reaction between these drugs may occur for the same CYP450 enzyme. If one of the drugs inhibits the CYP450 enzyme, an elevation in the systemic concentration of the other drug will occur. This elevation in drug concentration may be uneventful or undesirable depending on the magnitude of the change in concentration of the drug at the site of pharmacological action and the therapeutic index of the drug. The therapeutic index is defined as the ratio of the highest drug exposure that results in no toxicity divided by highest drug exposure that produces the desired efficacy. If the elevated drug concentration is undesirable leading to a toxicity end-point, then a drug-drug interaction (DDI) has occurred [257]. In other words, one drug has altered the PK and/or PD of another drug. Thus, drug candidates that are strong inhibitors of CYP450s have the potential of DDI. It should be kept in mind that drug candidates that are strong inhibitors of CYP450s must always be assessed in the context of the expected *in vivo* systemic exposure of the drug candidate before a critical go/no-go DDI decision is made concerning the elimination of the drug candidate. Thus, the assessment of the potential of a drug candidate to inhibit a specific or several CYPP450s is important to understand at the LO stage gate [49, 55, 258-265].

The eight major human CYP450 isoforms are CYP1A1, CYP2B6, CYP2C8, CYP2C9, CYP2C19, CYP2D6, CYP3A4, and CYP3A5. A human CYP450 inhibition assay is designed by incubating - for a fixed period of time - an isoform-specific substrate with a range of drug candidate concentrations (i.e., 0.1 – 25 μM) in an *in vitro* hepatic system including isolated hepatocytes [266], hepatic membrane subcellular fractions coupled with an electron-generation reaction [65, 267] or purified recombinant human CYP450 enzymes coupled with an electron-generation reaction [55, 268]. At the end of the incubation, the concentration of the metabolite formed from the isoform-specific substrate at each of the drug candidate concentrations is measured *via* fluorescence [55], or LC/MS/MS technique [259]. A decrease in the concentration of the metabolite formed from the isoform-specific substrate with increasing drug candidate concentration indicates the inhibition of CYP450 by the drug candidate, which can be calculated as an IC_{50} value; that is, the drug candidate concentration that produces 50% inhibition of the formation of the metabolite from the isoform-specific substrate [269]. At the LO stage gate, the *in vitro* CYP450 inhibition of a drug candidate is typically determined using human liver microsomes as the *in vitro* hepatic system. Using this "gold standard" assay, the drug candidate's inhibition is classified as weak or no inhibition ($IC_{50} > 10$ μM), moderate inhibition (1 μM $< IC_{50} <$ 10 μM) and strong inhibition ($IC_{50} <$ 1 μM) inhibitors.

For a complete assessment of a strong inhibitor drug candidate, reversible or irreversible inhibition should be distinguished [68, 257, 270]. Reversible CYP450 inhibition reactions are caused by competitive interactions at the same CYP450 from multiple substrates and typically occur over a short time period (hours to days). Irreversible binding (i.e., covalent binding) of the drug candidate to the apoprotein or the heme-prosthetic moieties of the CYP450 enzyme or a quasi-strong complexation between the drug candidate and the CYP450 results in the inactivation of the CYP450 over a long time period (days to weeks). These time-dependent reactions processes are referred to as time dependent-inhibition (TDI). If the TDI is caused by the reaction of metabolites formed during the biotransformation of the drug candidate, the time dependent-inhibition is termed a mechanism-based inhibition (MBI). Both reversible and irreversible inhibition may be caused by MBI. All of these TDI mechanisms result in loss of CYP450 activity over time and thus, changes in the clearance of the drug candidate from the body. While reversible inhibition of CYP450 enzymes causes short-term effects on the pharmacokinetic properties of the drug candidate, irreversible inactivation of CYP450s causes long-term effects on the pharmacokinetic properties of the drug candidate, as the inactivation CYP450 enzyme must be replaced by newly synthesized enzyme. Reversible *in vitro* TDI may or may not be important clinically, and careful assessments must be made to determine if a drug candidate should advance to another stage gate or be terminated. Drug candidates that are irreversible binders to CYP450s are typically eliminated unless this type of mechanism is desired to engage the target. At the LO stage gate, there are several different experimental designs including IC_{50} shift, progress curve, and kinetic determination to determine if a drug candidate undergoes TDI/MBI [99, 257, 270, 271]. Generally, these assays determine the apparent IC_{50} of the drug candidate toward a standard substrate with and without a preincubation period of the enzyme with the drug candidate. At present all these TDI/MBI assays are used at the LO stage gate.

In general, most CYP450 oxidation reactions show hyperbolic saturation kinetics and competitive inhibition between substrates. Therefore, some PK properties can be predicted with standard Michaelis-Menten kinetic analyzes; thus, DDI can be possibly predicted from inhibition studies. However, CYP3A4 enzyme reactions have been shown to exhibit unusual enzyme kinetics that complicates the interpretation of the experimental results including auto-activation, partial inhibition and substrate inhibition [260]. Auto-activation occurs when the activator is the substrate itself, which results in a sigmoidal saturation kinetic curve. In partial inhibition, the inhibitor does not completely inhibit substrate metabolism even at saturating concentrations. Substrate inhibition happens when an increase in substrate concentration beyond a certain value results in a decrease in the rate of metabolism. Also, CYP3A4 enzyme reactions show enzyme activation. Activation is an increase in the reaction rate of one substrate in the present of another substrate. Several studies suggest that CYP3A4 is an allosteric protein and allosteric binding at two distinct sites is a main factor leading to enzymatic activation [263], although the identity of the allosteric site has never been defined. The complexity of substrate binding to CYP3A4 also reflects on the dependence of the inhibition on the probe substrate used in the assay. Dramatic differences have been found in CYP3A4 inhibition assay using different probe compounds. All these unusual kinetics of CYP3A4 complicate the inhibition assay and the reliability of DDI prediction from *in vitro* data.

In Vitro CYPP450 Induction

It has been known for many years that some CYP450 enzymes are inducible by exogenous signaling molecules [166]. The result of CYP450 enzyme induction is an increased clearance of the drug that results in a reduction of pharmacological effects of a specific drug and the attenuation the efficacy of co-administered drugs in a time-dependent manner. Also, a drug that is metabolized by the induced CYP450 enzyme to a reactive metabolite may lead to a toxicity end-point. Unlike reversible CYP450 inhibition, the time course of CYP450 induction varies accordingly to different inducers and thus, the impact of CYP450 induction on drug metabolism may not be immediate. Although CYP450 induction is an undesirable property of a drug, the overall implications of induction to drug discovery are less important at the LO stage gate. In contrast to inhibition reactions, the DDI associated with CYP450 induction is not normally considered a safety issue in most cases. CYP450 induction does not necessarily prevent an NCE from entering into clinical development.

CYP450 induction resulting in an increase in enzymatic activity of CYP450 enzymes occurs at the transcriptional level of CYP450 enzymes through activation of respective nuclear receptors genes after chronic exposure to a drug. The nuclear receptors most commonly involved in regulation of CYP450 enzymes are the aryl hydrocarbon receptor (AhR) [272, 273], the constitutive androstane receptor (CAR) [274, 275] and the pregnane X receptor (PXR) [276, 277]. Activation of AhR results in enhanced expression of CYP1A and CYP1B gene products, while activation of CAR and/or PXR results in enhanced expression of CYP3A, CYP2B, CYP2C8 and UGT1A1 gene products. While all of the nuclear receptors could be assayed, the assessment of the potential of a drug candidate to activate PXR is the most important to determine at the LO stage gate since it regulates the majority of CYP450 enzymes including CYP2B6, CYP2C8, CYP2C9, CYP2C19, CYP3A4, and CYP3A5. In addition, PXR up-regulates the transporter MDR1 that encodes the broad-specificity P-glycoprotein (Pgp) efflux pump. Stable cell lines with reporter gene constructs for PXR are commercially available, and high throughput PXP assays have been developed [277]. Briefly, DPX2:HepG2 cells stably transfected with the PXRE and the CYP3A4 promoter fused to a luciferase gene are seeded in a 96-well plate and then treated 24 hours later with a drug candidate. These treated cells are incubated for 24 hours at 37°C and then assayed using the Bright-Glo reagent. The ability of a drug candidate to activate hPXR is expressed relative to its level of cytotoxicity [99, 276, 277].

The current "gold standard" for assessment of the potential of a drug candidate to induce CYP450 enzymes uses isolated human hepatocytes [278]. There has been substantial progress in human hepatocyte isolation and culture methods since the early reports published in the 1980s [166]. At the present, fresh primary human hepatocytes and cryopreserved primary human hepatocytes have been widely used for the evaluation of CYP induction in humans. Enzymatic activity assay has been a common detection method for CYP induction. Similar to the traditional inhibition assays described above, it measures the induction by quantifying the metabolite of a CYP-specific model substrate generated by the treated hepatocytes. In a typical assay, primary hepatocytes are first treated with drug candidates for 72 hours. Model substrates are then added to the treated cells. CYP induction leads to higher enzyme activity and generates more metabolites. Quantification of metabolites has been achieved by using LC/MS/MS. Cryo-preserved cells are less favorable in these assays due to the attachment of cells to plate surfaces and the loss of certain cellular functions [245, 246]. The scarcity of fresh human hepatocytes has demanded increasingly sensitive detection protocols using smaller numbers of cells in CYP induction assays.

In Vitro Assays for Membrane Transporters

The ATP-binding cassette (ABC) and the solute carrier (SLC) family of genes are responsible for a variety of brain, intestinal, hepatic, and renal membrane transporters that play a crucial role in drug efficacy and safety [279]. For example, the ABC superfamily is composed of ATP-driven pumps responsible for the efflux of many drugs across biological membranes including the blood-brain barrier where they act to prevent drugs from entering the brain, the intestinal epithelium where it limits absorption, the bile canalicular membrane of hepatocytes, and the brush border membrane of renal epithelial cells. P-glycoprotein (Pgp) [280] - the product of the ABCB1 gene - and breast cancer resistance protein (BCRP) [281] - the product of the ABCG1 gene - are expressed widely in these cell types, interacts with diverse classes of substrates and inhibitors and are overexpressed in many cancers. The multidrug resistance-associated protein 2 (MRP2) [282] - the product of the ABCC2 gene - is an organic anion and large conjugate compound transporter located in the intestine, liver, and kidney. The SLC superfamily is composed of over 300 proteins organized into 52 families. The organic anion transporting polypeptides (OATPs) family is responsible for the uptake of a wide range of compounds [283]. The OATP1B1 and the OATP1B3 transporters are predominately expressed at the sinusoidal membrane of hepatocytes. The renal organic anion transports (i.e., OAT1, OAT3) and the organic cation transporter 2 (OCT2) are primarily expressed in the proximal tubule epithelial cells of the kidney [284]. While all of the transporters could be assayed [99], the assessment of the potential of a drug candidate to be a substrate or inhibitor of P-gp is the most important to determine at the LO stage gate. The most widely used *in vitro* assays includes the Caco-2 cell line, the overexpressed P-gp porcine kidney cell line (LLC-K1), and the overexpressed P-gp MDCK cell line [99]. The *in vivo* P-gp assay uses P-gp knockout mice. At the LO stage gate, the *in vitro* P-gp results should be used in conjunction with TDI results; for example, if a compound demonstrated reversible TDI and were also a substrate or inhibitor of P-gp, these combined liabilities of the compound increases its potential for a DDI and the compound could be terminated for a compound without these liabilities.

"fast" In Vivo PK Screens

Pharmacokinetic profiles, in particular absolute bioavailability, elimination half-life, clearance, and distribution of volume, have become an integral part of many drug discovery programs [56, 64, 219, 119, 220, 279-298]. In orally dosed animal experiments, an effective approach for profiling a large number of drug candidates is to first minimize the in-life phase of the experiment by reducing the number of animals, the number of plasma collections and the overall collection time. For both orally and intravenously administered drugs, 6 animals and 5- to 6-time points per animal over a collection period of 6- to 8-hrs. are sufficient to profile most drug candidates. Secondly, the bioanalytical phase is minimized by the use of abbreviated method development schemes [57, 60, 105, 285, 286] and abbreviated calibration procedures. By using automated procedures to handle plasma preparations [287], and rapid LC/MS/MS techniques to perform *in vivo* studies, it is possible to measure the oral bioavailability and other pharmacokinetic parameters within a time frame that is consistent with the iterative cycle of a typical drug discovery research project [286-291]. At the LO stage gate, this is the preferred method to collect *in vivo* PK data.

Cassette dosing techniques [50, 176] and sample pooling methods [297, 298] have also been used to increase PK throughput by dosing multiple drug candidates to animals in a single dose or by combining single dosed plasma samples across similar or all time points, respectively. Retrospective investigations of cassette

(N-in-1) techniques and discrete sampling methods have been completed [220]. Potential drug-drug interactions that could result from dosing multiple drug candidates and potentially could hinder PK interpretations were found not to occur as frequently as had previously been expected. A comparison of 31 compounds in several species revealed that the systemic clearance, volume of distribution, oral bioavailability, and renal excretion were significantly correlated with N-in-1 dosing techniques and discrete studies.

In vitro Toxicity Assays

Drug-induced toxicity is another major factor contributing to high attrition rates of NCEs in clinical studies since predictions of risk in humans based upon preclinical *in vitro* or *in vivo* results obtained from cell lines and animal models, respectively, are not highly correlated, in many cases, to human toxicity [34, 54, 71, 81, 90, 93, 96, 279, 299]. Most *in vitro* and *in vivo* toxicity models have varying degrees of accuracy with the correlation between human and animal toxicities being best for genotoxicity, carcinogenicity, hematological, and gastrointestinal diseases; the poorest correlation being between human and animal toxicity models are for adverse drug reactions (ADRs) such as hypersensitivity, cutaneous reactions, and hepatotoxicity [300]. It should be remembered that *in vitro* cell toxicity assays are far removed from *in vivo* human toxicities and one should expect a significant amount of false-positive and false-negative data from *in vitro* toxicity assays. At the LO stage gate, the best strategy to uncover drug candidate toxicity liabilities for limited range of toxicities is to use a panel of well-characterized *in vitro* toxicity screens around a particular type of toxicity in parallel along with efficacy and ADME/PK optimization to identify drug candidates with the best overall profile. For example, accepting the fact that there is a poor correlation between human and animal models for ADR toxicity, a practical strategy for evaluating drug candidates for potential ADRs at the LO stage gate is the reactive intermediates panel screening approach outlined in Fig. (**2**). Here several *in vitro* assays are used for the detection of reactive intermediates that could be generated by biotransformation reactions of the drug candidate. The panel approach is proposed largely based on the hypothesis that reactive intermediates derived from drug candidates are mechanistically involved in ADRs by undefined mechanisms such as covalent binding of reactive intermediates to macromolecules and the impact of reactive intermediates on cellular functions. While this approach will be imperfect in detecting ADRs, it should help to minimize or to identify drug candidates that can be potentially bioactivated to reactive intermediates. A similar panel screening approach at the

LO stage gate is shown in Fig. (**2**) for other toxicities including genotoxicity, oxidative stress, and cytotoxicity.

In Vitro ADR: Reactive Intermediates (Hard and Soft Electrophiles, and Radials)

Reactive intermediates such as biotransformation derived electrophilic metabolites (Phase 1 or 2), and radicals may cause some types of hepatotoxicity, liver failure or other ADRs. To help medicinal chemists to avoid reactive intermediates, a comprehensive listing of many structural motifs used by medicinal chemists for drug design that potentially could form reactive intermediates has been published [301]. *In vitro* screening and identification of reactive metabolites are normally conducted using liver microsomes that contain both Phase 1 drug metabolizing enzymes such as cytochrome CYPP450s , FMO, dehydrogenases and oxidases and Phase 2 enzymes (i.e., UPD-glycosyltransferases) [302]. In microsomal incubations, drugs can be metabolized to stable metabolites or a variety of Phase 1 reactive metabolites. Phase 1 reactive metabolites can be broadly classified as either electrophiles or free radicals. Electrophiles are electron-deficient molecules and thus, covalently reactive with molecules that are electron-rich (i.e., nucleophiles). Electrophiles and nucleophiles can be broadly classified as "hard" or "soft". Hard electrophiles react with hard nucleophiles (i.e., basic groups in DNA and lysine residues in proteins). The cyanide anion (CN-), which is a "hard nucleophile, has been used to trap hard reactive metabolites [302]. This technique has been used to detect reactive iminium intermediates in the metabolism of alicyclic amines by trapping them with sodium or potassium cyanide. Soft electrophiles react with soft nucleophiles (i.e., the thiol groups in GSH and the thiol groups in cysteine residues in proteins). A majority of electrophilic metabolites can be classified as "soft" reactive metabolites that readily react with a sulfhydryl group of γ-glutamylcysteinylglycine (GSH). As an endogenous scavenger of chemically reactive species, GSH has been widely used in microsomal incubations to trap several structural classes of reactive metabolites, which include quinones, quinone imines, iminoquinone methides, epoxides, arene oxides and nitrenium ions. Formed GSH conjugates can be structurally characterized using tandem mass spectrometry, and a common MS method is neutral loss scan since all GSH adducts display a neutral loss of 129 Da in collision-induced dissociation. A highly efficient method is to use stable-isotope trapping combined with electrospray ionization (ESI)-tandem LC-MS/MS neutral loss scanning for detecting and identifying reactive intermediates [77, 81, 89, 99, 303-305]. Radicals are reactive metabolites that possess unpaired electrons. Formation of radicals can potentially lead to oxidative damages to DNA and proteins, and thus results in drug-induced toxicities. For drug candidates containing carboxylic acid groups, Phase 2 reactions to generate acyl glucuronides are common. There are several assays available to determine the reactivity of acyl glucuronides [99]. The use of electron spin resonance (ESR) spectroscopy and spin-trap reagents for trapping *in vitro* and *in vivo* reduced and oxidized radicals in biological fluids are well-established techniques [305]. Typically, radicals abstract hydrogen atoms from macromolecules rather than becoming covalently bound. It is clear from the above discussion that a combination of several trapping methods is necessary to evaluate the formation of reactive intermediates. While all of the reactive intermediate assays could be run, the assessment of the potential of a drug candidate to produce reactive metabolites using the stable-isotope trapping assay [77] is the most important to determine at the LO stage gate.

In Vitro Genotoxicity: DNA Binding

The covalent binding of the drug candidate or reactive metabolites to DNA can be assessed in microsomal incubations in the presence of DNA. Resulting DNA-drug complex can be detected by ^{32}P-postlabeling analysis [306]. The method includes enzymatic digestion of DNA sample to the adducted nucleoside 3'-monophosphates and partial purification of the adducted nucleotides, followed by the 5'-labeling with ^{32}P. For analysis of DNA adducts, polyethyleneimine-cellulose TLC plates are generally utilized to resolve ^{32}P-labeled DNA adducts (^{32}P-postlabeling/TLC analysis). This technique is reliable but time-consuming and labor intensive. An improved approach has been developed to increase the throughput of the post-labeling analysis. Non-denaturing 30% polyacrylamide gel electrophoresis (PAGE) has been adapted for the ^{32}P-postlabeling analysis (^{32}P-postlabeling/PAGE analysis) to expedite the analyses [307]. The major advantages of this technique are: (*a*) many DNA samples can be loaded concomitantly on the PAGE along with standard markers, (*b*) DNA adducts can be resolved in only a few hours, and (*c*) exposure to ^{32}P during the handling can be minimized. The detection limit for both ^{32}P-postlabeling/TLC and ^{32}P-postlabeling/PAGE analyses is approximately 7 adducts/10^9 nucleotides. HPLC in-line methods with a radioisotope detector can be used (^{32}P-postlabeling/HPLC analysis) to increase the resolution and detection limit (approximately 3 adducts/10^{10} nucleotides) of DNA adducts [308].

In Vitro Genotoxicity: Mutagens and Carcinogens Assays

Ames tests for mutagens are some of the most commonly performed toxicity assays in the pharmaceutical industry [309]. *Salmonella typhimurium* (*S.*

typhimurium) and *Escherichia coli* (*E. coli*) reverse-mutant bacterial strains are used to detect point mutations in DNA caused by the presence of a parent drug candidate or its metabolites. *In–vitro* metabolic activation of the parent drug candidate is usually achieved with Aroclor treated rat-liver S9 enzymatic fractions. Point mutations may take the form of base pair substitutions or from small insertions or deletions in the bacterial genome. In some cases, the mutations may be caused by the enzymatic biotransformation of the parent drug candidate to electrophilic species that are capable of covalently binding to DNA. The bacterial reverse-mutation test uses several strains carrying mutations in the operon coding for histidine biosynthesis. If the bacteria are exposed to a parent drug candidate or its metabolites that induce point mutations, reverse mutation from histidine auxotrophy to histidine prototrophy occurs and restores the functional capability of the bacterial to synthesize this essential amino acid. While there have been large amounts of data collected using the Ames test [310], the traditional full format Ames test requires a significant amount of drug substance and is resource intense. The Ames II assays is a liquid microtiter modification of the traditional Ames test that consumes less drug substance, is less resource intense and offers higher sample throughput [311]. The Ames II assay appears to be a reasonable alternative to the traditional Ames test.

It is important to remember that the various mutagenicity assays detect different genetic alterations, and, therefore, drug candidates can give uniformly positive and negative results in the various tests. Also, positive *in vitro* results may arise due to experimental artifacts such as changes in pH, or osmolality or impurities in the sample. Typically in drug discovery, the drug candidate is synthesized in small batches throughout the course of the project. It has been our experience that these batches can give different Ames results presumably due to different batch-to-batch levels of impurities. The interpretation of at least two different types of mutagenicity assays is important in making successful go/no-go decisions at the LO stage gate.

In Vitro Oxidative Stress: Reactive Oxygen Species

Reactive oxygen species (ROS) are by-products primarily from mitochondria cellular metabolism processes [312]. Typically, these ROS are free radicals that are removed by reactions with antioxidant enzymes including catalase, superoxide dismutase, and GSH peroxidase or non-enzymatic antioxidants including flavonoids, ubiquinone, GSH, and several vitamins (i.e., A, E and C). If the antioxidant defense system cannot remove these free radicals for some reason, then they can potentially react with cellular macromolecules such as lipids, proteins, and DNA to either cause damage or cell death. The present of ROS activity can be assumed from detecting oxidatively damaged lipids, proteins, or DNA as well as from ROS trapping experiments and by measuring the depletion of enzymatic/non-enzymatic antioxidants [313]. There is a variety of fluorometric assays with medium throughput screening capability that can be used to detect oxidative stress [314, 315]. At the LO stage gate, a widely used fluorometric probe is 2', 7'-dichloro dihydrofluorescein diacetate (H_2DCF) that can be used to detect ROS [313].

In Vitro Cytotoxicity: Hepatotoxicity Assays

A drug-induced liver injury has been the most common reason for the withdrawal of approved drugs. Therefore, evaluation of hepatotoxicity is very critical in drug discovery. A robust hepatotoxicity screen using a series immortalized cell lines and primary hepatocytes cultures have been described [316]. The immortalized cell system originally derived from normal human hepatocytes consisted of a series of four CYP450 enzyme-transfected cell lines (i.e., CYP3A4, CYP2C9, CYP2C19, and CYP2D6) and one non-metabolizing parental cell line. The CYP450-transfected cell lines were shown to metabolize specific substrates and generate metabolite profiles similar to those produced by human microsomes and supersomes while the parent cell line had no significant Phase 1 or 2 metabolizing capabilities. Drug candidates solubilized in DMSO were exposed for 20 hours to confluent cultures. After this time period, IC_{50} values were determine using a 12-point dose-response curve. The toxic endpoint was a decrease in the energy status of the cells measured by loss of mitochondrial function using an MTS tetrazolium salt assay [317] or an ATP assay [318]. Using a total of 679 marketed drugs, the assay correctly predicted 585 out of 587 non-hepatotoxic drugs, 15 out of 21 severely hepatotoxic drugs, and 51 out of 71 variably hepatotoxic drugs. From this validation study, a drug candidate with an IC_{50} values of <50 μM in any of the five cell lines was interpreted to have an increased potential risk for clinical hepatic liabilities.

NCE SELECTION STAGE GATE (LATE PHASE DRUG DISCOVERY)

The primary goal at the NCE selection stage gate is to ensure that the final NCE selected for a Phase I clinical trial will not expose humans to unreasonable health risks. Once the selection criteria for efficacy/potency, ADME, PK, and toxicity have been defined and accomplished at the LO stage gate, several drug candidates are promoted to the final NCE selection stage gate. Selected drug candidates will receive more extensive *in vivo* animal efficacy/PK, and toxicity/TK testing in order to select the final NCE for a Phase I clinical trial. Typically, the final NCE's preclinical IND package will include all the physicochemical characteristics of the NCE, manufacturing information to ensure that the NCE can be adequately synthesized with no batch-to-batch purity variations, *in vivo* efficacy data, *in vitro* ADME properties, *in vivo* metabolite profiles, *in vivo* drug and metabolite distribution, *in vivo* human PK predictions, *in vitro* toxicity profiles and *in vivo* acute and chronic toxicity evaluations of the NCE. However, the exact type of *in vitro* and *in vivo* data profiles necessary for the preclinical IND package varies depending on the therapeutic target [319]. For example, there are a large variety of *in vitro* and *in vivo* toxicity tests available in the literature including endogenous hormone response pathway screening, genotoxicity, oxidative stress, hERG inhibition, endoplasmic reticulum stress, mitochondrial toxicity, heat-shock response, apoptosis, cell growth and proliferation, DNA damage and repair, and screening panels of safety-related targets including GPCRs, transporters, ligand-gated ion channels and enzymes. Unfortunately, there is little consensus on what *in vitro* and *in vivo* toxicity assays should be tested. Keeping this in mind, a list of a few common toxicity assays used at the NCE selection stage gate will be discussed in the next sections (Figs. **1** and **2**).

Chemical Synthesis Scale-up

Since larger quantities of drug candidates are required for many of the *in vivo* and *in vitro* assays at the NCE selection stage gate, this is a good point to involve the development group that has the responsibility of scaling up drug substances for future clinical studies. Scale-up chemistry problems associated with the drug candidates at the NCE selection stage gate should be solved before the compound enters clinical development. It is extremely important that there are no batch-to-batch purity variations. Assuming scale-up and batch-to-batch purity issues will be corrected in clinical development can be risky.

It is important to determine the drug candidate's chemical state (i.e., solid or liquid), its *in vivo* route of administration, and the formulation that is necessary to obtain the maximum drug exposure for *in vivo* and *in vitro* toxicity studies [320-322]. At the final NCE selection stage gate, it is necessary to investigate a number of different formulation conditions and salt forms of the drug candidate. Typically, changes in the pH of the oral or intravenous (iv) formulations are tried first followed by solubilizing agents, such as solutol, methylcellulose, propylene glycol, or emulsion vehicles, such as sesame or corn oil or nano-suspension formulations [84, 231, 320-324]. The potential impact of a poor formulation on obtaining accurate PK parameters is well understood and appreciated since it directly affects efficacy and toxicity studies [325, 326]. Therefore, a good formulation must be established as early as possible at the NCE selection stage gate. In addition, the complete physicochemical characteristics for the solid form of the drug candidate including salt type, LogP, pKa, LogD, $LogS_0$, stability, solubility, polymorphism, hygroscopicity, melting point, and particle size, to name a few, or the liquid form including primarily solubility and shelf-life stability will also be required in the IND package of the final NCE.

In Vivo ADME/PK

A variety of *in vitro* ADME properties need to be re-investigated for drug candidates using larger panels of specific assays outlined above to estimate *in vivo* ADME properties. For example, *in vitro* drug inhibition, induction, CYP450 and UGT phenotyping, and major membrane transporters play a crucial role in drug efficacy and safety and needs to be fully investigated. Plasma protein/tissue binding studies are used many times to explain pharmacodynamic data since this type of binding effects drug concentrations at the active sites [190, 191, 327, 328]; for example, *in vivo* microdialysis sampling techniques have been used effectively to understand drug concentrations in the brain and to relates these concentrations to central nervous system (CNS) effects [329]. *In situ* absorption models have been shown to predict accurately human oral absorption for both passive and carrier-mediated transport mechanisms of drugs [218, 224, 239, 330, 331]. *In vivo* metabolite profiling can be used to establish appropriate toxicity species [65, 74, 77, 332-334]. *In vivo* drug and metabolite distribution studies need to be completed to help with the interpretation of efficacy and toxicity studies. In addition, single low and high dosed PK parameters are determined to understand exposure levels, to establish toxicity dosing levels, and to establish the therapeutic index of drug candidates. Information on pharmacokinetics-pharmacodynamic (PK/PD) relationships can be obtained during this process [335].

In estimating *in vivo* human PK parameters (%F, %Fa, CL_T, Vd_{ss}, and $t_{½}$) allometric scaling [76] and PBPK modeling [186-188, 336, 337] techniques are used at the NCE stage gate. The traditional approach uses allometric scaling techniques to estimate *in vivo* human PK parameters [76]. Allometric scaling technique assumes that PK parameters can be related to the body mass raised to a constant power (i.e., fixed allometric exponent). Thus, by measuring a PK parameter using different species with different body masses (i.e., mice, rat, dog, and monkey), the fixed allometric exponent can be determined from the animal studies, and the human PK parameter can be estimated by extrapolating the allometric equation to the body mass of a human using the fixed allometric exponent. Typically, three species are used to determine the fixed allometric exponent; however, allometric scaling has been investigated using only one species. With only *in vivo* rat PK data to predict *in vivo* human PK data, three simple allometric scaling relationships were derived [76].

- CL_T (Human) $\approx$ 40 CL_T (Rat) (L/hr)
- Vd_{ss} (Human) $\approx$ 200 Vd_{ss} (Rat) (L)
- $t_{½}$ (Human) $\approx$ 4 $t_{½}$ (Rat) (hr)

The average-fold error for human CL_T predictions for N = 176 drugs was 2.25-fold with 79% of the drugs having a fold error less than 3-fold. The average-fold error for human Vd_{ss} predictions for N = 144 drugs was 1.85-fold with 84% of the drugs having a fold error less than 3-fold. The average fold error for human $t_{½}$ predictions for N = 145 drugs was 2.05-fold with 76% of the drugs having a fold error less than 3-fold. The PBPK modeling technique can be considered a more mathematically advanced model of the allometric scaling technique and their *in vivo* human PK predictions are typically similar. Since PBPK modeling explicitly incorporates physiological data and *in vitro* ADME data, the average-fold error for *in vivo* human PK parameters is typically <2-fold for 70% to 90% for the drugs that have been tested [336, 337]. While both methods can be used to predict *in vivo* human PK data, the PBPK method is preferred at the NCE selection stage gate since it also provides information on the initial human dose for a Phase I study. Unfortunately, the PBPK method appears to be currently underutilized at the LO stage gate where it could have a greater impact on the selection of drug candidates.

In Vitro and *In Vivo* Toxicity/TK

The assessment of potential toxicity of drug candidates and its metabolites is accomplished by using a variety of *in vitro* cellular assays and *in vivo* animal studies (i.e., mice, rats, hamsters, guinea pigs, rabbits and dogs) to aid in the final NCE selection stage gate (Fig. **2**). Pivotal toxicities studies are typically conducted under good laboratory practices (GLP) with all animal studies being approved by local animal ethics committees. Toxicity studies can be broadly classified as genetic toxicology (i.e. gene mutation assays), general toxicology (i.e., single and repeated dose studies), safety pharmacology (i.e., central nervous, cardiovascular, and respiratory systems), developmental and reproductive toxicology (i.e., fertility and teratology), and clinical pathology (i.e., serum chemistry, hematology, histopathology, and urine analyses).

For genetic toxicology, the Ames test [309, 338] and the mouse lymphoma assay (MLA) [339, 340] for mutagens are routinely conducted at the LO and NCE selection stage gates. The Ames test is a *Salmonella*/microsome bacterial mutagenicity assay, which is used worldwide. When drug candidates are found to be Ames-positive mutagens, in many cases, they also are discovered to be mammalian carcinogens. While this correlation is very good, it is not absolute since the Ames test relies on metabolic activation. Many drug candidates remain Ames-negative until the enzymatic transformation of the compound to electrophilic metabolites capable of covalently binding to DNA. Thus, metabolic activation is a critical step in this assay. The MLA quantifies genetic alterations by monitoring changes in the expression of the thymidine kinase gene. In some cases, the MLA assays uses cell lines that are heterozygous and it detects forward mutation of the wild-type cell line. Also, the MLA assay can be used for photo-mutagenicity evaluations. There are several other *in vitro* mammalian cell gene tests that can be used to detect drug-induced gene alterations – for example, the *in vivo* and *in vitro* Comet assay and reporter gene systems [99]. Drug candidates that are found positive for the Ames test and/or the mouse lymphoma assay are usually terminated.

For general toxicology, single and repeated dose rodent (mouse, rat) PK studies are conducted to establish oral and iv exposures, CL_T, Vd_{ss}, and half-life values of the drug candidate. Typically, the repeat-dosed is selected to continue up to 10 or to 100 times the efficacy dose. In some laboratories, an abbreviated repeat-dosed PK study is performed in rats and dogs at the NCE selection stage gate (i.e., 3- to 5-days) [341]. These abbreviated repeat dosed studies help to establish the criteria for the 1 to 6 months toxicity studies required for the final NCE's IND package. From single and repeat PK studies, acute toxicity studies are established where a toxic dose of the drug candidate is administered in one or more doses within a 24-hour period. *In vivo* acute toxicity studies are required in at least one species of animal providing the TK parameters, preliminary identification of any target organs of toxicity, any delayed toxicity, and the estimation of the starting dose for Phase I human studies [342].

For safety pharmacology assays such as cardiovascular (CV) and central nervous system (CNS), the adverse effects of drug candidates on these systems at the highest possible dose are determined at the NCE selection stage gate [343-353]. Clinicians have used the QT interval of an electrocardiogram (ECG) as a surrogate marker in animals and humans to predict adverse CV effects. This measurement became extremely important to the pharmaceutical industry in the late 1990s when marketed drugs such as mibefradil, astemizole, cisapride, and terfenadine were withdrawn from the market due to a potentially fatal cardiac arrhythmia called Torsades de Pointes that was apparently associated with QT prolongation [343, 344]. The QT interval of duration, which is part of the repolarization of the heart, is measured in milliseconds. Typically, a 5 to 10 milliseconds prolongation of the QT interval is used as a cut-off to "red flag" potentially risky drug candidates. A major factor that influences the elongation of the QT interval is the drug-induced inhibition of a potassium channel encoded by the human ether-a-go-go-related gene (hERG) [345-350]. This potassium channel is essential to the maintenance of normal cardiac function. Due to the seriousness of the problem, the use of *in vitro* hERG and *in vivo* QT interval prolongation assays as a surrogate for Torsades-type dysrhythmias is now required by regulatory authorities worldwide. CNS testing is typically evaluated using blood-brain-barrier assays [351, 352] and the Irwin test [353]. The Irwin test is a battery of neurobehavioral observations that are used in rats to determine potential neurotoxicity. Both the hERG/QT and Irwin assays are run at the LO stage gate, and drug candidates that fail these tests are usually terminated.

For developmental and reproductive toxicology, there are several *in vitro* assays that have been validated for the assessment of embryotoxicity including the embryonic stem cell test, the rat whole embryo culture assay and the micromass assay [354]. These types of assays are performed on the NCE that is entering Phase I clinical trials. Readers are referred to the literature for more details on these assays [355, 356]. Drug candidates that fail these tests are usually terminated.

For clinical pathology, a variety of cellular toxicity [357, 358] and hepatotoxicity endpoints are evaluated [359, 360] depending on the project. The characterization of clinical chemistry endpoints in animal models is helpful in detecting acute hepatotoxicity [361-363]. In addition, the histological examination of biopsies from animal toxicity studies to shed light on toxicity mechanisms is typically completed prior to the final selection of the NCE. Depending upon the severity of the clinical pathology results, drug candidates may be terminated.

CONCLUDING REMARKS

Medicinal chemists are faced with the tremendous challenge of synthetically balancing ADME, toxicity, PK and PD deficiencies in drug candidates using modern organic chemistry strategies. It is well understood that this challenge is not a trivial process. A compromise between these properties will typically be the goal rather than obtaining optimal properties for each. In many cases, even a compromise will not be attainable. Thus, understanding the limitations of preclinical assays that are used for drug-design optimization strategies is a must for all medicinal chemists. This review has focused on examining ADME and toxicity assays used at the three main stage gates (i.e., H2L, LO, and the final NCE selection) in a target-based drug discovery screening approach to drive the overall process to produce superior Phase I clinical candidates. Drug discovery preclinical ADME and toxicity information produced at these stage gates utilize *in silico* computational methods, *in vitro* high/low throughput assays that are within a timeframe that is consistent with the iterative cycle of a typical research project, and *in vivo* animal models (Figs. **1** and **2**). Based on the author's experiences, *in silico* ADME and toxicity predictions vary greatly. Their use in making go/no-go decisions in drug discovery has been limited due to this lack of predictability. However, ML classification methods [197] and ML-QSAR models [146] whereby drug candidates are classified into high/low ADME and toxicity properties based on physiological relevant cutoff values appears to have significant potential in improving *in silico* predictability [200-204]. The major constraint of most *in silico* methods continues to be the selection of appropriate molecular descriptors of drug candidates that fully represent their ADME and toxicity properties.

The traditional *in vitro* assay and *in vivo* animal models that are used to predict human *in vivo* ADME/PK properties of drug candidates were reviewed. These ADME/PK assays work reasonably well but have limitations in predicting *in vivo* human ADME/PK properties. A recent publication on R&D metrics and trends, from four major pharmaceutical companies, showed the clinical outcomes for 812 oral small-molecule drug candidates that were nominated to NCE status between the years 2000 and 2010 [364]. From this study, about 16% of the NCEs failures in Phase I clinical studies were terminated due to ADME/PK issues and about 1% in Phase II. Thus, ADME/PK defects remain a significant contributor to the attrition rates of NCEs in clinical trials being approximately on the same order as the 11% observed in the year 2000 but much lower than the 40% observed in the year 1991 [21]. There continues to be considerable interest in the literature to improve ADME/PK predictions by developing superior human *in vitro* cell-based systems that maintain *in vivo*-like cell environments in regards to cell-cell signaling pathways. Examples of new *in vitro* cell-based system approaches include developing three-dimensional (3D) *in vitro* systems to mimic natural organ or tissues [365], 3D multilayer spheroid models that more accurately replicates *in vivo* barriers to drug distribution [366], and cell types derived from human induced pluripotent stem cells to overcome limitations of primary human cells that are utilized in *in vitro* assays [367].

A selected range of *in vitro* and *in vivo* toxicity assays that are used by most pharmaceutical companies at the LO and NCE selection stage gates were reviewed. These *in vitro* toxicity assays are primarily directed toward a specific anticipated toxicity endpoint, and they have prevented drug candidates with these types of deficiencies from advancing into clinical studies. However, a recent publication on R&D metrics and trends, from four major pharmaceutical companies, showed that about 38% of the NCEs failures in Phase I clinical studies were terminated due to toxicity events and about 25% in Phase II [364]. Thus, a major contributor to the attrition rates of NCEs clinical trials continues to be related to drug toxicity being approximately on the same order as the 35% observed in the year 2000 and the 21% observed in the year 1991 [21]. While specific anticipated toxicity endpoint assays need to be run at the LO and NCE selection stage gates, the continuing high attrition rates of NCEs in clinical studies due to toxicity events highlights the need for the pharmaceutical industry to move beyond this traditional approach to screening methods that are based on undefined toxicity endpoints. One *in vitro* approach toward this goal is the application of high-content cell-based screening to investigate the effects of drug candidate concentration levels on integrated cellular signaling [368]. However, detecting cellular signaling changes at low concentration levels of drug candidates will need to be validated to determine that these signaling changes represent a toxicity event in humans. One *in vivo* approach that has been suggested to reduce toxicity attrition rates in clinical trials involves using nano-suspension formulation techniques to evaluate more drug candidates using acute toxicity dosing animal models at the LO stage gate [369].

Finally, understanding the limitations of ADME/PK and toxicity assays that were developed in the past will assist current scientists in their correct utilization and future scientists in developing superior *in silico* computational methods, *in vitro* assays, and *in vivo* animal screens that accurately predict *in vivo* human data and hopefully, increases the overall pharmaceutical R&D productivity.

CONFLICT OF INTEREST

The author confirms that the chapter contents have no conflict of interest. The views expressed here are solely those of the author and do not reflect the opinions of Janssen Research & Development, LLC.

ACKNOWLEDGEMENTS

The author gratefully thanks Z. Yan, J.A. Masucci, G. Leo, W. Lang, D.M. Ritchie, W. Tang, M. Dasgupta, B. Hasting, W. Hageman, W.N. Wu, W.J. Jones, J. Hall, M. Noureddine, and C. Cotto for their excellent scientific insight and candid opinions.

DISCLOSURE

This is an updated Chapter of the earlier review article [by Caldwell, G. W.; Yan, Z.; Tang, W.; Dasgupta, M.; Hasting, B. "ADME Optimization and Toxicity Assessment in Early- and Late-Phase Drug Discovery" Curr. Topics. Med. Chem., 2009, 9(11), 965-980], with additional text and references.

REFERENCES

[1] Ng, R. *Drugs: From Discovery to Approval,* 3rd ed.; John Wiley & Sons Inc.: New York, **2015**.

[2] Berry, I.R.; Martin, R.P. *The Pharmaceutical Regulatory Process,* 2nd ed.; Informa Healthcare: New York, **2008**.

[3] Munos, B. Lessons for 60 years of pharmaceutical innovation. *Nat. Rev. Drug Discov.*, **2009**, 8(12), 959-968.

[4] Scannell, J.W.; Blanckley, A.; Boldon, H.; Warrington, B. Diagnosing the decline in pharmaceutical R&D efficiency. *Nat. Rev. Drug Discov.*, **2012**, 11(3), 191-200.

[5] Nicolaou, K.C. Advancing the drug discovery and development process. *Angewandte Chemie, Int. Ed.*, **2014**, 53(35), 9128-9140.

[6] Eder, J.; Sedrani, R.; Wiesmann, C. The discovery of first-in-class drugs: origins and evolution. *Nat. Rev. Drug Discov.*, **2014**, 13(8), 577-587.

[7] Paul, S.M.; Mytelka, D.S.; Dunwiddie, C.T.; *et al.* How to improve R&D productivity: the pharmaceutical industry's grand challenge. *Nat. Rev. Drug Discov.*, **2010**, 9(3), 203-214.

[8] US Department of Health and Human Services. US Food and Drug Administration (FDA) http://www.fda.gov/Drugs/DevelopmentApproval Process/HowDrugsareDevelopedandApproved/DrugandBiologicApprovalReports/ANDAGenericDrugApprovals/default.htm/ [Accessed on: 19th June, **2015**).

[9] Prentis, R.A.; Walker, S.R. Pharmaceutical innovation by the seven UK-owned pharmaceutical companies (1964-1985). *Br. J. Clin. Pharmacol.*, **1988**, 25, 387-396.

[10] DiMasi, J.A. Risks, regulation, and rewards in new drug development in the United States. *Reg. Tox. Pharmacol.*, **1994**, 19, 228-235.

[11] DiMasi, J.A. Success rates for new drugs entering clinical testing in the United States. *Clin. Pharmacol. Ther.*, **1995**, 58, 1-14.

[12] Kennedy, T. Managing the drug discovery/development interface. *Drug Discov. Today*, **1997**, 2, 436-444.

[13] DiMasi, J.A. New drug development in the United States from 1963 to 1999. *Clin. Pharmacol. Ther.*, **2001**, 69(5), 286-96.
[14] DiMasi, J.A.; Hansen, R.W.; Grabowski H.G. The price of innovation: new estimates of drug development costs. *J. Health Econ.*, **2003**, 22(2), 151-85.
[15] Kola, I.; Landis, J. Can the pharmaceutical industry reduce attrition rates? *Nat. Rev. Drug Discov.*, **2004**, 3(8), 711-715.
[16] Houston JG, Banks MN, Binnie A. Case study: impact of technology investment on lead discovery at Bristol-Myers Squibb, 1998-2006. *Drug Discov. Today*, **2008**, 13(1-2), 44-51.
[17] Pammolli, F.; Magazzini, L.; Riccaboni, M. The productivity crisis in pharmaceutical R&D. *Nat. Rev. Drug Discov.*, **2011**, 10(6), 428-438.
[18] Dimitri, N. An assessment of R&D productivity in the pharmaceutical industry. *Trends Pharmacol. Sci.*, **2011**, 32(12), 683-685.
[19] Bunnage, M.E.; Getting pharmaceutical R&D back on target. *Nat. Chem. Bio.,* **2011**, 7:335-339.
[20] Scannell, J.W.; Blanckley, A.; Boldon, H.; Warrington, B. Diagnosing the decline in pharmaceutical R&D efficiency. *Nat. Rev. Drug Discov.*, **2012**, 11(3), 191-200.
[21] Khanna, I. Drug discovery in pharmaceutical industry: productivity challenges and trends. *Drug Discov. Today*, **2012**, 17(19-20), 1088-1102.
[22] Southan C, Varkonyi P, Boppana K, *et al.* Tracking 20 Years of Compound-to-Target Output from Literature and Patents. *PLOS ONE* **2013**;8(10): e77142
[23] Ringel, M.; Tollman, P.; Hersch, G.; Schulze, U. Does size matter in R&D productivity? If not, what does? *Nat. Rev. Drug Discov.*, **2013**, 12(10), 901-902
[24] Cumming JG, Finlay MRV, Giordanetto F, *et al.* Potential strategies for increasing drug-discovery productivity. *Future Med. Chem.*, **2014**, 6(5), 515-527.
[25] Schulze, U.; Baedeker, M.; Chen, Y.T.; Greber, D. R&D productivity: on the comeback trail. *Nat. Rev. Drug Discov.*, **2014**, 13(4), 331-332.
[26] Cook D, Brown D, Alexander R, *et al.* Lessons learned from the fate of AstraZeneca's drug pipeline: a five-dimensional framework. Nat. Rev. Drug Discov., 2014, 13(6), 419-431.
[27] Lindborg, S.R.; Persinger, C.C.; Sashegyi, A.; *et al.* Statistical refocusing in the design of Phase II trials offers promise of increased R&D productivity. *Nat. Rev. Drug Discov.*, **2014**, 13 (8), 638-640.
[28] Owens, P.K.; Raddad, E.; Miller, J.W.; *et al.* A decade of innovation in pharmaceutical R&D: the Chorus model. *Nat. Rev. Drug Discov.*, **2015**, 14(1), 17-28.
[29] Roberts, R.A.; Kavanagh, S.L.; Mellor, H.R.; *et al.* Reducing attrition in drug development: smart loading preclinical safety assessment. *Drug Discov. Today,* **2014**, 19(3), 341-347.
[30] Caldwell, G.W.; Yan, Z.; Masucci, J.A.; Hageman, W.; Leo, G.; Ritchie, D.M. Applied Pharmacokinetics in Drug Development: An overview of *Drug Discov. Pharm. Dev. Regul.*, **2003**, 1(2), 117-132.
[31] Heykants, J.; Meuldermans, W. Nonclinical kinetics and metabolism studies in support of the safety assessment of drugs. *Drug Inf. J.*, **1994**, 28, 163-172.
[32] Rodrigues, A.D. Use of *in vitro* human metabolism studies in drug development. An industrial Perspective. *Biochem. Pharmacol.*, **1994**, 48(12), 2147-56.
[33] Welling, P.G. Changing status of pharmacokinetics in drug discovery and development. *Int. J. Pharm. Adv.*, **1995**, 1(2), 121-33.
[34] Ball, S.E.; Scatina, J.A.; Sisenwine, S.F.; Fisher, G.L. The application of *in vitro* models of drug metabolism and toxicity in drug discovery and drug development. *Drug Chem. Toxicol.*, **1995**, 18(1) 1-28.
[35] Caldwell, J. The role of drug metabolism in drug discovery and development: Opportunities to enhance time- and cost-efficiency. *Pharm. Sci.*, **1996**, 2, 117-119.
[36] Humphrey, M.J. Application of metabolism and pharmacokinetics studies to the drug discovery process. *Drug Metab. Rev.*, **1996**, 28(3), 473-489.
[37] Cashman, J.R. Drug discovery and drug metabolism. Drug Discov. *Today*, **1996**, 1(5), 209-216.
[38] Caldwell, J. The importance of drug metabolism studies for efficient drug discovery and development. *Yakubutsu Dotai*, **1996**, 11(1), 119-26
[39] Rodrigues, A.D. Preclinical drug metabolism in the age of high-throughput screening: an industrial perspective. *Pharm. Res.*, **1997**, 14(11), 1504-1510.

[40] Lipinski, C.A.; Lombardo, F.; Dominy, B.W.; Feeney, P.J. Experimental and computational approaches to estimate solubility and permeability in drug discovery and development setting. *Adv. Drug Del. Rev.*, **1997**, 23, 3-25.
[41] Grass, G.M. Simulation models to predict oral drug absorption from *in vitro* data. *Adv. Drug Del. Rev.*, **1997**, 23(1-3), 199-219.
[42] Rodrigues, A.D. Rational high-throughput screening in preclinical drug metabolism. *Med. Chem. Res.*, **1998**, 8(7/8), 422-433.
[43] Caldwell, G.W.; Masucci, J.A.; Evangelisto, M.; White R. Evaluation of the Immobilized Artificial Membrane. Phosphatidylcholine. Drug Discovery Column for HPLC Screening Drug-Membrane Interactions. *J. Chromatogr. A*, **1998**, 800(2), 161-169.
[44] Masucci, J.A.; Caldwell, G.W.; Foley, J.P. A Comparison of the Retention Behavior of β-Blockers Using Immobilized Artificial Membrane Chromatography and Lysophospholipid Micellar Electrokinetic Capillary Chromatography. *J. Chromatogr. A*, **1998**, 810(1+2), 95-103.
[45] Tarit, M.H.; Berman, J. High-throughput approaches for evaluating absorption distribution, metabolism and excretion properties of lead compounds. *Cur. Opin. Chem. Biol.*, **1998**, 2 411-416.
[46] Smith, D.A. High throughput drug metabolism. *Biomed. Health Res.*, **1998**, 25, 137-143.
[47] Caldwell, G.W.; Easlick, S.M.; Gunnet, J.; Demarest, K.;. Masucci, J.A. In-Vitro Permeability of Eight β-Blockers Through a Caco-2 Monolayer Utilizing Electrospray Ionization Mass Spectrometry. *J. Mass Spectrum.*, **1998**, 33, 607-614.
[48] Curatolo, W. Physical chemical properties of oral drug candidates in the discovery and exploratory development setting. *Pharm. Sci. Technol. Today*, **1998**, 1(9), 387-393.
[49] Caldwell, G.W.; Masucci, J.A.; Chacon E. High throughput liquid chromatography-mass spectrometry assessment of the metabolic activity of commercially available hepatocytes from 96-well plates. *Comb. Chem. High Throughput Screen*, **1999**, 2(1), 39-51.
[50] Bayliss, M.K.; Frick, L.W. High-throughput pharmacokinetics: cassette dosing. *Curr. Opin. Drug Discov. Dev.*, **1999**, 2(1), 20-25.
[51] Smith, D.A.; van de Waterbeemd H. Pharmacokinetics and metabolism in early drug discovery. *Cur. Opin. Chem. Biol.*, **1999**, 3, 373-378.
[52] Sinko, P.J. Drug selection in early drug development: screening for acceptable pharmacokinetic properties using combined *in vitro* and computational approaches. *Curr. Opin. Drug Discov. Dev.*, **1999**, 2(1), 42-48.
[53] Johnson, D.E.; Braeckman, R.A.; Wolfgang G.Hi. Practical aspects of assessing toxicokinetics and toxicodynamics. *Curr. Opin. Drug Discov. Dev.*, **1999**, 2(1), 49-57.
[54] Todd, M.D.; Ulrich, R.G. Emerging technologies for accelerated toxicity evaluation of potential drug candidates. *Curr. Opin. Drug Discov. Dev.*, **1999**, 2(1), 58-68.
[55] Crespi, C.L. Higher-throughput screening with human cytochromes P450. *Curr. Opin. Drug Discov. Dev.*, **1999**, 2(1), 15-19.
[56] Caldwell, G.W. Compound optimization in early- and late-phase drug discovery: acceptable pharmacokinetics properties utilizing combined physicochemical, *in vitro* and *in vivo* screens. *Curr. Opin. Drug Discov.*, **2000**, 3, 30-41.
[57] Watt, A.P.; Morrison, D.; Evans, D.C. Approaches to higher-through pharmacokinetics (HTPK) in drug discovery. *Drug Discov. Today*, **2000**, 5(1), 17-24.
[58] Robertson, D.G.; Bulera, S.J. High-throughput toxicology: practical considerations. *Curr. Opin. Drug Discov.*, **2000**, 3, 42-47.
[59] Bajpai, M.; Adkison, K.K. High-through screening for lead optimization: a rational approach. *Curr. Opin. Drug Discov.*, **2000**, 3, 63-71.
[60] White, R.E. High-through screening in drug metabolism and pharmacokinetics support of drug discovery. *Annu. Rev. Pharmacol. Toxicol.*, **2000**, 40, 133-157.
[61] Eddershaw, P.J.; Beresford, A.P.; Bayliss, M.K. ADME/PK as part of a rational approach to drug discovery. *Drug Discov. Today*, **2000**, 5(9), 409-414.
[62] Bertrand, M.; Jackson, P.; Walther, B. Rapid assessment of drug metabolism in the drug discovery process. *Eur. J. Pharm. Sci.*, **2000**, 11, S61-S72.
[63] Lakings, D.B. Nonclinical Drug Development: Pharmacology, Drug Metabolism, and Toxicology. *New Drug Approv.*, **2000**, 100, 17-54.

[64] Caldwell, G.W.; Ritchie, D.M.; Masucci, J.A.; Hageman, W.; Yan, Z. The new pre-preclinical paradigm: compound optimization in early and late phase drug discovery. *Curr. Topics Med. Chem.*, **2001**, 1(5), 353-366.
[65] Yan, Z.; Caldwell, G.W. Metabolism Profiling and Cytochrome P450 Inhibition & Induction in Drug Discovery. *Curr. Topics Med. Chem.*, **2001**, 1(5). 403-425.
[66] Masucci, J.A.; Caldwell, G.W.; Evangelisto, M.; Jones, W.J.; Juzwin, S.; Sasso, P. The Use of On-line and Off-line Chromatographic Extraction Techniques coupled with Mass Spectrometry for Support of *In Vivo* and *In Vitro* Assays in Drug Discovery. *Curr. Topics Med. Chem.*, **2001**, 1(5), 463-471.
[67] Kerns, E.H.; Di, L. Multivariate pharmaceutical profiling for drug discovery. *Curr. Top. Med. Chem.*, **2002**, 2(1), 87-98.
[68] Yan, Z.; Rafferty, B.; Caldwell, G.W. Masucci, J.A. Rapidly Distinguishing Reversible and Irreversible Cyp450 Inhibitors by Using Fluorometric Kinetic Measurements. *Eur. J. Drug Metabol. Pharmaco.*, **2002**, 27(4), 281-287.
[69] Kerns, E.H.; Di, L. Pharmaceutical profiling in drug discovery. *Drug Discov. Today*, **2003**, 8(7), 316-323.
[70] Roberts, S.A. Drug metabolism and pharmacokinetics in drug discovery. *Curr. Opin. Drug Discov. Dev.*, **2003**, 6(1), 66-80.
[71] Muskal, S.M.; Jha, S.K.; Kishore, M.P.; Tyagi, P. A Simple and Readily Integratable Approach to Toxicity Prediction. *J. Chem. Inform. Comp. Sci.*, **2003**, 43(5), 1673-1678.
[72] Czar, M.J. *In vitro* toxicogenomics to predict *in vivo* pathology. *New Drugs*, **2003**, (2), 20, 22-23.
[73] Kerns, E.H.; Di, L.; Petusky, S.; Farris, M.; Ley, R.; Jupp, P. Combined application of parallel artificial membrane permeability assay and Caco-2 permeability assays in drug discovery. *J. Pharm. Sci.*, **2004**, 93(6), 1440-1453.
[74] Yan, Z.; Caldwell, G.W. Metabolic assessment in liver microsomes by co-activating cytochrome P450s and UDP-glycosyltransferases. *Eur. J. Drug Metabol. Pharmacol.*, **2003**, 28(3), 223-232.
[75] Yan, Z.; Caldwell, G.W., Eds. Optimization in Drug Discovery: *In Vitro* Methods.; Humana Press: Totowa, New Jersey, **2004**.
[76] Caldwell, G.W.; Masucci, J.A.; Yan, Z.; Hageman, W. Allometric scaling of pharmacokinetic parameters in drug discovery: Can human CL, Vss and t1/2 be predicted from in-vivo rat data. *Eur. J. Drug Metabol. Pharmacol.*, **2004**, 29(2), 133-143.
[77] Yan, Z.; Caldwell, G.W. Stable-Isotope Trapping and Rapid Identification of Reactive Metabolites Using the Isotope MS Signature. *Anal Chem.*, **2004**, 76(23), 6835-6847.
[78] Thompson, T.N. Drug metabolism *in vitro* and *in vivo* results: how do these data support drug discovery? Using Mass Spectrometry for Drug Metabolism Studies, CRC Press: Boca Raton, FL, **2005**, pp. 35-81.
[79] Gomez-Orellana, I. Strategies to improve oral drug bioavailability. *Expert Opin. Drug Del.*, **2005**, 2(3), 419-433.
[80] Raub, T.J. P-Glycoprotein Recognition of Substrates and Circumvention through Rational Drug Design. *Mol. Pharm.*, **2006**, 3(1), 3-25.
[81] Caldwell, G.W.; Yan, Z. Screening for reactive intermediates and toxicity assessment in drug discovery. *Curr. Top. Med. Chem.*, **2006**, 9(1), 47-60.
[82] Wunberg, T.; Hendrix, M.; Hillisch, A.; Lobell, M.; Meier, H.; Schmeck, C; Wild, H.; Hinzen, B. Improving the hit-to-lead process: data-driven assessment of drug-like and lead-like screening hits. *Drug Discov. Today*, **2006**, 11(3 & 4), 175-180.
[83] Thomas, V.H.; Bhattachar, S.; Hitchingham, L.; Zocharski, P.; Naath, M.; Surendran, N.; Stoner, C.L.; El-Kattan, A. The road map to oral bioavailability: an industrial perspective. *Expert Opin. Drug Metabol. Toxicol.*, **2006**, 2(4), 591-608.
[84] Neervannan, S. Preclinical formulations for discovery and toxicology: physicochemical challenges. *Expert Opin. Drug Metabol. Toxicol.*, **2006**, 2(5), 715-731.
[85] Subramanian, G.; Mjalli, A.M.M.; Kutz, M.E. Integrated approaches to perform *in silico* drug discovery. *Curr. Drug Discov. Technol.*, **2006**, 3(3), 189-197.
[86] Komura, H.; Iwaki, M. Pitfalls in high throughput screening for drug absorption optimization in drug discovery. *Curr. Anal. Chem.*, **2007**, 3(4), 302-309.

[87] Thakker, D.R. Strategic use of preclinical pharmacokinetic studies and *in vitro* models in optimizing ADME properties of lead compounds. *Biotechnol. Pharm. Aspects*, **2006**, 4, 1-23.

[88] Fortuna, A.; Alves, G.; Falcao, A. The importance of permeability screening in drugs discovery process: PAMPA, Caco-2 and rat everted gut assays. *Curr. Top. Pharmacol.*, **2007**, 11(2), 63-86.

[89] Yan, Z.; Caldwell, G.W.; Maher, N. Unbiased High-Throughput Screening of Reactive Metabolites on the Linear Ion Trap Mass Spectrometer Using Polarity Switch and Mass Tag Triggered Data-Dependent Acquisition", *Anal. Chem.*, **2008**, 80(16), 6410-6422.

[90] Schoonen, W.G.E.J.; Westerink, W.M.A.; Horbach, G.J. High-throughput screening for analysis of *in vitro* toxicity. *EXS*, **2009**, 99, 401-452.

[91] Yeon, J.H.; Park, J.-K. Drug Permeability Assay Using Microhole-Trapped Cells in a Microfluidic Device. *Anal. Chem.*, **2009**, 81(5), 1944-1951.

[92] Di, L.; Kerns, E.H.; Carter, G.T. Drug-like property concepts in pharmaceutical design. *Curr. Pharm. Des.*, **2009**, 15(19), 2184-2194.

[93] Caldwell, G.W.; Yan, Z. ADME optimization and toxicity assessment in early- and late-phase drug discovery. *Curr. Topics Med. Chem.*, **2009**, 9(11), 965-980.

[94] Dasgupta, M.; Tang, W.; Caldwell, G.W.; Yan, Z. Use of Stable Isotopic-labeled Probes to Facilitate LC/MS-based High Throughput Screening of Time-dependent CYP Inhibitors" *Rapid Commun. Mass Spectrom.*, **2010**, 24, 2177-2185.

[95] Yan, Z.; Caldwell, G.W. The Current Status of Time Dependent CYP Inhibition Assay and *In Silico* Drug-drug Interaction Predictions. *Curr. Topics. Med. Chem.*, **2012**, 12(11), 1291-1297.

[96] O'Brien, P.J. High-Content Analysis in Toxicology: Screening Substances for Human Toxicity Potential, Elucidating Subcellular Mechanisms and *In Vivo* Use as Translational Safety. *Basic Clinical Pharmacol. Toxicol.*, **2014**, 115(1), 4-17.

[97] Bal-Price, A.; Jennings, P., Eds.; *In vitro toxicology systems*; Humana Press: Totowa, New Jersey, **2014**.

[98] Keohavong, P.; Grant, S.G., Eds.; *Molecular Toxicology Protocols*. Humana Press: Totowa, New Jersey, **2014**.

[99] Caldwell, G.W.; Yan, Z., Eds.; *Optimization in Drug Discovery: In Vitro Methods* 2nd ed.; Humana Press: Totowa, New Jersey, **2014**.

[100] Caldwell, G.W. *In silico* tools used for compound selection during target-based drug discovery/ development. *Expert Opin. Drug Discov.*, **2015**, 10(7), 1-23.

[101] Kopec, K.K.; Bozyczko-Coyne, D.; Williams, M. Target identification and validation in drug discovery: the role of proteomics. *Biochem. Pharmacol.*, **2005**, 69(8), 1133-1139.

[102] Squarize CH, Castilho RM, Leopoldino AM. Drug target validation. *Curr. Methods Med. Chem. Bio. Phys.*, **2008**, 2, 167-185.

[103] Titov DV, Liu JO. Identification and validation of protein targets of bioactive small molecules. *Bioorg. Med. Chem.*, **2012**, 20(6), 1902-1909.

[104] Sharma N, Harikumar SL. Use of genomics and proteomics in pharmaceutical drug discovery and development: a review. *Inter. J. Pharm. Pharma. Sci.*, **2013**, 5(3), 24-28.

[105] Chen, K.-J.; Lee, J.-C. Establishment of high throughput screening (HTS) for drug discovery. *Huaxue*, **2008**, 66(4), 269-277.

[106] Macarron, R.; Banks, M.N.; Bojanic, D.; *et al.* Impact of high-throughput screening in biomedical research. *Nat. Rev. Drug Discov.*, **2011**, 10(3), 188-195.

[107] Shun, T.Y.; Lazo, J.S.; Sharlow, E.R.; Johnston, P.A. Identifying actives from HTS data sets: practical approaches for the selection of an appropriate HTS data-processing method and quality control review. *J. Biomol. Screen.*, **2011**, 16, 1-14.

[108] Janzen, W.P. Screening Technologies for Small Molecule Discovery: The State of the Art. *Chem. Bio.*, **2014**, 21(9), 1162-1170.

[109] Janzen, W.P.; Popa-Burke, I.G. Advances in improving the quality and flexibility of compound management. *J. Biomol. Screen.*, **2009**, 14(5), 444-451.

[110] Huggins, D.J.; Venkitaraman, A.R.; Spring, D.R. Rational Methods for the Selection of Diverse Screening Compounds. *ACS Chem. Bio.*, **2011**, 6(3), 208-217.

[111] Ramsay, D.L.; Kwasnoski, J.D.; Caldwell, G.W. Small molecule compound logistics outsourcing - going beyond the "thought experiment". *Curr. Topics Med. Chem.*, **2012**, 12(11), 1298-1308.

[112] Sink, R.; Gobec, S.; Pecar, S.; Zega, A. False positives in the early stages of drug discovery. *Curr. Med. Chem.*, **2010**, 17(34), 4231-4255.
[113] Lipinski, C.A. Lead- and drug-like compounds: the rule-of-five revolution. *Drug Discov. Today: Technol.*, **2004**, 1(4), 337-341.
[114] Colmenarejo G. *In silico* ADME prediction Data sets and models. *Curr. Comput.-Aided Drug Design*, **2005**, 1(4) 365-376.
[115] Van de Waterbeemd H. From *in vivo* to *in vitro/in silico* ADME: progress and challenges. *Expert Opin. Drug Metabol. Toxicol.*, **2005**, 1(1), 1-4.
[116] Narayanan R.; Gunturi S.B. *In silico* ADME modeling: prediction models for blood-brain barrier permeation using a systematic variable selection method. *Bioorg. Med. Chem.*, **2005**, 13(8), 3017-3028.
[117] Helma C. *In silico* predictive toxicology: The state-of-the-art and strategies to predict human health effects. *Curr. Opin. Drug Discov. Dev.*, **2005**, 8(1), 27-31.
[118] Subramanian, G.; Mjalli, A.M.M.; Kutz, M.E. Integrated approaches to perform *in silico* drug discovery. *Curr. Drug Discov. Technol.*, **2006**, 3(3), 189-197.
[119] Bidault, Y. A flexible approach for optimising *in silico* ADME /Tox characterization of lead candidates. *Expert Opin. Drug Metab. Toxicol.*, **2006**, 2(1), 157-168.
[120] Pang, Y. P. *In Silico* drug discovery: solving the "target-rich and lead-poor" imbalance using the genome-to-drug-lead paradigm. *Clin. Pharm. Ther.*, **2007**, 81(1), 30-34.
[121] Singh, B.; Dhake, A.S.; Sethi, D.; Paul, Y. *In silico* ADME predictions using quantitative structure pharmacokinetic relationships. Part I: fundamental aspects. *Pharm. Rev.*, **2007**, 5(29), 93-100.
[122] Singh, B.; Parle, M.; Paul, Y.; Khurana, L. *In silico* ADME predictions using quantitative structure pharmacokinetic relationships. Part II: descriptors. *Pharm. Rev.*, **2007**, 5(30), 63-68.
[123] Hop, C.E.C.A.; Cole, M.J.; Davidson, R.E.; *et al.* High throughput ADME screening: practical considerations, impact on the portfolio and enabler of *in silico* ADME models. *Curr. Drug Metab.*, **2008**, 9(9), 847-853.
[124] Moda, T.L.; Torres, L.G.; Carrara, A.E.; Andricopulo, A.D. PK/DB: database for pharmacokinetic properties and predictive *in silico* ADME models. *Bioinformatics*, **2008**, 24(19), 2270-2271.
[125] De Lisle, R.K.; Diller, D.J. *In silico* ADME/tox models: progress and challenges. *Curr. Comput.-Aided Drug Design.*, 2009, 5(2), 69-70.
[126] Gleeson, M.P.; Hersey, A.; Hannongbua, S. In-silico ADME models: a general assessment of their utility in drug discovery applications. *Curr. Topics Med. Chem.*, **2011**, 11(4), 358-381.
[127] Ou-Yang, Si.-S.; Lu, J.-Y.; Kong, X.-Q.; *et al.* Computational Drug Discovery. *Acta Pharmacol. Sinica*, **2012**, 33(9), 1131-1140.
[128] Ortwine, D.F.; Aliagas, I. Physicochemical and DMPK *In Silico* Models: Facilitating Their Use by Medicinal Chemists. *Mol. Pharma.*, **2013**, 10(4), 1153-1161.
[129] Cheng, F.; Li, W.; Liu, G.; Tang, Y. *In Silico* ADMET Prediction: Recent Advances, Current Challenges and Future Trends. *Curr. Topics Med. Chem.*, **2013**, 13(11), 1273-1289.
[130] Beck, B.; Geppert, T. Industrial applications of *in silico* ADMET. *J. Mole. Modeling*, **2014**, 20(7), 1-15.
[131] Ripphausen, P.; Nisius, B.; Peltason, L.; Bajorath, J. Quo Vadis, Virtual Screening? A Comprehensive Survey of Prospective Applications. *J. Med. Chem.*, **2010**, 53(24), 8461-8467.
[132] Ripphausen, P.; Nisius, B.; Bajorath, J. State-of-the-art in ligand based virtual screening. *Drug Discov. Today*, **2011**, 16, 372-376.
[133] Ortega, S.S.; Cara, L.C.L.; Salvador, M.K. *In silico* pharmacology for a multidisciplinary drug discovery process. *Drug Metabol. Drug Interact.*, **2012**, 27(4), 199-207.
[134] Stumpfe, D.; Ripphausen, P.; Bajorath, J. Virtual compound screening in drug discovery. *Future Med. Chem.*, **2012**, 4, 593-602.
[135] Zhu, T.; Cao, S.; Su, P.-C.; *et al.* Hit Identification and Optimization in Virtual Screening: Practical Recommendations Based on a Critical Literature Analysis. *J. Med. Chem.*, **2013**, 56(17), 6560-6572.
[136] Hoi, P.M.; Li, S.; Vong, C.T.; *et al.* Recent advances in structure-based drug design and virtual screening of VEGFR tyrosine kinase inhibitors. *Methods*, **2015**, 71, 85-91.
[137] Petrauskas, A.A.; Kolovanov, E.A. ACD/Log P method description. *Perspect. Drug Discov. Design.*, **2000**, 99-116.

[138] Leo, A.J.; Hoekman, D. Calculating log P(oct) with no missing fragments; the problem of estimating new interaction parameters. *Perspect. Drug Discov. Design.*, **2000**, 19-38.
[139] Yalkowsky, S.H.; He, Y. *Handbook of Aqueous Solubility Data.* CRC Press: Boca Raton, **2003**.
[140] Machatha, S.G.; Yalkowsky, S.H. Comparison of the octanol/water partition coefficients calculated by ClogP ACDlogP and KowWin to experimentally determined values. *Int. J. Pharm.*, **2005**, 294(1-2), 185-192.
[141] Avdeef, A. Drug ionization and physicochemical profiling. Methods Princ. *Med. Chem.*, **2008**, 37, 55-83.
[142] Mannhold, R.; Poda, G.I.; Ostermann, C.; Tetko, I.V. Calculation of molecular lipophilicity: state-of-the-art and comparison of logP methods on more than 96,000 compounds. *J. Pharm. Sci.*, **2009**, 98(3), 861-893.
[143] Advanced Chemistry Development, Inc., 110 Yonge Street, 14th floor, Toronto, Ontario, Canada M5C 1T4., http://www.acdlabs.com [Accessed on: 19th June, **2015**).
[144] Dearden, J.C. *In silico* prediction of aqueous solubility. *Expert Opin. Drug Discov.*, **2006**, 1(1), 31-52.
[145] Lipinski, C.A. Drug solubility in water and dimethylsulfoxide. *Methods Princ. Med. Chem.*, **2008**, 37, 257-282.
[146] Schroeter, T.S.; Schwaighofer, A.; Mika, S.; Ter Laak, A.; Suelzle, D.; Ganzer, U.; Heinrich, N.; Mueller, K.-R. Estimating the domain of applicability for machine learning QSAR models: a study on aqueous solubility of drug discovery molecules. *J. Comput.-Aided Mol. Design.*, **2007**, 21(9), 485-498.
[147] Hopfinger, A.J.; Esposito, E.X.; Llina, A.; Glen, R.C.; Goodman, J.M. Findings of the challenge to predict aqueous solubility. *J. Chem. Informa. Model.*, **2009**, 49, 1-5.
[148] Cherkasov, A.; Muratov, E.N.; Fourches, D.; *et al.* QSAR Modeling: Where Have You Been? Where Are You Going? *J. Med. Chem.*, **2014**, 57(12), 4977-5010.
[149] McDonagh, J.L.; Nath, N.; De Ferrari, L.; van Mourik, T.; Mitchell, J.B.O. Uniting Cheminformatics and Chemical Theory to Predict the Intrinsic Aqueous Solubility of Crystalline Druglike Molecules. *J. Chem. Informa. Model.*, **2014**, 54(3), 844-856.
[150] Palmer, D.S.; Mitchell, J.B.O. Is Experimental Data Quality the Limiting Factor in Predicting the Aqueous Solubility of Druglike Molecules? *Mol. Pharma.*, **2014**, 11(8), 2962-2972.
[151] Chevillard, F.; Lagorce, D.; Reynes, C.; *et al. In Silico* Prediction of Aqueous Solubility: A Multimodel Protocol Based on Chemical Similarity. *Mol. Pharma.*, **2012**, 9(11), 3127-3135.
[152] Bains, W.; Gilbert, R.; Sviridenko, L.; Gascon, J.-M.; Scoffin, R.; Birchall, K.; Harvey, I.; Caldwell, J. Evolutionary computational methods to predict oral bioavailability QSPRs. *Curr. Opin. Drug Discov. Dev.*, **2002**, 5(1), 44-51.
[153] Yoshida, F.; Topliss, J.G. QSAR Model for Drug Human Oral Bioavailability. *J. Med. Chem.*, **2000**; 43(13), 2575-2585.
[154] Van de Waterbeemd, H. *In silico* models to predict oral absorption. *Compr. Med. Chem. II*, **2006**, 5, 669-697.
[155] Linnankoski, J.; Maekelae, J.M.; Ranta, V.-P.; Urtti, A.; Yliperttula, M. Computational Prediction of Oral Drug Absorption Based on Absorption Rate Constants in Humans. *J. Med. Chem.*, **2006**, 49(12), 3674-3681.
[156] Gozalbes, R.; Jacewicz, M.; Annand, R.; Tsaioun, K.; Pineda-Lucena, A. QSAR-based permeability model for drug-like compounds. *Bioorg. Med. Chem.*, **2011**, 19(8), 2615-2624.
[157] Tetko, I.V.; Oprea, T.I. Early ADME/T predictions: toy or tool? *Chemoinformat. Appro. Virtual Screen.*, **2008**, 240-267.
[158] Lombardo, F.; Obach, R.S.; DiCapua, F.M.; Bakken, G.A.; Lu, J.; Potter, D.M.; Gao, F.; Miller, M.D.; Zhang, Y. A Hybrid Mixture Discriminant Analysis-Random Forest Computational Model for the Prediction of Volume of Distribution of Drugs in Human. *J. Med. Chem.*, **2006**, 49(7), 2262-2267.
[159] Ghafourian, T.; Barzegar-Jalali, M.; Dastmalchi, S.; Khavari-Khorasani, T.; Hakimiha, N.; Nokhodchi, A. QSPR models for the prediction of apparent volume of distribution. *Int. J. Pharm.*, **2006**, 319(1-2), 82-97.
[160] Sui, X.; Sun, J.; Wu, X.; Li, H.; Liu, J.; He, Z. Predicting the volume of distribution of drugs in humans. *Curr. Drug Metabol.*, **2008**, 9(6), 574-580.

[161] Obach, R.S. Prediction of human volume of distribution using *in vivo*, *in vitro*, and *in silico* approaches. *Ann. Rep. Med. Chem.*, **2007**, 42, 469-488.
[162] Berellini, G.; Springer, C.; Waters, N.J.; Lombardo, F. *In Silico* Prediction of Volume of Distribution in Human Using Linear and Nonlinear Models on a 669 Compound Data Set. *J. Med. Chem.*, **2009**, 52(14), 4488-4495.
[163] Stepensky, D. Prediction of Drug Disposition on the Basis of its Chemical Structure. *Clin. Pharmacol.*, **2013**, 52(6), 415-431.
[164] Louis, B.; Agrawal, V.K. Prediction of human volume of distribution values for drugs using linear and nonlinear quantitative structure pharmacokinetic relationship models. *Interdisciplinary Sci.: Comput. Life Sci.*, **2014**, 6(1), 71-83.
[165] Freitas, A.A.; Limbu, K.; Ghafourian, T. Predicting volume of distribution with decision tree-based regression methods using predicted tissue: plasma partition coefficients. *J. Cheminformatics*, **2015**, 7, 1-17.
[166] Lewis, D.F.V. *Cytochromes P450*, Taylor & Francis Inc.: Bristol, PA, **1996**.
[167] Benedetti, M.S.; Tipton, K.F. Monoamine oxidases and related amine oxidases as phase I enzymes in the metabolism of xenobiotics. *J. Neural Transm. Suppl.*, **1998**, 52, 149-171.
[168] Beedham, C. The role of non-P450 enzymes in drug oxidation. *Pharm. World Sci.*, **1997**, 19(6), 255-263.
[169] Cashman, J.R. Human flavin-containing monooxygenase: substrate specificity and role in drug metabolism. *Curr. Drug Metabol.*, **2000**, 1(2), 181-191.
[170] Kwon, Y. *Handbook of Essential Pharmacokinetics. Pharmacodynamics and Drug Metabolism for Industrial Scientists.* Kluwer Academic/Plenum Publishers: New York, NY, **2001**.
[171] Greene, N.; Judson, P.N.; Langowski, J.J.; Marchant, C.A. Knowledge-based expert systems for toxicity and metabolism prediction: DEREK, StAR and METEOR. *SAR and QSAR Environ. Res.*, **1999**, 10(2-3) 299-314.
[172] Testa, B.; Balmat, A.-L.; Long, A.; Judson, P. Predicting drug metabolism - an evaluation of the expert system METEOR. *Chem. Biodivers.*, **2005**, 2(7) 872-885.
[173] Zheng, M.; Luo, X.; Shen, Q.; Wang, Y.; Du, Y.; Zhu, W.; Jiang, H. Site of metabolism prediction for six biotransformations mediated by cytochromes P450. *Bioinformatics*, **2009**, 25(10), 1251-1258.
[174] Tarcsay, A.; Kiss, R.; Keseru, Gyoergy M. Site of metabolism prediction on cytochrome P450 2C9: a knowledge-based docking approach. *J. Comput.-Aided Mol. Design*, **2010**, 24(5), 399-408.
[175] Valerio, L.G., Jr.; Long, A. The *in silico* prediction of human-specific metabolites from hepatotoxic drugs. *Curr. Drug Discov. Technol.*, **2010**, 7(3), 170-187.
[176] CompuDrug International, Inc., 286 Balbay Dr. # 1A, Bal Harbor, Florida, FL 33154, U.S.A. http://www.compudrug.com [Accessed on: 19th June, **2015**].
[177] Vermeulen, N.P.E. Prediction of drug metabolism: The case of cytochrome P450 2D6. *Curr. Topics Med. Chem.*, **2003**, 3(11), 1227-1239.
[178] Li, H.; Sun, J.; Fan, X.; Sui, X.; Zhang, L.; Wang, Y.; He, Z. Considerations and recent advances in QSAR models for cytochrome P450-mediated drug metabolism prediction. *J. Comput.-Aided Mol. Design.*, **2008**, 22(11), 843-855.
[179] Speck-Planche, A.; Dias Soeiro Cordeiro, M.N. Review of Current Chemoinformatic Tools for Modeling Important Aspects of CYPs mediated Drug Metabolism. Integrating Metabolism Data with Other Biological Profiles to Enhance Drug Discovery. *Curr. Drug Metabol.*, **2014**, 15(4), 429-440.
[180] Lin, T.-H.; Tsai, T.-L. Constructing a linear QSAR for some metabolizable drugs by human or pig flavin-containing monooxygenases using some molecular features selected by a genetic algorithm trained SVM. *J. Theoretical. Bio.*, **2014**, 356, 85-97.
[181] Karlsson, F.H.; Bouchene, S.; Hilgendorf, C.; Dolgos, H.; Peters, S.A. Utility of *in vitro* systems and preclinical data for the prediction of human intestinal first-pass metabolism during drug discovery and preclinical development. *Drug Metabol. Dispos.*, **2013**, 41(12), 2033-2046.
[182] Di, L.; Obach, R.S. Addressing the Challenges of Low Clearance in Drug Research. *AAPS J.*, **2015**, 17(2), 352-357.
[183] Willmann, S.; Lippert, J.; Schmitt, W. From physicochemistry to absorption and distribution: predictive mechanistic modelling and computational tools. *Expert Opin. Drug Metab. Toxicol.*, **2005**, 1(1), 159-168.

[184] Emoto, C.; Murayama, N.; Rostami-Hodjegan, A.; Yamazaki, H. Utilization of estimated physicochemical properties as an integrated part of predicting hepatic clearance in the early drug-discovery stage: impact of plasma and microsomal binding. *Xenobiotica*, **2009**, 39(3), 227-235.

[185] Nikolic, K.; Agababa, D. Prediction of hepatic microsomal intrinsic clearance and human clearance values for drugs. *J. Mol. Graph. Model.*, **2009**, 28(3), 245-252.

[186] Simulations Plus, Inc., 42505 10th Street West Lancaster, CA 93534-7059, U.S.A., http://www.simulations-plus.com [Accessed on: 19th June, **2015**].

[187] Certara USA, Inc., 100 Overlook Center, Suite 101, Princeton, NJ 08540, U.S.A., http://www.simcyp.com [Accessed on: 19th June, **2015**].

[188] Jamei, M.; Marciniak, S.; Feng, K.; Barnett, A.; Tucker, G.; Rostami-Hodjegan, A. The Simcyp Population-based ADME Simulator. *Expert Opin. Drug Metab. Toxicol.*, **2009**, 5(2), 211-223.

[189] Veselinovic, J.B.; Toropov, A.A.; Toropova, A.P.; Nikolic, G.M.; Veselinovic, A.M. Monte Carlo Method-Based QSAR Modeling of Penicillins Binding to Human Serum Proteins, *Archiv der Pharmazie*, **2015**, 348(1), 62-67.

[190] Cohen, L.H. Plasma Protein-Binding Methods in Drug Discovery. In: *Optimization in Drug Discovery: In Vitro Methods*; Yan, Z.; Caldwell, G.W., Eds.; Humana Press Inc.: Totowa, New Jersey, **2004**, pp. 163-184.

[191] Caldwell, G.W.; Yan, Z. Isothermal titration Calorimetry Characterization of Drug-Binding Energetics to Blood Proteins. In: *Optimization in Drug Discovery: In Vitro Methods*; Yan, Z.; Caldwell, G.W., Eds.; Humana Press Inc.:Totowa, New Jersey, **2004**, pp. 163-184.

[192] Gleeson, M. P. Plasma protein binding affinity and its relationship to molecular structure: an in-silico analysis. *J. Med. Chem.*, **2007**, 50, 101-112.

[193] Rodgers, A.L.; Davis, A.M.; van de Waterbeemd, H. Time series QSAR analysis of human plasma protein binding data. *QSAR Comb. Sci.*, **2007**, 26, 511-521.

[194] Wood, D.J.; Buttar, D.; Cumming, J.G.; Davis, A.M.; Norinder, U.; Rodgers, S.L. Automated QSAR with a Hierarchy of Global and Local Models. *Mole. Informa.*, **2011**, 30(11-12), 960-972.

[195] Trotter, M.; Holden, S. Support vector machines for ADME property classification. *QSAR Comb. Sci.*, **2003**, 22, 533-548.

[196] Sakiyama,Y. The use of machine learning and nonlinear statistical tools for ADME prediction. *Expert Opin. Drug Metab. Toxicol.*, **2009**, 5, 149-169.

[197] Tao, L.; Zhang, P.; Qin, C.; Chen, S. Y.; Zhang, C.; Chen, Z.; Zhu, F.; Yang, S. Y.; Wei, Y. Q.; Chen, Y. Z. Recent progresses in the exploration of machine learning methods as in-silico ADME prediction tools. *Adv. Drug Delivery Rev.*, **2015**, [Online] http://dx.doi.org/10.1016/ j.addr.2015.03.014/ [Accessed on: 19th June, **2015**]

[198] Wegner, J.K.; Frohlich, H.; Zell, A. Feature selection for descriptor based classification models. 2. Human intestinal absorption (HIA), *J. Chem. Inf. Comput. Sci.*, **2004**, 44, 931-939.

[199] Yap, C. W.; Li, Z. R.; Chen, Y. Z. Quantitative structure pharmacokinetic relationships for drug clearance by using statistical learning methods. *J. Mol. Graph. Model.*, **2006**, 24, 383-395.

[200] Hsiao, Y.-W.; Fagerholm, U.; Norinder, U. *In Silico* Categorization of *in Vivo* Intrinsic Clearance Using Machine Learning. *Mole. Pharma.*, **2013**, 10(4), 1318-1321.

[201] Sakiyama, Y.; Yuki, H.; Moriya, T.; Hattori, K.; Suzuki, M.; Shimada, K.; Honma, T. Predicting human liver microsomal stability with machine learning techniques. J. *Mole. Graphics Model.*, **2008**, 26(6), 907-915.

[202] Cheng, F.; Yu, Y.; Shen, J.; Yang, L.; Li, W.; Liu, G.; Lee, P.W.; Tang, Y. Classification of cytochrome P450 inhibitors and noninhibitors using combined classifiers. *J. Chem. Informa. Model.*, **2011**, 51, 996-1011.

[203] Zsila, F.; Bikadi, Z.; Malik, D.; Hari, P.; Pechan, I.; Berces, A.; Hazai, E. Evaluation of drug-human serum albumin binding interactions with support vector machine aided online automated docking. *Bioinformatics*, **2011**, 27, 1806-1813.

[204] Chen, H.; Winiwarter, S.; Friden, M.; Antonsson, M.; Engkvist, O. *In silico* prediction of unbound brain-to-plasma concentration ratio using machine learning algorithms. *J. Mol. Graph. Model.*, **2011**, 29, 985-995.

[205] Tan, W.; Mei, H.; Chao, L.; Liu, T.; Pan, X.; Shu, M.; Yang, L. Combined QSAR and molecule docking studies on predicting P-glycoprotein inhibitors, *J. Comput. Aided Mol. Design.*, **2013**, 27, 1067-1073.

[206] Kumar, Rajnish; Sharma, Anju; Varadwaj, Pritish Kumar A prediction model for oral bioavailability of drugs using physicochemical properties by support vector machine. *J. Natural Sci., Bio. Med.*, **2011**, 2(2), 168-173.

[207] Cronin M.T.D.; Madden, J.C.; Eds.; *In silico toxicology*; RSC Publishing: Cambridge (UK), **2010**.

[208] Ekins, S. Progress in computational toxicology. *J. Pharmacol. Toxicol. Methods.*, **2014**, 69(2), 115-140.

[209] Pradeep, P.; Povinelli, R.J.; Merrill, S.J.; Bozdag, S.; Sem, D.S. Novel Uses of *In Vitro* Data to Develop Quantitative Biological Activity Relationship Models for *in Vivo* Carcinogenicity Prediction. Mole. *Informa.*, **2015**, 34(4), 236-245.

[210] Madden, J.C. Sources of chemical information, toxicity data and assessment of their quality. *Issues in Toxicol.*, **2013**, 17, 98-126.

[211] Omer, A.; Singh, P.; Yadav, N.K.; Singh, R.K. An Overview of Data Mining Algorithms in Drug Induced Toxicity Prediction. *Mini-Rev. Med. Chem.*, **2014**, 14(4), 345-354.

[212] Song, M.; Clark, M. Development and Evaluation of an *in Silico* Model for hERG Binding. *J. Chem. Informa. Model.*, **2006**, 46(1), 392-400.

[213] Dearden, J.C. Expert Systems for Toxicity Prediction.; In: *In silico Toxicology*; Cronin M.T.D.; Madden, J.C.; Eds.; RSC Publishing: Cambridge (UK), **2010**, pp. 478-507.

[214] Klon, A.E. Machine learning algorithms for the prediction of hERG and CYP450 binding in drug development. *Expert Opin. Drug Metabol. Toxicol.*, **2010**, 6(7), 821-833.

[215] Dobchev, D.A.; Pillai, G.G.; Karelson, M. *In Silico* Machine Learning Methods in Drug Development. *Curr. Topics Med. Chem.*, **2014**, 14(16), 1913-1922.

[216] Seal, A.; Passi, A.; Abdul Jaleel, U.C.; Wild, D.J. In-silico predictive mutagenicity model generation using supervised learning approaches. *J. Cheminformatics*, **2012**, 4, 1-10.

[217] Alelyunas, Y.W.; Liu, R.; Pelosi-Kilby, L.; Shen, C. Application of a dried-DMSO rapid throughput 24-h equilibrium solubility in advancing discovery candidates. *Eur. J. Pharm. Sci.*, **2009**, 37(2), 172-182.

[218] van de Waterbeemd, H. Physicochemical approaches to drug absorption. *Methods Princ. Med. Chem.*, **2009**, 40, 71-99.

[219] Liu, B.; Chang, J.; Gordon, W.P.; Isbell, J.; Zhou, Y.; Tuntland, T. Snapshot PK a rapid rodent *in vivo* preclinical screening approach. *Drug Discov. Today*, **2008**, 13(7-8), 360-7.

[220] He, K.; Qian, M.; Wong, H.; Bai, S. A.; He, B.; Brogdon, B.; Grace, J. E.; Xin, B.; Wu, J.; Ren, S. X.; Zeng, H.; Deng, Y.; Graden, D. M.; Olah, T. V.; Unger, S. E.; Luettgen, J. M.; Knabb, R. M.; Pinto, D. J.; Lam, P. Y. S.; Duan, J.; Wexler, R. R.; Decicco, C. P.; Christ, D. D.; Grossman, S. J. N-in-1 dosing pharmacokinetics in drug discovery: experience, theoretical and practical considerations. *J. Pharm. Sci.*, **2008**, 97(7), 2568-2580.

[221] Cox, K.A.; White, R.E.; Korfmacher, W.A. Rapid determination of pharmacokinetic properties of new chemical entities: *In vivo* approaches. *Comb. Chem. High Throughput Screen.*, **2002**; 5(1), 29-37.

[222] Kerns, EH.; Di, L.; Carter, G.T. *In vitro* solubility assays in drug discovery. *Curr. Drug Metabol.*, **2008**, 9(9), 879-885.

[223] Guo, Y.; Shen, H. pKa, Solubility, and Lipophilicity. In: *Optimization in Drug Discovery: In Vitro Methods*; Yan, Z.; Caldwell, G.W., Eds.; Humana Press Inc.: Totowa, New Jersey, **2004**, pp. 1-18.

[224] Avdeef, A. Absorption and Drug Development: Solubility, Permeability and Charge State, Wiley-Interscience: New York, New York, **2003**.

[225] Baka, E.; Comer, J.E.A.; Takacs-Novak, K. Study of equilibrium solubility measurement by saturation shake - flask method using hydrochlorothiazide as model compound. *J. Pharm. Biomed. Anal.*, **2008**, 46(2), 335-341.

[226] Yamashita, T.; Dohta, Y.; Nakamura, T.; Fukami, T. High-speed solubility screening assay using ultra-performance liquid chromatography/mass spectrometry in drug discovery. *J. Chromatogr. A*, **2008**, 1182(1), 72-76.

[227] Hageman, M.J. Solubility, solubilization and dissolution in drug delivery during lead optimization. *Biotechnol. Pharm. Aspects*, **2006**, 4, 99-130.

[228] Bard, B.; Martel, S.; Carrupt, P.-A. High throughput UV method for the estimation of thermodynamic solubility and the determination of the solubility in biorelevant media. *Eur. J. Pharm. Sci.*, **2008**, 33(3), 230-240.

[229] Zhou, L.; Yang, L.; Tilton, S.; Wang, J. Development of a high throughput equilibrium solubility assay using miniaturized shake - flask method in early drug discovery. *J. Pharm. Sci.*, **2007**, 96(11), 3052-3071.

[230] Glomme, A.; Maerz, J.; Dressman, J.B. Comparison of a miniaturized shake - flask solubility method with automated potentiometric acid/base titrations and calculated solubilities. *J. Pharm. Sci.*, **2005**, 94(1), 1-16.

[231] Caldwell, G.W.; Hasting, B.; Masucci, J.A.; Yan, Z. Small molecule formulation screening strategies in drug discovery. In: *Optimization in Drug Discovery: In Vitro Methods* 2nd ed.; Caldwell, G.W.; Yan, Z.; Eds. Humana Press Inc.: Totowa, New Jersey, **2014**, pp. 1-20.

[232] Uchida, M.; Fukazawa, T.; Yamazaki, Y.; Hashimoto, H.; Miyamoto, Y. A modified fast (4 day) 96-well plate Caco-2 permeability assay. *J. Pharm. Toxic. Methods*, **2009**, 59(1), 39-43.

[233] Fortuna, A.; Alves, G.; Falcao, A. The importance of permeability screening in drugs discovery process: PAMPA, Caco-2 and rat everted gut assays. *Curr. Top. Pharmacol.*, **2007**, 11(2), 63-86.

[234] Thomas, S.; Brightman, F.; Gill, H.; Lee, S.; Pufong, B. Simulation modelling of human intestinal absorption using Caco-2 permeability and kinetic solubility data for early drug discovery. *J. Pharm. Sci.*, **2008**, 97(10), 4557-4574.

[235] Go, W.-J.; Cheon, E.-pa; Han, H.-K. Comparison of Caco - 2 and MDCK cells as an in-vitro ADME screening model. *Yakche Hakhoechi*, **2008**, 38(3), 183-189.

[236] Sun, H.; Chow, E.C.Y.; Liu, S.; Du, Y.; Pang, K.S. The Caco-2 cell monolayer: usefulness and limitations. *Expert Opin. Drug Metab. Toxicol.*, **2008**, 4(4), 395-411.

[237] Avdeef, A.; Bendels, S.; Di, L.; Faller, B.; Kansy, M.; Sugano, K.; Yamauchi, Y. Parallel artificial membrane permeability assay (PAMPA)-critical factors for better predictions of absorption. *J. Pharm. Sci.*, **2007**, 96(11), 2893-2909.

[238] Irvine, J.D.; Takahashi, L.; Lockhart, K.; Cheong, J.; Tolan, J.W.; Selick, H.E.; Grove, J.R. MDCK (Madin-Darby canine kidney) cells: a tool for membrane permeability screening. *J. Pharm. Sci.*, **1999**, 88(1), 28-33.

[239] Hidalgo, Ismael J. Assessing the absorption of new pharmaceuticals. *Curr. Topics Med. Chem.*, **2001**, 1(5), 385-401.

[240] Caldwell, G.W.; Ferguson, C.; Buerger, R.; Kulp, L.; Yan, Z. Permeability assessment using 5-day cultured Caco-2 cell monolayers. In: *Optimization in Drug Discovery: In Vitro Methods* 2nd ed.; Caldwell, G.W.; Yan, Z. Eds.; Humana Press Inc.: Totowa, New Jersey, **2014**, pp. 49-76.

[241] Chen, M.-L.; Amidon, G.L.; Benet, L.Z.; Lennernas, H.; Yu, L.X. The BCS, BDDCS, and Regulatory Guidances. *Pharma. Res.*, **2011**, 28(7), 1774-1778.

[242] Volpe, D.A. Application of Method Suitability for Drug Permeability Classification. *AAPS J.*, **2010**, 12(4), 670-678.

[243] Vanhulle, V.P.; Martiat, G.A.; Verbeeck, R.K.; Horsmans, Y.; Calderon, P.B.; Eeckhoudt, S.L.; Taper, H.S.; Delzenne, N. Cryopreservation of rat precision-cut liver slices by ultrarapid freezing. Influence on phase I and II metabolism and on cell viability upon incubation for 24 hours. *Life Sci.* **2001**, 68(21), 2391-2403.

[244] Ekins, S.; Ring, B.J.; Grace, J.; McRobie-Belle, D.J.; Wrighton, S.A. Present and future *in vitro* approaches for drug metabolism. *J. Pharmacol. Toxicol. Methods*, **2001**, 2000, 44(1), 313-324.

[245] Guillouzo, A.; Rialland, L.; Fautrel, A.; Guyomard, C. Survival and function of isolated hepatocytes after cryopreservation. *Chem. Biol. Interact.*, **1999**, 121(1), 7-16.

[246] Li, A.P. Overview: hepatocytes and cryopreservation-a personal historical perspective. *Chem. Biol. Interact.*, **1999**, 121(1), 1-5.

[247] Umehara, K.; Kudo, S.; Hirao, Y.; Morita, S.; Uchida, M.; Odomi, M.; Miyamoto, G. Oxidative cleavage of the octyl side chain of 1-(3,4-dichlorobenzyl)-5-octylbiguanide (OPB-2045) in rat and dog liver preparations. *Drug Metab. Dispos.*, **2000**, 28(8), 887-894.

[248] Wu, W.-N.; McKown, L.A. *In vitro* drug metabolite profiling using hepatic S9 and human liver microsomes. In: *Optimization in Drug Discovery: In Vitro Methods*; Yan, Z.; Caldwell, G.W., Eds.; Humana Press Inc.: Totowa, New Jersey, **2004**, pp. 163-184.

[249] Caldwell, G.W.; Yan, Z. Metabolic assessment in alamethicin-activated liver microsomes: Co-activating CYPs and UGTs. In: *Optimization in Drug Discovery: In Vitro Methods* 2nd ed.; Caldwell, G.W.; Yan, Z. Eds.; Humana Press Inc.: Totowa, New Jersey, **2014**, pp. 101-115.

[250] Shou, W.Z.; Magis, L.; Li, A.C.; Weng, N.; Bryant, M.S. A novel approach to perform metabolite screening during the quantitative LC-MS/MS analyses of *in vitro* metabolic stability samples using a hybrid triple-quadrupole linear ion trap mass spectrometer. *J. Mass Spectrom.*, **2005**, 40(10), 1347-1356.

[251] Fonsi M.; Orsale M.V.; Monteagudo E. High-throughput microsomal stability assay for screening new chemical entities in drug discovery. *J. Soc. Biomol. Screen.*, **2008**, 13(9), 862-9.

[252] Di, L.; Kerns, E.H.; Ma, X.J.; Huang, Y.; Carter, G.T. Applications of high throughput microsomal stability assay in drug discovery. *Comb. Chem. High Throughput Screen.*, **2008**, 11(6), 469-476.

[253] Di, L; Kerns, E.H.; Li, S.Q.; Petusky, S.L. High throughput microsomal stability assay for insoluble compounds. *Int. J. Pharm.*, **2006**, 317(1), 54-60.

[254] Sensenhauser, C. *In Vitro* CYP/FMO Reaction Phenotyping. In: *Optimization in Drug Discovery: In Vitro Methods* 2nd ed.; Caldwell, G.W.; Yan, Z. Eds.; Humana Press Inc.: Totowa, New Jersey, **2014**, pp. 137-169.

[255] Court, M.H. Phenotyping UDP-glucuronosyltransferases (UGTs) involved in human drug metabolism: An update. In: *Optimization in Drug Discovery: In Vitro Methods* 2nd ed.; Caldwell, G.W.; Yan, Z. Eds.; Humana Press Inc.: Totowa, New Jersey, **2014**, pp. 117-136.

[256] Yan, Z.; Caldwell, G.W.; Wu, W.N.; McKown, L.A.; Rafferty, B.; Jones, W.; Masucci, J.A. *In vitro* identification of metabolic pathways and cytochrome P450 enzymes involved in the metabolism of Etoperidone. *Xenobiotica*, **2002**, 32(11), 949-962.

[257] Obach, R.S. Inhibition of drug-metabolizing enzymes and drug - drug interactions in drug discovery and development. *Drug Drug Interac. Pharm. Development*, **2008**, 75-93.

[258] Shou, M.; Dai, R. Analysis of *in vitro* cytochrome P450 inhibition in drug discovery and development. *Drug Metabol. Drug Des. Dev.*, **2008**, 513-544.

[259] Smith, D.; Sadagopan, N.; Zientek, M.; Reddy, A.; Cohen, L. Analytical approaches to determine cytochrome P450 inhibitory potential of new chemical entities in drug discovery. *J. Chromatogr. B: Anal. Technol. Biomed. Life Sci.*, **2007**, 850(1-2), 455-463.

[260] Di, L.; Kerns, E.H.; Li, S.Q.; Carter, G.T. Comparison of cytochrome P450 inhibition assays for drug discovery using human liver microsomes with LC-MS, rhCYP450 isozymes with fluorescence, and double cocktail with LC-MS. *Int. J. Pharm.*, **2007**, 335(1-2), 1-11.

[261] Peng, S.X.; Barbone, A.G.; Ritchie, D.M. High-throughput cytochrome p450 inhibition assays by ultrafast gradient liquid chromatography with tandem mass spectrometry using monolithic columns. *Rapid Commun. Mass Spectrom: RCM*, **2003**, 17(6), 509-18.

[262] Houston, J.B.; Kenworthy, K.E. *In vitro-in vivo* scaling of CYP kinetic data not consistent with the classical Michaelis-Menten model. *Drug Metab. Dispos.*, **2000**, 28(3), 246-254.

[263] Emoto, C.; Yamazaki, H.; Iketaki, H.; Yamasaki, S.; Satoh, T.; Shimizu, R.; Suzuki, S.; Shimada, N.; Nakajima, M.; Yokoi, T. Cooperativity of α-naphthoflavone in cytochrome P450 3A-dependent drug oxidation activities in hepatic and intestinal microsomes from mouse and human. *Xenobiotica*, **2001**, 31(5), 265-275.

[264] Furuta, S.; Kamada, E.; Suzuki, T.; Sugimoto, T.; Kawabata, Y.; Shinozaki, Y.; Sano, H. Inhibition of drug metabolism in human liver microsomes by nizatidine, cimetidine and omeprazole. *Xenobiotica*, **2001**, 31(1), 1-10.

[265] McGinnity, D.F.; Waters, N.J.; Tucker, J.; Riley, R.J. Integrated *in vitro* analysis for the *in vivo* prediction of cytochrome P450-mediated drug-drug interactions. *Drug Metab. Dispos.*, **2008**, 36(6), 1126-1134.

[266] Zhang, J.G.; Stresser, D.M. Assessment of CYP3A4 time-dependent inhibition in plated and suspended human hepatocytes. In: *Optimization in Drug Discovery: In Vitro Methods* 2nd ed.; Caldwell, G.W.; Yan, Z. Eds.; Humana Press Inc.: Totowa, New Jersey, **2014**, pp. 255-268.

[267] Yan, Z.; Caldwell, G.W. Evaluation of cytochrome P450 inhibition in human liver microsomes. In: *Optimization in Drug Discovery: In Vitro Methods*; Yan, Z.; Caldwell, G.W., Eds.; Humana Press Inc.: Totowa, New Jersey, **2004**, pp. 231-244.

[268] Stresser, D.M. High-throughput screening of human cytochrome P450 inhibitors using fluorometric substrates: methodology for 25 enzymes/substrate pairs. In: *Optimization in Drug Discovery: In Vitro Methods*; Yan, Z.; Caldwell, G.W., Eds.; Humana Press Inc.: Totowa, New Jersey, **2004**, pp. 231-244.

[269] Caldwell, G.W.; Yan, Z.; Lang, W.; Masucci, J.A. The IC50 concept revisited. *Curr. Topics Med. Chem.*, **2012**, 12(11), 1282-1290.

[270] Caldwell, G.W.; Yan, Z. Rapidly distinguishing reversible and time-dependent CYP450 inhibition using human liver microsomes, co-incubation, and continuous fluorometric kinetic analyses. In: *Optimization in Drug Discovery: In Vitro Methods* 2nd ed.; Caldwell, G.W.; Yan, Z. Eds.; Humana Press Inc.: Totowa, New Jersey, **2014**, pp. 101-115.

[271] Caldwell, G.W.; Yan, Z. CYP Time-Dependent Inhibition (TDI) Using an IC50 Shift Assay with Stable Isotopic Labeled Substrate Probes to Facilitate Liquid Chromatography/Mass Spectrometry Analyses In: *Optimization in Drug Discovery: In Vitro Methods* 2nd ed.; Caldwell, G.W.; Yan, Z. Eds.; Humana Press Inc.: Totowa, New Jersey, 2014, pp. 315-336.

[272] Flaveny, C.A.; Murray, I.A.; Chiaro, C.R.; Perdew, G.H. Ligand selectivity and gene regulation by the human aryl hydrocarbon receptor in transgenic mice. *Mol. Pharmacol.*, **2009**, 75(6), 1412-20.

[273] Soshilov, A.A.; Denison, M.S. DNA binding (gel retardation assay) analysis for identification of aryl hydrocarbon (Ah) receptor agonists and antagonists. In: *Optimization in Drug Discovery: In Vitro Methods* 2nd ed.; Caldwell, G.W.; Yan, Z. Eds.; Humana Press Inc.: Totowa, New Jersey, **2014**, pp. 207-219.

[274] Qatanani, M.; Moore, D.D. CAR, the continuously advancing receptor, in drug metabolism and disease. *Curr. Drug Metab.*, **2005**, 6(4), 329-339.

[275] Lynch, C.; Li, H.; Wang, H. Characterization of constitutive androstane receptor (CAR) activation. In: *Optimization in Drug Discovery: In Vitro Methods* 2nd ed.; Caldwell, G.W.; Yan, Z. Eds.; Humana Press Inc.: Totowa, New Jersey, **2014**, pp. 195-205.

[276] Cui, X.; Thomas, A.; Gerlach, V.; White, R.E.; Morrison, R.A.; Cheng, K.-C. Application and interpretation of hPXR screening data: Validation of reporter signal requirements for prediction of clinically relevant CYP3A4 inducers. *Biochem. Pharm.*, **2008**, 76(5), 680-689.

[277] Raucy, J.L. Human pregnane x receptor (hPXR) activation assay in stable cell lines. In: *Optimization in Drug Discovery: In Vitro Methods* 2nd ed.; Caldwell, G.W.; Yan, Z. Eds.; Humana Press Inc.: Totowa, New Jersey, **2014**, pp. 171-193.

[278] Singer, M.; Sensenhauser, C.; Dallas, S. *In vitro* CYP induction using human hepatocyte. In: *Optimization in Drug Discovery: In Vitro Methods* 2nd ed.; Caldwell, G.W.; Yan, Z. Eds.; Humana Press Inc.: Totowa, New Jersey, **2014**, pp. 237-253.

[279] DeGorter, M.K., *et al.* Drug transporters in drug efficacy and toxicity. *Annu. Rev. Pharmacol. Toxicol.*, **2012**, 52, 249-273.

[280] Thiebaut, F.; Tsuruo, T.; Hamada, H.; Gottesman, M.M.; Pastan, I.; Willingham, M.C. Cellular localization of the multidrug-resistance gene product P-glycoprotein in normal human tissues. *Proc. Natl. Acad. Sci. USA.*, **1987**, 84, 7735-7738.

[281] Doyle, L.A.; Yang, W.; Abruzzo, L.V.; Krogmann, T.; Gao, Y.; Rishi, A.K.; Ross, D.D. A multidrug resistance transporter from human MCF-& breast cancer cells. *Proc. Natl. Acad. Sci. USA.*, **1998**, 95, 15665-15670.

[282] Kruh, G.D.; Belinsky, M.G. The MRP family of drug efflux pumps. *Oncogene*, **2003**, 22(47), 7537-7552.

[283] Kalliokoski, A.; Niemi, M. Impact of OATP transporters on pharmacokinetics. *Br. J. Pharmacol.*, **2009**, 158(3), 693-705.

[284] Lepist, E.I.; Ray, A.S. Renal drug-drug interactions: what we have learned and where we are going. *Expert Opin. Drug Toxicol.*, **2012**, 8(4), 433-448.

[285] Hoke, S.H.II; Tomlinson, J.A.; Bolden, R.D.; Morand, K.L.; Pinkston, J.D.; Wehmeyer, K.R. Increasing Bioanalytical Throughput Using pcSFC-MS/MS: 10 Minutes per 96-Well Plate. *Anal. Chem.*, **2001**, 73(13), 3083-3088.

[286] Tong, X.; Wang, J.; Ita, I.E.; Zheng, S.; Pivnichny, J.V.; Griffin, P. *High throughput in vivo pharmacokinetics studies and in vitro assays by using Beckman liquid handling system*, ANYL-077; Am. Chem. Soc. 221st National Meeting, San Diego, CA, April 1-5, **2001**.

[287] Gao, V.C.X.; Luo, W.C.; Ye, Q.; Thoolen, M. Column switching in high-performance liquid chromatography with tandem mass spectrometric detection for high-throughput preclinical pharmacokinetic studies. *J. Chromatogr. A*, **1998**, 828(1 + 2), 141-148.

[288] Muck, W. Quantitative analysis of pharmacokinetic study samples by liquid chromatography coupled to tandem mass spectrometry (LC-MS/MS). *Pharmazie*, **1999**, 54(9), 639-644.

[289] Poon, G.K.; Kwei, G.; Wang, R.; Lyons, K.; Chen, Q.; Didolkar, V.; Hop, C.E.C.A. Integrating qualitative and quantitative liquid chromatography/ tandem mass spectrometric analysis to support drug discovery. *Rapid Commun. Mass Spectrom.*, **1999**, 13(19), 1943-1950.

[290] Romanyshyn, L.; Tiller, P.R.; Hop, C.E.C.A. Bioanalytical applications of "fast chromatography" to high-throughput liquid chromatography/ tandem mass spectrometric quantitation. *Rapid Commun. Mass Spectrom.*, **2000**, 14(18), 1662-1668.

[291] Wu, J.-T. The development of a staggered parallel separation liquid chromatography/tandem mass spectrometry system with on-line extraction for high-throughput screening of drug candidates in biological fluids. *Rapid Commun. Mass Spectrom.*, 2001, 15(2), 73-81.

[292] Korfmacher, W. Strategies and techniques for higher throughput ADME/ PK assays. High-Throughput *Anal. Pharm. Ind.*, **2009**, 205-231.

[293] Musteata, F.M.; de Lannoy, I.; Gien, B.; Pawliszyn, J. Blood sampling without blood draws for *in vivo* pharmacokinetic studies in rats. *J. Pharm. Biomed. Anal.*, **2008**, 47(4-5), 907-912.

[294] Xu, X. HPLC method development for drug discovery LC-MS assays in rapid pharmacokinetics applications. *Separation Sci. Technology*, **2007**, 8, 317-351.

[295] Bock, M.J.; Neilson, K.L.; Dudley, A. Use of trifluoroacetic acid to quantify small, polar compounds in rat plasma during discovery-phase pharmacokinetic evaluation. J. Chromatogr. B: *Analyt. Technol. Biomed. Life Sci.*, **2007**, 856(1-2), 165-170.

[296] White, R.E.; Manitpisitkul, P. Pharmacokinetic theory of cassette dosing in drug discovery screening. *Drug Metab. Dispos.*, **2001**, 29(7), 957-966.

[297] Kuo, Be.-S.; Van Noord, T.; Feng, M.R.; Wright, D.S. Sample pooling to expedite bioanalysis and pharmacokinetic research. J. *Pharm. Biomed. Anal.*, **1998**, 16(5), 837-846.

[298] Wu, J.-T.; Zeng, H.; Qian, M.; Brogdon, B.L.; Unger, S.E. Direct Plasma Sample Injection in Multiple-Component LC-MS-MS Assays for High-Throughput Pharmacokinetic Screening. *Anal. Chem.*, **2000**, 72(1), 61-67.

[299] Johnson, D.E.; Wolfgang, G.H.I. Assessing the potential toxicity of new pharmaceuticals. *Curr. Topics Med. Chem.*, **2001**, 1(4), 233-245.

[300] Olson, H.; Betton, G.; Robinson, D.; Thomas K.; Monro A.; Kolaja G.; Lilly P.; Sanders J.; Sipes G.; Bracken W.; Dorato M.; Van Deun K.; Smith P.; Berger B.; Heller A. Concordance of the Toxicity of Pharmaceuticals in Humans and in Animals. *Regul. Toxicol. Pharmacol.*, **2000**, 32(1), 56-67.

[301] Kalgutkar, A.S.; Gardner, I.; Obach, R.S.; Shaffer, C.L.;, Callegari, E.; Henne, K.R.; Mutlib, A.E.; Dalvie, D.K.; Lee, J.S.; Nakai, Y.; O'Donnell, J.P.; Boer, J.; Harriman, S.P. A comprehensive listing of bioactivation pathways of organic functional groups. *Curr. Drug Metab.*, **2005**, 6, 161-225.

[302] Argoti, D.; Liang, L.; Conteh, A.; Chen, L.; Bershas, D.; Yu, C.-P.; Vouros, P.; Yang, E. Cyanide Trapping of Iminium Ion Reactive Intermediates Followed by Detection and Structure Identification Using Liquid Chromatography-Tandem Mass Spectrometry (LC-MS/MS). *Chem. Res. Toxicol.*, **2005**, 18 (10), 1537-1544.

[303] Yan, Z.; Zhong, H.M.; Maher, N.; Torres, R.; Leo, G.C.; Caldwell, G.W.; Huebert, N. Bioactivation of 4-methylphenol (p-cresol) *via* cytochrome P450-mediated aromatic oxidation in human liver microsomes. *Drug Metab. Dispos.*, **2005**, 33 (12), 1867-1876.

[304] Yan, Z.; Li, J.; Huebert, N.; Caldwell, G.W.; Du, Y.; Zhong, H.M. Detection of a novel reactive metabolite of diclofenac: evidence for CYP2C9-mediated bioactivation *via* arene oxides. *Drug Metab. Dispos.*, **2005**, 33 (6), 706-713.

[305] Dikalova, A.E.; Kadiiska, M.B.; Mason, R.P. An *in vivo* ESR spin-trapping study: free radical generation in rats from formate intoxication-role of the Fenton reaction. *Proc. Nat. Acad. Sci. USA*, **2001**, 98(24), 13549-13553.

[306] Pluim, D.; Maliepaard, M.; van Waardenburg, R.C.; Beijnen, J.H.; Schellens, J.H. ^{32}P-postlabeling assay for the quantification of the major platinum-DNA adducts. *Anal. Biochem.*, **1999**, 275, 30-38.

[307] Terashima, I.; Suzuki, N.; Shibutani, S. ^{32}P-Postlabeling/polyacrylamide gel electrophoresis analysis: application to the detection of DNA adducts. *Chem. Res. Toxicol.*, **2002**, 15, 305-311.

[308] Shibutani, S.; Suzuki, N.; Laxmi, Y.R.S.; Schild, L.J.; Divi, R.L.; Grollman, A.P.; Poirier, M.C. Identification of Tamoxifen-DNA Adducts in Monkeys Treated with Tamoxifen. *Cancer Res.*, **2003**, 63(15), 4402-4406.

[309] Maron, D.M.; Ames, B.N. Revised methods for the Salmonella mutagenicity test. *Mutat. Res.*, **1983**, 113, 173-215.

[310] Judson, P.N.; Cooke, P.A.; Doerrer, N.G.; Greene, N.; Hanzlik, R.P.; Hardy, C.; Hartmann, A.; Hinchliffe, D.; Holder, J.; Mueller, L.; Steger-Hartmann, T.; Rothfuss, A.; Smith, M.; Thomas, K.; Vessey, J.D.; Zeiger, E. Towards the creation of an international toxicology information centre. *Toxicology*, **2005**, 213(1-2) 117-128.

[311] Fluckiger-Isler, S.; Baumeister, M.; Braun, K.; Gervais, V.; Nguyen-Hasler, N.; Reimann, R.; Van Gompel, J.; Wunderlich, H.-G.; Engelhardt, G. Assessment of the performance of the AmesII™ assay: a collaborative study with 19 coded compounds. *Mutat. Res.*, **2004**, 558(3), 181-197.

[312] Miyamoto, Y.; Koh, Y.H.; Park, Y.S.; Fujiwara, N.; Sakiyama, H.; Misonou, Y.; Ookawara, T.; Suzuki, K.; Honke, K.; Taniguchi, N. Oxidative stress caused by inactivation of glutathione peroxidase and adaptive responses. *Biol. Chem.*, **2003**, 384(4), 567-574.

[313] Lautraite, S.; Lasserre-Bigot, D.; Bars, R.; Carmichael, N. Optimization of cell-based assays for medium throughput screening of oxidative stress. *In Vitro*, **2003**, 17, 207-220.

[314] Loetchutinat, C.; Kothan, S.; Dechsupa, S.; Meesungnoen, J.; Jay-Gerin, J.-P.; Mankhetkorn, S. Spectrofluorometric determination of intracellular levels of reactive oxygen species in drug-sensitive and drug-resistant cancer cells using the 2', 7' -dichlorofluorescein diacetate assay. *Radiat. Phys. Chem.*, 2004, Volume Date **2005**, 72(2-3), 323-331.

[315] Schoonen, W.G.E.J.; de Roos, J.A.D.M.; Westerink, W.M.A.; Debiton, E. Cytotoxic effects of 110 reference compounds on HepG2 cells and for 60 compounds on HeLa, ECC-1 and CHO cells. *Toxicol. In Vitro*, **2005**, 19(4), 491-503.

[316] Dambach, D.M.; Andrews, B.A.; Moulin, F. New technologies and screening strategies for hepatotoxicity: Use of *in vitro* models. *Toxicol. Pathol.*, **2005**, 33, 17-26.

[317] York, J.L.; Maddox, L.C.; Zimniak, P.; McHugh, T.E.; Grant, D.F. Reduction of MTT by glutathione S-transferase. *BioTechniques*, **1998**, 25(4), 626-628.

[318] Riss, T.L.; Moravec, R.A. Use of Multiple Assay Endpoints to Investigate the Effects of Incubation Time, Dose of Toxin, and Plating Density in Cell-Based Cytotoxicity Assays. *Assay Drug Dev. Technol.*, **2004**, 2(1), 51-62.

[319] US Department of Health and Human Services. US Food and Drug Administration (FDA) http://www.fda.gov/Drugs/DevelopmentApproval Process/HowDrugsareDevelopedandApproved/ApprovalApplications/InvestigationalNewDrugINDApplication/ucm176522.htm [Accessed 19th June, **2015**].

[320] Higgins, J.D. Perspectives on preformulation programs at the discovery -development inferface. *Am. Pharm. Rev.*, **2009**, 12(3), 42-46.

[321] Merisko-Liversidge, E.M.; Liversidge, G.G. Drug nanoparticles: formulating poorly water-soluble compounds. *Toxicolo. Pathol.*, **2008**, 36(1), 43-48.

[322] Huang, L.-F.; Dong, J. Formulation strategies and practice used for drug candidates with water-insoluble properties for toxicology, biology, and pharmacology studies in discovery support. *Water-Insoluble Drug Formul.* 2nd ed., **2008**, pp. 113-132.

[323] Wahlstrom, J.L.; Chiang, P.-C.; Ghosh, S.; Warren, C.J.; Wene, S.P.; Albin, L.A.; Smith, M.E.; Roberds, S.L. Pharmacokinetic evaluation of a 1,3-dicyclohexylurea nanosuspension formulation to support early efficacy assessment. *Nanoscale Res. Lett.*, **2007**, 2(6), 291-296.

[324] Li, P.; Zhao, L. Developing early formulations: Practice and perspective. *Int. J. Pharm.*, **2007**, 341(1-2), 1-19.

[325] Bittner, B.; Mountfield, R.J. Formulations and related activities for the oral administration of poorly water-soluble compounds in early discovery animal studies: an overview of frequently applied approaches: part 2. *Pharm. Ind.*, **2002**, 64(9), 985-991.

[326] Bittner, B.; Mountfield, R. J. Intravenous administration of poorly soluble new drug entities in early drug discovery: the potential impact of formulation on pharmacokinetic parameters. *Curr. Opin. Drug Discov. Dev.*, **2002**, 5(1), 59-71.

[327] Fessey, R.E.; Austin, R.P.; Barton, P.; Davis, A.M.; Wenlock, M.C. The role of plasma protein binding in drug discovery. *Pharmacokinetic Profiling in Drug Research: Biological, Physicochemical, and Computational Strategies*, [LogP2004, Lipophilicity Symposium], 3rd, Zurich, Switzerland, Feb. 29-Mar. 4, 2004 (**2006**), pp. 119-141.

[328] Zeitlinger, M.A.; Derendorf, H.; Mouton, J.W.; Cars, O.; Craig, W.A.; Andes, D.; Theuretzbacher, U. Protein Binding: Do We Ever Learn? Antimicrob. *Agents Chemother.*, 2011, 55(7), 3067-3074.

[329] Crick, E.W.; Osorio, I.; Bhavaraju, N.C.; Linz, T.H.; Lunte, C.E. An investigation into the pharmacokinetics of 3-mercaptopropionic acid and development of a steady-state chemical seizure model using *in vivo* microdialysis and electrophysiological monitoring. *Epilepsy Res.*, **2007**, 74(2-3), 116-125.

[330] Salphati, L.; Childers, K.; Pan, L.; Tsutsui, K.; Takahashi, L. Evaluation of a single-pass intestinal-perfusion method in rat for the prediction of absorption in man. *J. Pharm. Pharmacol.*, *2001*, 53(7), 1007-1013.

[331] Letendre, L.; Scott, M.; Dobson, G.; Hidalgo, I.; Aungst, B. Evaluating barriers to bioavailability *in vivo*: Validation of a technique for separately assessing gastrointestinal absorption and hepatic extraction. *Pharm. Res.*, **2004**, 21(8), 1457-1462.

[332] Chen, Y.; Monshouwer, M.; Fitch, W.L. Analytical Tools and Approaches for Metabolite Identification in Early Drug Discovery. *Pharm. Res.*, **2007**, 4(2), 248-257.

[333] Baranczewski, P.; Stanczak, A.; Kautiainen, A.; Sandin, P.; Edlund, P.-O. Introduction to early *in vitro* identification of metabolites of new chemical entities in drug discovery and development. *Pharmacol. Rep.*, **2006**, 58(3), 341-352.

[334] Beattie, I.; Joncour, K.; Lawson, K. Ultra performance liquid chromatography coupled to orthogonal quadrupole TOF-MS(MS) for metabolite identification. *LC-GC Europe*, **2005**, 19-21, 24-27.

[335] Mager, D.E.; Woo, S.; Jusko, W.J. Scaling pharmacodynamics from *in vitro* and preclinical animal studies to humans. *Drug Metab. Pharm.*, **2009**, 24(1), 16-24.

[336] Parrott, N.; Paquereau, N.; Coassolo, P.; Lave, T. An evaluation of the unility of physiologically based models of pharmacokinetics in early drug discovery. *J. Pharm. Sci.*, **2005**, 94, 2327-2343.

[337] Jones, H.; Parrott, N.; Jorga, K.; Lave, T. A novel strategy for physiologically based predictions of human pharmacokinetics. *Clin. Pharmacokinet.*, **2006**, 45, 511-542.

[338] Hakura, A. Improved AMES test for genotoxicity assessment of drugs: preincubation assay using a low concentration of dimethyl sulfoxide. In: *Optimization in Drug Discovery: In Vitro Methods* 2nd ed.; Caldwell, G.W.; Yan, Z. Eds.; Humana Press Inc.: Totowa, New Jersey, **2014**, pp. 545-559.

[339] Applegate, M.L.; Moore, M.M.; Broder, C.B.; Burrell, A.; Hozier, J.C. Molecular dissection of mutations at the heterozygous thymidine kinase locus in mouse lymphoma cells. *Proc. Natl. Acad. Sci.*, 1990, 87 51-55.

[340] Mei, N.; Guo, X.; Moore, M.M. Methods for using the mouse lymphoma assay to screen for chemical mutagenicity and photo-mutagenicity. In: *Optimization in Drug Discovery: In Vitro Methods* 2nd ed.; Caldwell, G.W.; Yan, Z. Eds.; Humana Press Inc.: Totowa, New Jersey, **2014**, pp. 561-592.

[341] Korfmacher, W.; Fitzgerald, M.; Luo, Y.; Ho, S.; Wang, J.; Wu, Z.; Snow, G.; O'Shea, T. Capillary microsampling of whole blood for mouse PK studies: an easy route to serial blood sampling. *Bioanalysis*, **2015**, 7(4), 449-461.

[342] Liu, S.-Y.; Chang, C.-H.; Lee, T.-W. Single dose acute toxicity testing for N,N-bis(2-mercaptoethyl)-N',N' diethylethylenediamine in beagles. *Reg. Toxicol. Pharmacol.*, **2014**, 69(2), 217-225.

[343] Gralinski, M.R. The assessment of potential for QT interval prolongation with new pharmaceuticals. Impact on drug development. *J. Pharmacol. Toxicol. Methods*, **2000**, 43(2), 91-99.

[344] January, C.T.; Gong, Q.; Zhou, Z. Defective protein trafficking of HERG K+ channels in human congenital long QT syndrome. *Pharm. News*, **2000**, 7(4), 27-34.

[345] Abrahamsson, C. Neuropeptide Y1- and Y2-receptor-mediated cardiovascular effects in the anesthetized guinea pig, rat, and rabbit. *J. Cardiovasc. Pharmacol.*, **2000**, 36(4), 451-458.

[346] Pourrias, B.; Porsolt, R.D.; Lacroix, P. QT interval prolongation by noncardiovascular drugs: a proposed assessment strategy. *Drug Dev. Res.*, **1999**, 47, 55-62.

[347] Fermini, B.; Fossa, A.A. Preclinical assessment of drug-induced QT interval prolongation. Current issues and impact on drug discovery. *Ann. Rep. Med. Chem.*, **2004**, 39, 323-334.

[348] Guo, L.; Guthrie, H. Automated electrophysiology in the preclinical evaluation of drugs for potential QT prolongation. *J. Pharmacol. Toxicol. Methods*, **2005**, 52(1) 123-135.

[349] Hammond, T.G.; Pollard, C.E. Use of *in vitro* methods to predict QT prolongation. *Toxicol. Appl. Pharmacol.*, **2005**, 207(Suppl.), 446-450.

[350] Finlayson, K.; Witchel, H.J.; McCulloch, J.; Sharkey, J. Acquired QT interval prolongation and HERG: implications for drug discovery and development. *Eur. J. Pharmacol.*, **2004**, 500(1-3), 129-142.

[351] Abbott, N.J.; Dolman, D.E.M.; Patabendige, A.K. Assays to predict drug permeation across the blood - brain barrier, and distribution to brain. *Curr. Drug Metab.*, **2008**, 9(9), 901-910.
[352] Kortagere, S.; Chekmarev, D.; Welsh, W.J.; Ekins, S. New Predictive Models for Blood - Brain Barrier Permeability of Drug -like Molecules. *Pharm. Res.*, **2008**, 25(8), 1836-1845.
[353] Blokland, A.; Hinz, V.; Schmidt, B.H. Effects of metrifonate and tacrine in the spatial Morris task and modified Irwin Test: evaluation of the efficacy/safety profile in rats. *Drug Dev. Res.*, **1995**, 36(4), 166-179.
[354] Hood, R.D., Ed.; *Developmental and Reproductive Toxicology: A Practical Approach.*, CRC Press: Boca Raton, FL, **2014**.
[355] Le Coz, F.; Suzuki, N.; Nagahori, H.; Omori, T.; Saito, K. Hand1-Luc embryonic stem cell test (Hand1-Luc EST): a novel rapid and highly reproducible *in vitro* test for embryotoxicity by measuring cytotoxicity and differentiation toxicity using engineered mouse ES cells. *J. Toxicol. Sci.*, **2015**, 40(2), 251-261.
[356] Zou, P.; Xing, L.; Tang, Q.; Liu, R.; Hao, W. Comparative evaluation of the teratogenicity of genistein and genistin using rat whole embryo culture and limbud micromass culture methods. *Food Chem. Toxicol.*, **2012**, 50(8), 2831-2836.
[357] Riss, T.L.; Moravec, R.A. Use of Multiple Assay Endpoints to Investigate the Effects of Incubation Time, Dose of Toxin, and Plating Density in Cell-Based Cytotoxicity Assays. *Assay and Drug Develop. Technol.*, **2004**, 2(1), 51-62.
[358] Mikus, J.; Steverding D. A simple colorimetric method to screen drug cytotoxicity against Leishmania [by] using the dye Alamar Blue. *Parasitol. Int.*, **2000**, 48(3), 265-269.
[359] McMillian, M.K.; Grant, E.R.; Zhong, Z.; Parker; J.B.; Li, L.; Zivin, R.A.; Burczynski, M.E.; Johnson, D. Nile Red binding to HepG2 cells: an improved assay for *in vitro* studies of hepatosteatosis. *In Vitro Mol. Toxicol.*, **2001**, 14(3), 177-190.
[360] Xu, J.J.; Diaz, D.; O'Brien, P.J. Applications of cytotoxicity assays and pre-lethal mechanistic assays for assessment of human hepatotoxicity potential. *Chem.-Biol. Interact.*, **2004**, 150(1), 115-128.
[361] Alarcon-Corredor, O.M.; Alfonso, R. Clinical and biochemical alterations in rats treated with high doses of vitamin A. *Arch. Latinoamericanos de Nut.*, **2007**, 57(3), 224-230.
[362] Bandyopadhyay, S.; Varshney, J.P.; Hoque, M.; Ghosh, M.K. Clinical, serum bio-chemical and enzymatic features of dogs with cholecystic disorders. *Indian Vet. J.*, **2008**, 85(4), 368-370.
[363] Schuurman, H.-J.; Smith, H.T. Reference values for clinical chemistry and clinical hematology parameters in cynomolgus monkeys. *Xenotransplantation*, **2005**, 12(1), 72-5.
[364] Waring, M.J.; Pairaudeau, G.; Arrowsmith, J.; Mandrell, S.; Weir, A.; Leach, A.R.; Pickett, S.D.; Leeson, P.D.; Owen, R.M.; Pennie, W.D.; *et al.* An analysis of the attrition of drug candidates from four major pharmaceutical companies. *Nat. Rev. Drug Discov.*, **2015**, 14(7), 475-86.
[365] Hosseinkhani, H. 3D *in vitro* technology for drug discovery. *Curr. Drug Safety*, **2012**, 7(1), 37-43.
[366] Achilli, T.-M.; McCalla, S.; Meyer, J.; Tripathi, A.; Morgan, J.R. Multilayer Spheroids To Quantify Drug Uptake and Diffusion in 3D. *Molecular Pharma.*, **2014**, 11(7), 2071-2081.
[367] Sengupta, S.; Johnson, B.P.; Swanson, S.A.; Stewart, R.; Bradfield, C.A.; Thomson, J.A. Aggregate Culture of Human Embryonic Stem Cell-Derived Hepatocytes in Suspension Are an Improved *In Vitro* Model for Drug Metabolism and Toxicity Testing. *Toxicol. Sci.*, **2014**, 140(1), 236-245.
[368] Chen, M.; Tung, C.-W.; Shi, Q.; Guo, L.; Shi, L.; Fang, H.; Borlak, J.; Tong, W. A testing strategy to predict risk for drug-induced liver injury in humans using high-content screen assays and the 'rule-of-two' model. *Arch. Toxicol.*, **2014**, 88(7), 1439-1449.
[369] Komasaka, T.; Fujimura, H.; Tagawa, T.; Sugiyama, A.; Kitano, Y. Practical method for preparing nanosuspension formulations for toxicology studies in the discovery stage: formulation optimization and *in vitro/in vivo* evaluation of nanosized poorly water-soluble compounds. *Chem. Pharma. Bulletin*, **2014**, 62(11), 1073-1082.

CHAPTER 2

The Oxidative Stress Menace to Coronary Vasculature: Any Place for Antioxidants?

Alexandros Briasoulis, Charalampos Antoniades, George-Angelo Papamikroulis, Evangelos Oikonomou and Dimitris Tousoulis*

1st Cardiology Department, University of Athens Medical School, "Hippokration" Hospital, Athens, Greece

Abstract: Oxidative stress occurs due to the combination of excess reactive oxygen species and insufficient antioxidant capacity. Oxidative stress has been correlated with endothelial dysfunction, the pathogenesis of atherosclerosis as well as with high incidence of cardiovascular disease. A variety of antioxidants has been studied, during the past few years, for the prevention and treatment of atherosclerosis. Early observational studies, focusing on dietary antioxidants, demonstrate an inverse association between antioxidant intake and major cardiovascular events, and supported a number of large-scale, randomized, placebo-controlled clinical trials, which investigated the effect of selected antioxidant therapies on primary and secondary cardiovascular prevention. However, the findings appear controversial, since disappointing results have been reported from many studies with little or no decrease in cardiovascular risk while others showed significant reduction of the oxidative stress and improvement of endothelial function. A few rational explanations of these controversial findings have been proposed and should be taken into account in future clinical studies. This chapter provides contemporary data concerning pathophysiology of oxidative stress, its relation to atherogenesis and the potential role of antioxidant therapy in reducing cardiovascular risk *via* primary and secondary prevention.

Keywords: Antioxidants, atherosclerosis, cardiovascular risk, clinical trial, Coronary Artery Disease, diet, endothelial dysfunction, free radicals, hyper-lipidemia, hypertension, inflammatory markers, nitric oxide, oxidase enzymes, oxidative markers, oxidative stress, patients, primary prevention, reactive oxygen species, secondary prevention, supplementation, vitamins

INTRODUCTION

The vascular endothelium consists of a thin semipermeable layer of cells upon the basal membrane covering the internal surface of vessels and forming a boundary

***Corresponding author Dimitris Tousoulis:** University of Athens Medical School, Vas.Sophias 114, Athens, Greece; Tel: 30-210-7782446; Fax: 30-210-7784590; E-mail: drtousoulis@hotmail.com

Atta-ur-Rahman, M. Iqbal Choudhary and Allen B. Reitz (Eds.)

between the vessel wall and the blood flow. Its structural and functional integrity is vital for the protection of the vessel wall and circulatory function [1]. Over the last decade, the vascular endothelium has emerged mostly as a paracrine organ responsible for the secretion of several beneficial substances with anti-atherogenic effects [2]. One of these chemical molecules, nitric oxide (NO) is widely known for its anti-thrombotic, anti-apoptotic, anti-inflammatory and antioxidant effects. However, in pathophysiological conditions NO production is reduced, Reactive Oxygen Species (ROS) and adhesion molecules are accumulated, insulin resistance is observed and oxidative status is impaired while oxidative stress constitute an essential link from endothelial dysfunction to progression of coronary artery disease (CAD) progression [3].

Several pro-oxidant pathways related to endothelial injury, have been identified, and these oxidation mechanisms may be important in the structural and functional changes that occur during the initiation and progression of CAD. This consistency has provided the rationale for the study of antioxidant strategies in clinical trials, with the currently available data being rather controversial. In general, secondary prevention trials have shown more consistent benefit than primary prevention trials.

Therefore in this chapter we present current information regarding molecular and biochemical aspects of oxidative pathways and we focus on pharmacokinetic and pharmacodynamic aspects of antioxidants and its therapeutic potential in clinical practice.

PHYSIOLOGY OF REACTIVE OXYGEN SPECIES METABOLISM

ROS, are highly reactive oxygen-based metabolic products with unpaired electrons. Physiologically ROS are signaling molecules that influence many processes, by serving as an integral component of cellular signaling pathways [4]. Under basal conditions, a healthy endothelium maintains its functional and structural vascular properties and keeps the balance between vasoconstriction and vasodilation, growth promotion and inhibition, pro-thrombotic and anti-thrombotic mechanisms, pro-inflammatory and anti-inflammatory status as well as oxidative and anti-oxidative mechanisms [5, 6]. To protect against cellular oxidant stress and its adverse effects, adaptive enzymatic mechanisms have evolved to metabolize ROS, into less reactive forms and minimize their potential to produce oxidative damage and cellular dysfunction. The balance between ROS production and their removal by antioxidant systems is called “redox state”. Oxidative stress is defined as an excess production of ROS relative to the levels of antioxidants which can modify endothelial function, cell and organism phenotype [7].

The forms of ROS that are relevant in biological systems include free radicals, such as superoxide (O_2^-) and hydroxyl radical (OH•), and non-radicals capable of generating free radicals, such as hydrogen peroxide (H_2O_2). ROS are produced by various oxidase enzymes, including nicotinamide-adenine dinucleotide phosphate (NADPH) oxidase, xanthine oxidase, uncoupled endothelial NO synthase (eNOS), cyclooxygenase, glucose oxidase, and lipooxygenase, and mitochondrial electron transport [8, 9].

On the other hand, a number of key antioxidant enzymes such as superoxide dismutases, catalases, glutathione peroxidases, hemeoxygenase, and glucose-6-phosphate dehydrogenase, limit intracellular and extracellular accumulation of deleterious ROS [10, 11].

O_2^- is a primary radical that could lead to the formation of other ROS, such as OH• and H_2O_2. A small amount of O_2^- is normally produced as a byproduct of molecular oxygen during mitochondrial oxidative phosphorylation. O_2^- gets inactivated by either NO or superoxide dismutase (SOD). SOD enzymes, rapidly converts O_2^- to H_2O_2, which is itself broken down by glutathione peroxidase (GSHPx) and catalase to water. Under pathological conditions, the single-electron reduction of H_2O_2 may lead to formation of highly reactive OH radicals, either *via* the Fenton reaction in the presence of endogenous iron or *via* Haber-Weiss reaction by reacting with O_2^- (Fig. **1**).

REACTIVE OXYGNE SPECIES AND ENDOTHELIAL DYSFUNCTION

A major mechanism by which ROS contribute to vascular disease is *via* the superoxide-mediated inactivation of NO. This reaction, which is essentially diffusion-limited, not only results in reduced NO bioavailability, and thus a loss of its vasoprotective actions, but also in the formation of peroxynitrite ($ONOO^-$). Through its powerful oxidizing properties, peroxynitrite causes irreversible damage to macromolecules including proteins, lipids and DNA, thereby disrupting crucial cell signaling pathways and promoting cell death. Peroxynitrite can also react with smaller cellular molecules, including the key eNOS cofactor, tetrahydropterin. In the absence of tetrahydropterin, eNOS becomes enzymatically uncoupled and generates superoxide instead of NO, thus exacerbating oxidative stress within the cell [7, 12]. Superoxide may also cause lipid peroxidation

(LOOH) while myeloperoxidase (MPO), in turn, catalyzes the reaction of hydrogen peroxide and chloride (Cl^-) to form hypochlorous acid (HOCl) [11].

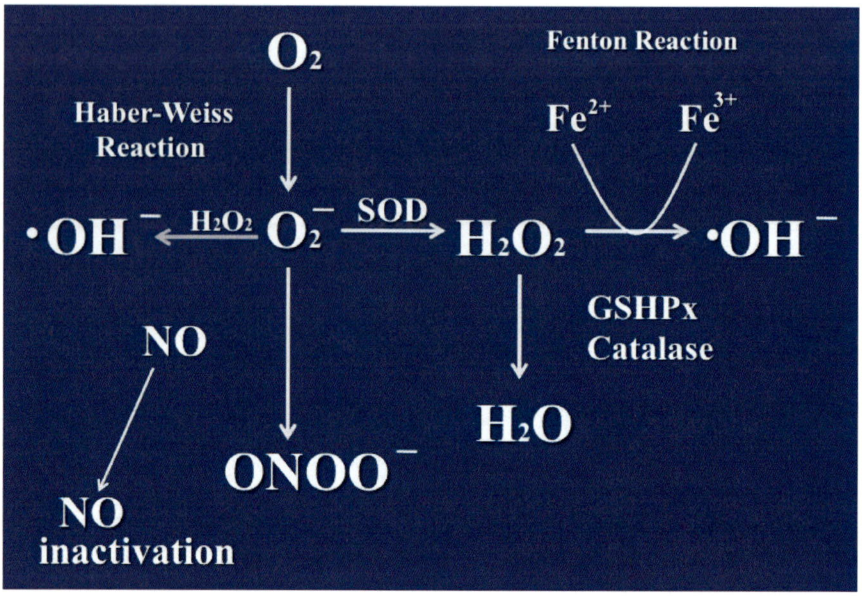

Figure 1: Reactive oxygen species metabolism. O_2^- is a primary radical that could lead to the formation of other ROS, such as OH• and H_2O_2. O_2^- gets inactivated by either nitric oxide or superoxide dismutase. Glutathione peroxidase catalyze H_2O_2 to H_2O. O_2^-: Superoxide; ROS: Reactive oxygen species; OH•: Hydroxyl radical; H_2O_2: hydrogen peroxide; NO: Nitric oxide; SOD: Superoxide dismutase; GSHPx: Glutathione peroxidase catalyse.

In addition cardiovascular risk factors increase the expression and/or activity of NADPH oxidases (Nox) in the vascular wall [12]. NOXs are the only enzymes with the primary function of generating ROS [13, 14]. Other enzymes (cyclooxygenases, cytochrome P450, enzymes of the mitochondrial electron transport chain) can produce ROS, but only as a byproduct of their normal function. Moreover, abnormal or non- physiological sources of ROS exist, including 'uncoupled' eNOS and xanthine oxidase, which are dysfunctional variants of eNOS and xanthine dehydrogenase, respectively [15-17]. Of note, these latter enzymes only become ROS-generators in the presence of an external source of ROS (such as NOXs) and can thus be considered as secondary sources of ROS. NOXs catalyze the transfer of electrons from cytosolic NADPH to molecular O2 *via* their membrane-bound catalytic NOX or DUOX subunit to generate the ROS, super-oxide or H_2O_2 [18].

NOX1, NOX2, NOX4, and NOX5 have emerged as major sources of ROS production in the endothelium. Under physiological conditions these enzymes are expressed at relatively low levels on membranes of the endoplasmic reticulum and nucleus where they contribute to intracellular redox signaling processes. However, in cardiovascular risk/disease settings such as hyperlipidemia, hypertension, diabetes, and atherosclerosis, endothelial expression of NOX2, and to a lesser extent of NOX1, appears to be enhanced, resulting in excessive superoxide generation – especially in the extracellular compartment. This ultimately leads to loss of vasoprotective NO, damage to macromolecules including extracellular matrix proteins, and activation of pro-inflammatory signaling pathways, all of which contribute to vessel disease [12].

Xanthine oxidase is another potential source of ROS associate with vascular disease [11]. It is an iron–sulfur molybdenum flavoprotein enzyme that catalyzes the last steps of purine metabolism, the transformation of hypoxanthine and xanthine to uric acid, with O_2^- or H_2O_2 generation as byproducts. In addition to being a housekeeping enzyme, xanthine oxidase is known as the rate-limiting enzyme in purine catabolism [19]. The enzyme readily donates electrons to molecular oxygen, *via* an one-electron reduction pathway producing O_2^- and a two-electron reduction pathway generating H_2O_2. Both ex vivo experiments and patient studies show beneficial effects of xanthine oxidase inhibitors in post-myocardial infarction ventricular function and endothelial function respectively [20, 21].

THE IMPACT OF OXIDATIVE STRESS IN ATHEROSCLEROSIS

ROS lead to NO inactivation and impaired endothelium-dependent vasorelaxation, while they also cause vascular smooth muscle cell hypertrophy, inflammatory molecule expression, platelet activation and/or adhesion [12], endothelial cells apoptosis [22], increase monocyte adhesion [23] and play an important role in angiogenesis [24].

Risk factors for atherosclerosis induce coronary atherogenesis, partly through the increased free radicals production [25, 26]. Hypertension is associated with enhanced angiotensin II-induced O_2^- production and increased oxidative stress in humans [27], while diabetes mellitus increases free radicals production through the increased formation of advanced glycation end products [25]. More specifically, hyperglycemia stimulates cellular ROS production by different sources, with the mitochondrial electron transport chain acting as the initial O_2^- producer. Overproduction of mitochondrial derived O_2^- leads to activation of protein kinase C (PKC) and formation of advanced glycation end-products (AGEs) [28]. PKC and advanced glycation end-products can, on the one hand, activate NADPH oxidase and, on the other hand, inhibit eNOS activity through post-translational modifications [29].

Smoking also leads to endothelial dysfunction, possibly by a non-specific attenuation of the vessel at the level of vascular smooth muscle and by an enhanced degradation of nitric oxide as a result of endothelial and/or smooth muscle production of oxygen derived free radicals. O_2^- from cigarette smoking may reach the vascular endothelium and react with NO [30], reducing the vasoactive levels of NO, and diminishing the response to endothelium-dependent vasodilators *via* formation of peroxynitrite anion ($ONOO^-$) a highly reactive intermediate with strong cytotoxic potency [30, 31]. The damaging free radicals in cigarette smoke may cause either direct arterial wall injury or may initiate and/or accelerate secondary processes including depletion of antioxidants, lipid peroxidation [32] and activation of phagocyte-platelet endothelial cell interactions. Oxygen free radicals in the vascular wall may initiate a vicious cycle by oxidizing lipids or lipoproteins, which in fact may inhibit endothelium-dependent vasodilation directly or indirectly by the formation of superoxide anions [33].

Another major involvement of free radicals in atherogenesis is the oxidation of LDL, which leads to the formation of oxidized-LDL, a highly reactive free radical with direct cytotoxic effects on endothelial cells, while it is chemoattractant for circulating monocytes and it induces vascular smooth muscle cells proliferation and cytokine production [34]. Furthermore, oxidized-LDL is taken up by monocytes to form foam cells, triggering atherogenesis [25].

STUDIES OF ANTIOXIDANTS AS PRIMARY OR SECONDARY PREVENTION MEASURES IN CORONARY ARTERY DISEASE

Nowadays, there is a large body of experimental evidence in animals and epidemiologic data in humans, obtained from studies that examined plasma levels of antioxidants, supporting the oxidation hypothesis. Unfortunately limited data exist concerning the impact of antioxidant regimens.

Antioxidants are divided into three major categories of molecules: a) antioxidant proteins that bind metal ions, reducing the hydroxyl radical production (e.g., transferrin for iron (Fe) ion, ceruloplasmin and albumin for copper); b) intracellular antioxidant enzymes such as superoxide dismutase, catalase and glutathione peroxidase (whose enzymatic activity is selenium-dependent). The above enzymes are responsible for the trapping of free radicals produced in situ, as by-products of cellular metabolism; c) extracellular antioxidant substances such as α-tocopherol, β-carotene, vitamin C and others.

Beyond these natural antioxidants, there is also a large number of pharmaceutical antioxidant agents used in clinical practice. The best-studied antioxidants used for the treatment of cardiovascular disease are vitamin E and vitamin C. However, the antioxidant capacity of these "antioxidants" cannot easily be determined, since the existent markers of oxidative stress such as malonyldialdehyde (MDA), lipid hydroperoxides (LPO), or F2-isoprostanes are rather unreliable [25].

Observational Studies on Antioxidants and Risk of Coronary Artery Disease (Summarized on Table 1)

Large-scale population-based studies were designed to investigate the association between decreased plasma antioxidant levels and risk for atherosclerosis. In these early studies, research was focused principally on the antioxidants β-carotene, ascorbic acid and α-tocopherol and how their intake was related to atherogenesis and major adverse events. This effort aims to provide information whether dietary antioxidants could explain the different cardiovascular risk observed between patient populations, as well as to provide ample evidence to introduce these compounds as dietary supplements in order to decrease risk of CAD.

Optimism and enthusiasm for trials on antioxidant supplementation came from studies like the Health Professionals Follow-up Study [35], the Nurses Health Study [36], the Eastern Finland Study [37], the ARIC trial [38], the Rotterdam study[39], the Iowa Women's Health Study [40] and the National Health and Nutritional Examination Study (NHANES III) [41, 42].

Findings from the NHANES III, Eastern Finland Study, ARIC trial and Rotterdam study suggested that ascorbic acid was associated with a decreased risk for CAD [37-39, 42]. On the other hand the Health Professionals Follow-up Study, Nurses Health Study and Iowa Women's Study all reported that Vitamin C had no protective role in the prevention of CAD [43].

Despite the great excitement derived from these epidemiological studies concerning the protective profit of antioxidants, when considered together, it becomes definite that they have a number of serious limitations, which may have influenced the results. The metric by which antioxidant intake was assessed,

Table 1. Observational studies on antioxidants and coronary artery disease risk.

Study	Study Type	Vitamin Type	Number of Patients	Age	Population Characteristics	Duration (Years)	Outcome	RR
Health Professionals Follow-up Study [35]	Prospective cohort study	B-carotene	39,910	40-75	Male professionals free of CAD, diabetes or hyper-cholesterolemia	4	β-carotene decreases cardiovascular events Greater risk reduction for former or never smokers	0.71
Nurses Health Study [36]	Prospective cohort study	Carotenoids	73,286	30-55	Healthy female nurses	12	Inverse association between dietary β-carotenoid levels and incident CAD	0.74
Eastern Finland Study [37]	Prospective cohort study	Vitamin C	1,605	42-60	Men without either symptomatic CAD or ischemia on exercise testing at entry	6	Vitamin C deficiency is a risk factor for CAD	3.5
ARIC trial [38]	Prospective cohort study	Dietary and supplemental antioxidants (Vitamin C, α-tocopherol and provitamin A carotenoids)	6,318 females and 4,989 males	45-64	Dietary criteria	4	Limited support for the hypothesis that dietary vitamin C and alpha-tocopherol may protect against atherosclerotic disease, especially in individuals > 55 years old.	
Rotterdam study [39]	Prospective cohort study	Dietary beta-carotene, vitamin C, and vitamin E	4,367	55-94	No previous cardiovascular disease	4	Vitamin C intake was significantly inversely associated with peripheral arterial disease in women In men, vitamin E intake was inversely associated with peripheral arterial disease	0.64 and 0.67
Iowa Women's Health Study [40]	Prospective cohort study	Dietary antioxidant vitamins	34,486	55-69	Postmenopausal women with no cardiovascular disease	7	In postmenopausal women the intake of vitamin E from food is inversely associated with the risk of death from CAD. Intakes of vitamins A and C was not associated with lower CAD mortality	0.42 for the highest quartile of vitamin E consumption
National Health and Nutritional Examination Study (NHANES III) [41]	Prospective cohort study	Serum carotenoids	13,293	(35-90)	Adults enrolled in the Third National Health and Nutrition Examination Survey, 1988 to 1994, with measured serum carotenoids and mortality follow-up through 2006	18	Higher α and β-carotene levels were associated with lower rates of cardiovascular mortality	1.38

CAD: Coronary artery disease; RR: Relative risk; In this column we present the relative risks where available.

the lack of measures of plasma antioxidant levels, and other potential confounders such as healthy lifestyle behaviors that correlate with antioxidant intake may have played a role in the positive outcomes [44].

Early Clinical Trials on Antioxidants and Cardiovascular Disease Prevention (Summarized on Table 2)

The encouraging findings from the aforementioned observational studies set up the basis to support a number of randomized primary and secondary prevention clinical trials.

The primary prevention Alpha-Tocopherol, Beta-Carotene Cancer Prevention (ATBC) study examined cardiovascular disease as a secondary outcome in a total of 29,133 male smokers randomized to α-tocopherol, β-carotene, a combination of the two, or placebo. At eight years of follow-up, neither antioxidant was found to decrease significantly cardiovascular disease [45, 68, 69]. Similar results were observed in the Beta-Carotene and Retinol Efficacy Trial (CARET), which was early terminated after finding that β -carotene and retinol (vitamin A) had no considerable or statistically important effect on incident cardiovascular disease [47, 70]. Negative findings were also reported in the Vitamin E Atherosclerosis Prevention Study (VEAPS) [71].

Vitamin C has been studied in combinations with vitamin E, beta carotene, and other vitamins and minerals in large scale clinical trials that reported no significant cardiovascular effects. Vitamin C alone has only been evaluated among 8171 women at high risk for cardiovascular disease [53], and 13887 men at low-intermediate risk for cardiovascular disease [55] and no effect on cardiovascular disease was found with use of 500 mg/day of vitamin C. Therefore, the clinical utility of vitamin C alone in preventing cardiovascular disease among those at low initial risk of cardiovascular disease remains uncertain

Despite the negative or neutral results reported in the majority of prevention studies, the Women's Health Study in 40.000 women had favorable results [51]. Furthermore, the Cambridge Heart Antioxidant Study (CHAOS) [48] a secondary prevention trial observed a reduced cardiovascular risk in patients with established disease owed to the use of antioxidants. An interesting study which was carried out to determine whether vitamin C infusion before elective percutaneous coronary intervention is able to affect the incidence of periprocedural myocardial injury revealed that a 3 gm vitamin C infusion within 6 hours before intervention can reduce periprocedural myocardial injury [66].

Table 2. Prospective clinical trials on antioxidants and cardiovascular disease prevention.

Study	Prevention	Vitamin Type	Vitamin Dose	Number of Patients	Age	Population Characteristics	Duration (Years)	Outcome	RR
Alpha-Tocopherol–Beta-Carotene Cancer Prevention Study [45]	primary	Vitamin E	50 mg	22,269	50–69	Male smokers	4.7	Decreases angina pectoris events	0.91 (0.83–0.99)
Alpha-Tocopherol–Beta-Carotene Cancer Prevention Study [46]	secondary	Vitamin E	50 mg	1,862	50–69	Male smokers who had a history of MI	5.3	38% reduction in the occurrence of MI	0.62 (0.41–0.96)
Alpha-Tocopherol–Beta-Carotene Cancer Prevention Study [45]	primary	b-Carotene	20 mg	27,271	50–69	Male smokers with no history of MI	6.1	No effect on CAD	1.03 (0.91–1.16)
Alpha-Tocopherol–Beta-Carotene Cancer Prevention Study [45]	primary	Vitamin E+ b-Carotene	50mg of Vitamin E and 20mg of b-carotene	27,271	50–69	Male smokers with no history of MI	6.1	No effect on CAD	0.97 (0.86–1.09)
Beta-Carotene and Retinol Efficacy Trial [47]	primary	b-carotene +retinol	30 mg b-carotene, 25,000 IU retinol	18,314	45–69	Heavy smokers and asbestos workers	4	Borderline increases CAD mortality	1.26 (0.99–1.61)
Cambridge Heart Antioxidant Study (CHAOS) [48]	secondary	Vitamin E	400 or 800 IU	2,002	62	Subjects with CAD	1.4	Decreases nonfatal acute MI	0.23 (0.11–0.47)
Heart Outcomes Prevention Evaluation trial [49]	primary	Vitamin E	400 IU	9,541	>55	High risk for CAD subjects	4.5	No effect on MI, stroke	1.05 (0.95–1.16)
Heart Protection Study [50]	secondary	Vitamin C+ Vitamin E+ b-caroten	600 mg vitamin E, 250 mg vitamin C, 20 mg b-carotene	20,536	40–80	Subjects at high CAD risk	5	No effect on CAD mortality	1.05 (0.95–1.15)
Women's Health Study [51]	Primary	α-tocopherol	600IU of natural-source vitamin-E on alternate days	39,876	≥45	Healthy women	10.1	Decrease in cardiovascular mortality in healthy women, but no effect on the incidence of myocardial infarction	0.76

Table 2. contd....

Study	Prevention	Vitamin Type	Vitamin Dose	Number of Patients	Age	Population Characteristics	Duration (Years)	Outcome	RR
SU.VI.MAX [52]	primary	Vitamin C+ Vitamin E+ b-caroten+ zinc+ selenium	120mg+ 300mg+ 6mg+20mg+ 100µg	1,162	53	Healthy Subjects	7.2	No effect on carotid atherosclerosis and arterial stiffness	-
Women's Antioxidant Cardiovascular Study [53]	primary	b-carotene	50 mg on alternate days	8,171	Post-menopausal women >40	Subjects with CAD history/risk	9,4	No effect MI, Stroke, CAD	1.02 (0.92–1.13)
Women's Antioxidant Cardiovascular Study [53]	secondary	Vitamin C	500 mg	8,171	Post-menopausal women >40	Subjects with CAD history/risk	9.4	No effect on MI, Stroke, CAD	1.02 (0.92–1.13)
Women's Antioxidant Cardiovascular Study [53]	primary	Vitamin E	600 IU	8,171	Post-menopausal women >40	Subjects with CAD history/risk	9.4	No effect MI, Stroke, CAD	0.94 (0.85–1.04)
Physicians' Health Study [54]	primary	b-Carotene	50 mg on alternate days	22,071	40–84	Healthy male physicians	12	No effect on MI	0.96 (0.84–1.09)
Physicians' Health Study II [55]	primary	Vitamin E	400IU	13,887	>50	Males at low/medium risk	10	No effect on cardiovascular mortality	1.07 (0.90-1.28)
Physicians' Health Study II [55]	primary	Vitamin C	500mg/day	13887	>50	Males at low/medium risk	10	No effect on cardiovascular mortality	1.02 (0.85-1.21)
Physicians' Health Study II [55]	secondary	Vitamin E	400IU	754	>50	Males with cardiovascular disease	10	No effect on cardiovascular mortality	0.82 (0.62-1.09)
Physicians' Health Study II [55]	secondary	Vitamin C	500mg/day	754	>50	Males with cardiovascular disease	10	No effect on cardiovascular mortality	0.96 (0.72-1.27)
Skin Cancer Prevention Study [56]	primary	b-Carotene	50 mg	1,805	<85	Subjects with skin cancer patients	8.2	No effect on CAD mortality	1.16 (0.82–1.64)
Gruppo Italiano per lo Studio della Sopravvivenza nell'Infarto Miocardico-Prevention Trial [57]	secondary	Vitamin E	300 mg	11,324	No limits	Post-MI adults subjects	3.5	No effect on MI, stroke	0.98 (0.87–1.10)

Table 2. contd....

Study	Prevention	Vitamin Type	Vitamin Dose	Number of Patients	Age	Population Characteristics	Duration (Years)	Outcome	RR
Secondary Prevention with Antioxidants of Cardiovascular Disease in End-stage Renal Disease [58]	secondary	Vitamin E	800 IU	196	40–75	Hemodialysis patients	2	Decreases CVD	0.46 (0.27–0.78)
Microalbuminuria Cardiovascular Renal Outcomes/HOPE [59]	secondary	Vitamin E	400 IU	3,654	65	Diabetic subjects	4.5	No effect on MI, stroke	1.03 (0.88–1.21)
Antioxidant Supplementation in Atherosclerosis Prevention Study [60]	secondary	Vitamin C+ Vitamin E	182IU a-tocopherol, 500 mg vitamin C	520	45–69	Hypercholesterolemic subjects	3	Slow the progression of IMT thickening in men but not in women	-
Primary Prevention Project [61]	Primary	Vitamin E	300mg	4,495	Mean age 64years	Subjects at CAD risk	3.6	No effect on MI, stroke	1.07 (0.74–1.56)
Probucol and multivitamins in the prevention of restenosis [62]	Secondary	Probucol or Vitamin C+ Vitamin E+ b-caroten or both	500mg, 500mg, 700IU, 30,000IU	317	Mean age 58.5 year	Subjects before and after coronary angioplasty	6 months	No beneficial effect of antioxidant	-
HOPE-TOO [63]	Secondary	Vitamin E	400IU	2007	Mean age 66 years	Subjects with vascular disease or diabetes	4	No effect on cardiovascular mortality	1.04 (0.96-1.14)
Ujeyl A *et al.* [64]	Primary	Vitamins C and E	500-1000mg 400-800IU Twice/day	40	-	Adults 2 years post-heart transplantation	10	Delay of allograft vasculopathy with antioxidants	0.19 (0.05-0.76)
Devaraj S *et al* [65]	Secondary	Vitamin E	1200IU/day	90	52-68	Patients with CAD	2	α-tocopherol supplementation in patients with CAD was safe and significantly reduced plasma biomarkers of oxidative stress and inflammation but had no significant effect on carotid intima-medial thickness during 2 years of follow-up	0.5
Wanq ZJ *et al* [66]	Primary	Vitamin C	3gr	532	-	Patients scheduled for elective PCI for de novo lesions in native coronary arteries	1	Vitamin C infusion before PCI was associated with a lower incidence of peri-procedural MI	-
Women's Angiographic Vitamin and Estrogen (WAVE) Trial [67]	secondary	Vitamin C+ Vitamin E+ estrogen	0.625 equine estrogen/ day, 400 IU vitamin E and 500 mg vitamin C bid	423	Postmenopausal	Post-menopausal women with CAD	5	No effect on MI, stroke	

MI, myocardial infarction; CAD, Coronary artery disease; RR, relative risk; IMT; Intima-media thickness; PCI: Percutaneous coronary intervention. In the RR column we present the relative risk with the 95% confidence intervals where available.

To elucidate the complex associations between antioxidant and cardiovascular risk several metanalysis have been conducted. In a pooled analysis of 9 cohorts, vitamin C supplement use exceeding 700 mg/d was significantly associated with a 25% reduction in coronary heart disease risk [72]. However in one analysis included data from 188,209 individuals who participated in 15 placebo-controlled trials antioxidants failed to favorably influence the cardiovascular events (RR=1.00, 95% CI: 0.96-01.03). This meta-analysis, however had a few limitations, such as the use of pooled and non patient-level data and could not account for potential differences as a result of supplement dose and follow up times [73]. Another meta-analysis identified 50 randomized controlled trials including 294,478 in total (156,663 in the treatment or intervention group and 137,815 in the control or referent group) to evaluate the effect of antioxidants on cardiovascular disease. Using a fixed effects model, the primary analysis found that vitamins or antioxidant supplements did not reduce the risk for major cardiovascular events (RR=1.00; 95% CI:0.98-1.02), leading to the conclusion that antioxidant vitamins or supplements had no beneficial effect in the primary or secondary prevention of cardiovascular disease [74].

Other Antioxidants and Related Clinical Trials (Summarized on Table 3)

Coenzyme Q10 (CoQ10)

Coenzyme Q10 (2,3 dimethoxyl-5 meth-6-decaprenyl benzoquinone) is a key component of the electron transport chain. It can be supplemented orally and is derived from mevalonic acid and phenylalanine [96]. The mechanism of CoQ10 action is not likely to be secondary to a superoxide scavenger effect given CoQ10's hydrophobic properties. CoQ10 increases the ability of Complexes I and II to transfer electron down the mitochondrial electron transport chain and therefore can reduce mitochondrial superoxide production [97]. CoQ10 may also exert an antioxidant effect by reducing lipid peroxidation at the level of the plasma membrane [98]. Lee BJ *et al.* have shown that higher plasma levels of CoQ10 were significantly associated with reduced risk of cardiovascular disease, indicating its potential cardioprotective role [99].

A few studies have demonstrated favorable results regarding the correlation between CoQ10 supplementation, leading to higher serum levels of CoQ10 and decrease in inflammatory molecules [75-77]. These findings advocate the role of CoQ10 as an antioxidant.

Table 3. Other antioxidants and relevant clinical trials.

Study	Prevention	Antioxidant Type	Antioxidant Dose	Number of Patients	Age	Population Characteristics	Duration	Outcome
Lee BJ *et al.* [75]	Secondary	CoQ10	300mg/day	51	55.4-83.2	Patients with CAD	12 weeks	Significantly lower levels of inflammatory markers (TNF-a and IL-6)
Lee BJ *et al.* [76]	Secondary	CoQ10	60mg/day or 150mg/day	51	69.8-84.6	Patients with CAD	12 weeks	CoQ10 supplementation at a dosage of 150 mg decreases C-reactive protein, IL-6, homocysteine, malondialdehyde and superoxidase dismutase
Tiano L *et al.* [77]	Secondary	CoQ10	300mg/day	38	Mean age 55±4	CAD patients	1 month	Improvement in the endothelium-dependent vasodilation, in the endothelium extracellular superoxide dismutase activity
Tome-Carneiro *et al.* [78]	Secondary	Resveratrol	8mg/day	35	47-75	Hypertensive male patients with type II diabetes mellitus	1 year	Treatment decreases inflammatory cytokines, such as C-C motif chemokine ligand 3, Tumor Necrosis Factor-α, IL-6, -8 -1β and C-X-C motif chem0kine ligand 2
Tome-Carneiro *et al.* [79]	Secondary	Resveratrol	8 mg/day	75	48-72	Stable-CAD patients	1	Resveratrol increases the anti-inflammatory serum adiponectin and decreases the thrombogenic plasminogen activator inhibitor type 1
Maqyar K. *et al.* [80]	Secondary	Resveratrol	10 mg/day	40	-	Post-infarction Caucasian patients	3 months	Improves left ventricle diastolic function, endothelial function, reduces LDL-cholesterol levels
Potsdam arm of the European Prospective Investigation into Cancer and Nutrition study [81]	Primary	Chocolate	Mean intake ~6gr/day	19,357	35-65	Healthy adults	8 years	Chocolate consumption appears to lower cardiovascular disease risk, in part through reducing blood pressure. The inverse association may be stronger for stroke than for myocardial infarction.
Zomer E. *et al.* [82]	Primary	Chocolate	100gr/day	2013	-	People with hypertension who met the criteria for metabolic syndrome, with no history of cardiovascular disease and not receiving antihypertensive therapy	10 years	The blood pressure and cholesterol lowering effects of dark chocolate consumption are beneficial in the prevention of cardiovascular events in a population with metabolic syndrome..
Inami S. *et al.* [83]	Primary	Polyphenol 70S	500mg/day of catechin, which is equivalent to 6 cups of green tea per day	40	28.6-49.7	Healthy adult volunteers	4 weeks	reduces circulating Ox-LDL

Table 3. contd….

Study	Prevention	Antioxidant Type	Antioxidant Dose	Number of Patients	Age	Population Characteristics	Duration	Outcome
Widlansky ME *et al.* [84]	Secondary	Epigallocatechin gallate (EGCG)	300mg/day	42	48-68	Clinically stable CAD patients	2 weeks	Epigallocatechin gallate acutely improves endothelial function
Siani A *et al.* [85]	Primary	L-arginine	-	6	39±4	Healthy adults	1 week	Reduces blood pressure
Siasos G *et al.* [86]	Primary	L-arginine	21gr/day	12	-	Healty smokers	2 days	Improves endothelial function
Kelly BS *et al.* [87]	Primary & Secondary	L-arginine	9 gr/day for 9 days following by 18 gr/day for 9 more days	16	-	6 normotensive volunteers and 10 kidney transplant patients	18 days	Restores NO-mediated vasidilatory response and reduction in blood pressure
Yin WH *et al.* [88]	Secondary	L-arginine and Vitamin C	10 gr/day L-arginine or 500 mg/day Vitamin C, as active control	31	-	Stable CAD patients	6 weeks	Consumption of either L-arginine or vitamin C significantly increases brachial artery flow-mediated dilatation
Lee BJ *et al.* [89]	Secondary	Acetyl-L-Carnitine	1000 mg/day	47	61.3-83.8	CAD patients	12 weeks	increases activity of antioxidant enzymes
Selenium Therapy in Coronary Artery disease Patients (SETCAP) Study [90]	Secondary	Sodium selenite	200 or 500 μg/day	465	61-70	CAD patients	12 weeks	Dose-dependent increase in Gltutathione peroxide-1 (GPx-1) protein expression and activity
Mohammadpour AH *et al.* [91]	secondary	Pentoxifylline (PTX)	400 mg x 3 / day	40	44.43-63.15	CAD patients	2 months	Decreases serum levels of Intercellular adhesion molecule 1 and Vascular cell adhesion molecule 1
Fernandes JL *et al.* [92]	Secondary	Pentoxifylline (PTX)	400 mg /day	64	18-80	Patients with Acute Coronary Syndromes (ACS)	6 months	Significant reduction of CRP and TNF-alpha
Sobeninin IA *et al.* [93]	Secondary	Garlic powder	150 mg x 2 / day	51	40-55	CAD patients	12 months	Decreases LDL cholesterol levels and cardiovascular risk
Wu YZ *et al.* [94]	Secondary	Ginkgo biloba extract (GBE)	87.5 mg/day	80	45-74	CAD patients	2 weeks	Increases left anterior descending coronary artery blood flow
Enseleit F *et al.* [95]	secondary	Pycnogenol	200 mg/day	23	≥18	CAD patients	8 weeks	Improves flow-mediated dilatation and decreases 15-F_{2t}-Isoprostane

CAD: Coronary artery disease; IL: Interleukin; ox-LDL: oxidized low density lipoprotein; CRP: C reactive protein; TNF-alpha: Tumor necrosis factor alpha; LDL: Low density lipoprotein.

On the other hand, there is very limited information to determine whether CoQ10 would prevent cardiovascular disease; however, a meta-analysis of 12 trials including 362 patients in total, found that CoQ10 supplements decreased blood pressure with a drop in systolic blood pressure of 11-17mmHg and diastolic blood pressure of 8-10mmHg, compared with placebo. Despite these optimistic findings, due to the unreliability of some of the studies included, no recommendations could be made [100]. Another meta-analysis including 194 patients found that CoQ10 improved endothelial function, as assessed by flow-mediated dilatation [101].

Although these early studies are encouraging, this antioxidant remains to be studied in adequately powered large-scale randomized trials.

Resveratrol (RES)

Resveratrol (3,5,4′ -trihydroxystilbene) is a polyphenolic compound found in the skin of many berries and grapes. Its reported benefits are wide ranging, and include anti-oxidant, anticancer, cardioprotective, and anti-aging effects, and it is thought to be primarily responsible for the beneficial effects of red wine [102]. Specifically in regard to cardioprotection, the drug has been shown to be anti-atherogenic [102], reduces low density lipoprotein (LDL) while increases high-density lipoprotein (HDL) [103-105], reduces oxidative stress by upregulating antioxidant enzymes such as superoxide dismutase, glutathione reductase and catalase [106, 107], improves glucose homeostasis [105, 108], prevent platelet aggregation [109-111] and monocyte recruitment [112], promote coronary vasorelaxation [113, 114]. As resveratrol has so many actions it is an intriguing preventive and therapeutic option for the treatment of CAD, which itself is an extremely complex disease.

Early studies concerning the supplementation of diet with resveratrol exhibit promising results [78-80]. More specifically, dietary intake of resveratrol decreases inflammatory cytokines and LDL and improves left ventricle diastolic and endothelial function.

Flavonoids

Flavonoids are constitutes from several polyphenols. They are found in multiple fruits, vegetables, and beverages such as red wine, tea, cocoa, and dark chocolate. [115]. Flavonoids have been tested as antioxidant therapy for cardiovascular disease since epidemiological data suggest improved cardiovascular outcomes in individuals with high intake of dietary flavonoid [115-117]. Moreover, *in vitro* and cellular studies suggest the antioxidant efficacy of these substances [118-121]. Nevertheless, flavonoids have limited bioavailability after oral consumption which suggests that beyond their scavenging activity they have an additional role in cell to cell interactions which may be the key mechanism mediating beneficial cardiovascular effects [121]. This hypothesis is further confirmed by evidence showing that inhibition of NADPH oxidase activity occurs through molecular pathways that restrain angiotensin converting enzyme [122, 123], increase eNOS-specific NO production through the estrogen receptor [124], and alter COX-2 expression [125]. Recently we have shown that concord grape juice, a rich source of flavonoids, ameliorates endothelial function and imroves vascular elasticity of the arterial circuit in healthy smokers and attenuates acute smoking-induced impairment of arterial wall properties [126]. However, studies investigating the anti-hypertensive effects of flavonoids are inconclusive. In several small studies of short duration dark chocolate therapy have demonstrated blood pressure lowering effects in hypertensive patients [127-130]. Nevertheless, studies in normotensive and pre-hypertensive individuals have demonstrated no benefit [131, 132]. The specific flavonoids and combination of flavonoids that exert the largest beneficial effects remain unknown.

Dark Chocolate/Cocoa

Dark chocolate/cocoa has been consumed for at least 2500 years thanks to its good taste, and for long time it has been regarded as a medicine [133]. It is naturally polyphenol-rich food. The favorable cardiovascular effects of cocoa have been shown by several randomized studies. These effects includes amelioration of systemic hypertension, enhancement of endothelial function, improvement of arterial stiffness, glycemic control and platelet adhesiveness [134, 135] which are mediated by polyphenols [81, 82]. Hence, direct dietary supplementation with flavonol instead of whole chocolate consumption deserves further study.

Tea and Catechins

Lower cardiovascular risk following tea consumption has been observed for many years, but the molecular pathways remain unidentified [136]. Tea, especially black and green tea, contains catechins, which are monomeric flavanols that include epicatechin, epigallocatechin, epigallocatechingallate and epicatechingallate. *In vitro* studies have confirmed that polyphenolic compounds and tea catechins can effectively scavenge ROS,by enhancing production and//or bioavailability of endothelium-derived NO, or have indirect antioxidant activityby effecting transcription factors and enzyme activities [137-139]. Inami S. *et al.* administered 500mg of Polyphenon 70S, extracted from green tea leaves with hot water. Its compounds include epigallocatechingallate (35.6%), epicatechingallate (10.2%), gallocatechingallate (3.8%), catechingallate (0.6%), epigallocatechin (15.0%), and epicatechin (8.7%). In the catechin group, each individual received Polyphenon 70S after a meal once a day *per os* capsules containing 500 mg of catechin. 500 mg of catechin is equal to the dose containing 6 or 7 cups of green tea [83]. Widlansky ME *et al.* administered 300mg of epigallocatechingallate EGCG [84]. Both studies showed favorable results regarding cardiovascular risk, such as reduction of oxidized LDL and improvement in endothelial function respectively. In a meta-analysis of 10 cohort studies and 7 case-control studies, Peters U *et al.* demonstrated a decrease by 11% in the incidence rate of myocardial infarction with an increase in tea consumption by 3 cups per day [115]

L-arginine

L-arginine is an amino acid as well as the main substrate of eNOS in a reaction whose end-product is NO and depends on tetrahydrobiopterin [140]. Potential dietary sources include milkproducts, beef, wheat germ, nuts, and soybeans. Reduction in intracellular levels of L-arginine has been observed in hypertensive subjects [141, 142].While L-arginine deficiency itself does not appears to lead to uncoupling of eNOS [143], it may lead to lower bioavailability of NO, probably contributing to hypertension. Interestingly, short term treatment with L-arginine improves endothelial function and vascular elastic properties of healthy smokers [126, 144], supposing that L-arginine supplementation might limit hypertension by re-establishing physiological NO bioavailability. This concept is supported by studies demonstrating the anti-hypertensive effect of L-arginine supplementation in salt-sensitive rats [145], healthy human subjects [85], hypertensive diabetics [146] patients with chronic kidney disease [87] and diabetic patients in combination with N-acetylcysteine, a precursor of glutathione [146]. L-arginine's anti-hypertensive response may be mediated in part by its suppressive effects on angiotensin II and endothelin-1, and its potentiating effects on insulin [147]. Findings from another study suggest that oral L-arginine supplement improved endothelial function and reduced LDL oxidation in stable CAD patients [88].

Nevertheless, potential toxic side-effects of homocysteine due to L-arginine supplementation have raised new concerns [148]. The majority of L-arginine is converted to creatine, leading to increased homocysteine levels [149].

Homocysteine can increase oxidative stress [150]. According to a recent study this mechanism correlates with L-arginine metabolism in humans [151], suggesting a potential pathway for neutralizing the eNOS-related anti-oxidant effects of L-arginine.

Acetyl-L-Carnitine and α-Lipoic Acid (αLA)

Acetyl-L-Carnitine is a key compound in the transport of fatty acids into mitochondria for beta-oxidation. The antioxidant mechanism of Acetyl-L-Carnitine supplementation appears to be secondary to reductions in mitochondrial ROS production in synergy with concomitant α-Lipoic Acid therapy [152]. The exact intra-mitochondrial mechanism of Acetyl-L-Carnitine's effects are not clear, and prior work in older rats demonstrates Acetyl-L-Carnitine potential to be pro-oxidative when used alone [153]. The effect of Acetyl-L-Carnitine was investigated by Lee BJ *et al.* [89], who conclude that CAD patients might benefit from using Acetyl-L-Carnitine supplements to increase their anti-oxidation capacity, since Acetyl-L-Carnitine supplementation at a dose of 1000 mg/day was found to be associated with a significant reduction in oxidative stress and an increase in antioxidant enzymes activities (Catalase, Superoxide dismutase).

α-Lipoic acid is a dithiol compound synthesized from octanoic acid in mitochondria. The *in vivo* and *in vitro* effects of α -Lipoic Acid have been thoroughly reviewed elsewhere [154, 155]. α -Lipoic Acid has moderate oral bioavailability [156]. While α -Lipoic Acid is a potent *in vitro* antioxidant, the limited plasma concentrations achievable with supplementation and rapid clearance of α-Lipoic Acid suggest that free radical scavenging and anti-oxidant recycling are less likely to be the primary *in vivo* activity of α -Lipoic Acid. Participation in mitochondrial-associated metabolic pathways, in cell signaling that may improve coupling of eNOS, and anti-inflammatory responeses are among the potential beneficial effects of α -Lipoic Acid supplementation [155, 157].

Human data with respect to the anti-hypertensive effects of these compounds is limited to two small studies which have shown some promising results. Consistent with animal data,combined Acetyl-L-Carnitine and α-Lipoic Acid therapy reduced systolic blood pressure in coronary arterydisease patients with hypertension and/or metabolic syndrome at the time of enrollment [158].Also consistent with prior cell culture and animal work [159], type 2 diabetic subjectssupplemented with 2 grams/day of acetyl-L-carnitine showed significantly lowered bloodpressure and improved insulin sensitivity [160].

Selenium

Selenium is a nonmetal trace element indispensable for the formation of selenoproteins, such as the antioxidant enzymes glutathione preoxidase and thioredoxin reductase. Dietary sources of selenium include nuts, particularly Brazilian nuts, cereals, meat, tuna, eggs and mushrooms. Clinical trials on selenium supplementation for the primary prevention of cardiovascular disease have been conducted. Twelve randomized controlled trials with totally 19,715 patients who were followed up for three months or longer were included in a meta-analysis. According to this analysis, selenium supplementation had no effect on all-cause or cardiovascular mortality and no effect on any major adverse cardiovascular event. Taking into deep account all the available data, it is clear that there is not sufficient data to justify the dietary supplementation with selenium for primary prevention of cardiovascular disease [161]. As far as the secondary prevention of coronary artery disease is concerned Schnabel R. *et al.* found that Sodium selenite supplementation increases Glutathione peroxidase-1 activity in endothelial cells and in CAD patients [90]. Future studies though have to demonstrate whether long-term CAD outcome can be improved.

Pentoxifylline (PTX)

Pentoxyfilinit is a phosphodeisterase inhibitor with immune-modulating properties and ability to inhibit T helper 1 lymphocytes activity. In animal models pharmacological treatment with pentoxyfilin has resulted in reduced atherosclerosis and modification in plaque morphology [162-164]. In humans only few studies have focused specifically in modulating this particular immune response. Pentoxyfilin has been shown to increase Inteleukin 10 (IL-10), reduce IL-6 and inhibit CRP expression and activity in peripheral blood cells [165, 166] and has been extensively used in patients with peripheral vascular disease [167]. Fernandes JL *et al.* demonstrated a beneficial effect of pentoxyfilin in the immunomodulation of the inflammatory activity in patients with CAD with a significant decrease in pro-inflammatory markers and increase in anti-inflammatory cytokines towards a more stable immune phenotype [92].

Garlic (Allium sativum L. fam. Alliaceae)

Garlic (Allium sativum L. fam. Alliaceae) is one of the mostly-studied herbal remedies. Traditionally, it was considered as first line treatment of infections, rheumatic disorders, heart problems, diabetes mellitus, and many other pathologies.

Several experiments have confirmed its beneficial effects on hypertension, bacterial infections and hyperglycemia [168]. Clinically, garlic has been evaluated for a number of conditions, including hypertension, hypercholesterolemia, diabetes, and for the prevention of arteriosclerosis and cancer [168]. Several studies have indicated that garlic and its constituents inhibit key enzymes involved in cholesterol and fatty acid synthesis [169-172]. However, the clinical evidence is far from compelling. Garlic contains a number of different biologically active compounds and sulfur-containing amino acids, such as allicin, ajoene, S-allylcysteine, S-methylcysteine, diallyl disulfide and sulfoxides, that may be responsible for its antiatherosclerotic activity [171, 173]. In recent years, anti-atherosclerotic and cardiovascular protective effects of garlic have been extensively evaluated. Evidence from numerous studies demonstrate that garlic-based preparations can enhance fibrinolytic activity, inhibite platelet aggregation and reduce blood pressure [174] and normalize plasma lipids by inhibiting human squalene monooxygenase and HMG-CoA reductase, as well as other key enzymes involved in cholesterol biosynthesis [175]. Sobenin IA *et al.* demonstrated that a 12-month treatment with garlic tablets results in a significant reduction of 1.5 and 1.3 times in cardiovascular risk in men and women respectively. Cardiovascular risk reduction was mainly mediated by a decrease in LDL cholesterol (32.9mg/dl in men and 27.3 mg/dl in women). Interestingly, the most outstanding effects were observed in men. On the other hand, the insufficient sample size of the female group may have played a critical role in the result [93].

Ginkgo Biloba Extract (GBE)

Ginkgo biloba extract is a specific and complex product prepared from the dried leaves of *Ginkgo biloba*that has been used therapeutically in traditional Chinese medicine for 5000 years.The pharmacologically active constituents, flavonol glycosides and the terpene lactones have been identified. The terpene lactones comprise of ginkgolides A, B, C and bilobalides. The extract scavenges excess free radicals and pretreatment with ginkgo biloba extract reduces damage by free radicals in patients undergoing coronary bypass surgery. Release of NO and cytokines, such as prostacyclines was shown to be stimulated. Recently, Wu YZ *et al.* suggested that Ginkgo biloba extract treatment in CAD patients led to an increase of left anterior descending coronary artery blood flow, as it was assessed by significant improvement in maximal diastolic peak velocity, maximal systolic peak velocity and diastolic time velocity integral, which might be correlated to the restoration of the delicate equilibrium between NO and endothelin 1 [94].

Pycnogenol (Pinuspinaster ssp. atlantica)

Pycnogenol (Pinuspinaster ssp. atlantica) is a bark extract of the French maritime pine tree that has been used in traditional medicine. Pycnogenol exerts anti-inflammatory effects, inhibits platelet aggregation LDL peroxidation and lipid peroxidation [176]. Enseleit F. *et al.* reported thatpycnogenol improved the endothelial function of CAD patients by reducing oxidative stress [95].

Intervention *via* Mediterranean Diet

Based on positive findings from observational cohort studies of higher dietary intake or supplementation of antioxidants [177-180], the American Heart Association recommended the consumption of a balanced diet with emphasis on antioxidant-rich fruits, vegetables and whole grains [181]. The findings from clinical trials on antioxidant supplements have led to the resurgence of the interest on studying dietary antioxidants that are consumed through fruit, vegetable and nut intake. The beneficial antioxidant effects of diet have been examined extensively in the PREvention con DIetaMEDiterranea (PREDIMED) study, a multicenter randomized controlled clinical trial that evaluated the effects of a Mediterranean diet on cardiovascular outcomes in individuals without cardiovascular disease. Participants were subgrouped in a randomized way to a Mediterranean diet containing nuts or extra-virgin olive oil, or a control low-fat diet. After a median period of 4.8 years, there was a 28–30%reduction in major adverse cardiovascular events in individuals randomized to the supplemented Mediterranean diet [182]. These findings were attributed, in part, to a reduction in oxidant stress, as markers of oxidant stress (i.e. oxidized low-density lipoproteinsand malondialdehyde) were decreased in individuals assigned to the Mediterranean diet [183]. Although itis not clear what dietary antioxidants are responsible for the cardiovascular risk reduction, it is evident that dietary intervention with the Mediterranean diet offered the best possible outcomes of all the antioxidants studied.

FAILURE OF ANTIOXIDANTS? WHY?

Pathophysiologically, oxidative stress has been suggested to contribute to the central mechanism by which all clinical risk factors (hypertension, hyperlipidemia, smoking and/or diabetes mellitus) lead to endothelial dysfunction. Understandably, given this view, the reduction of oxidative stress would be anticipated to have a potentially favorable impact on CAD.

Despite the abundance of laboratory and observational evidence indicating that antioxidants should prevent CAD, the results from large-scale, randomized, placebo-controlled clinical trials have been rather controversial. There are many potential and speculative explanations for this.

First of all, an obvious reason could be that the wrong antioxidant or combination of antioxidant has been studied. Perhaps the vitamins used in the antioxidant trials were selected because they were available and not ideal or equivalent to the natural source. The synthetic versions of the antioxidants tested in clinical intervention trials may not completely mimic the natural forms of antioxidants and as a result the cardiovascular risk reduction seen in dietary studies may be associated with other ingredients in the foods consumed [184]. For example, potential additive or antagonistic effects of other antioxidants administered concomitantly with vitamin E have been reported. In particular, vitamin C is thought tobe important in the regeneration of the original form of vitamin E, acting as a co-antioxidant. Another issue associated with the use of vitamin supplementation in clinical trials is their metabolism. Vitamin E is a fat-soluble vitamin and its absorption occurs in the small intestine along with other fat-soluble nutrients. Consequently, the bioavailability of vitamin E is impacted by the consumption of dietary fats and plant sterols which share similar digestion and absorption pathways. In contrast a-tocopherol also acts as a pro-oxidant molecule by its reaction with lipid peroxyl radicals and conversion into an oxidized product, a-tocopheroxyl free radical. High levels of this vitamin may displace other fat solubleantioxidants (i.e. g-tocopherol), disrupting the naturalbalance of antioxidant system.The decreased absorption ofnatural g-tocopherol caused by α-tocopherol supplements canresult in a significant reduction of plasma g-tocopherol levels. γ-Tocopherol can provide greater antioxidant protection againstspecific oxidants. Vitamin E may also inhibit human cytosolic glutathione S-transferases, which are involved in the detoxification of drugs and endogenous toxins.

Secondly, the dosage of these vitamins varies among different studies. Not all antioxidants are functionally similar and those included in clinical trials may have different threshold effects for retracting cellular response related to oxidative stress and the pathogenesis of atherosclerosis. The optimal dose of antioxidants remains controversial; 800 units of vitamin E and 500 mg of vitamin C have been suggested to be the effective threshold doses. As it has also been proposed with other agents, there is individual variability in metabolism and in response to vitamin supplementation as well, meaning that same doses of vitamins to everybody will not lead to the same antioxidant effect.

Moreover, although there are several markers of oxidative stress (F2-isoprostanes, thiobarbituric acid reactive substances, oxidized LDL and free oxygen radicals), so far, none of the candidate markers or methods has been widely accepted as the gold standard one to use in clinical trials. Interestingly, many studies, in which positive results of antioxidant supplementation were observed, reported a significant increase in plasma antioxidant levels, while other studies did not include a metric system to determine whether the dose of antioxidant given actually reduced oxidant stress and did not examine dose-response [185].

Another potential explanation may be the inadequate duration of the antioxidant therapy. When considering adequate duration therapy, it is important to keep in mind that in animal studies antioxidants were proved to be effective through the prevention of the earliest forms of atherosclerotic lesion formation [186]. Therefore, even clinical trials lasting years may be too short to show any benefit. A trial several decades long may be required to see some impact of vitamin E supplementation. When primary or secondary prevention in adults is studied, we are looking at the effects of 40 or 50 years of oxidative stress on the endothelium, and it might be unrealistic to expect that we can turn that around with 1 or 2 years of antioxidant therapy. There may be an impact, but follow-up has not been long enough to prove it.

A final possible explanation for the incapacity of clinical trials to highlight the profits of antioxidants would be that the wrong patients may have been studied. The majority of randomized clinical trials enrolled patients with pre-existing cardiovascular event or at risk of cardiovascular disease. Although the selection of high-risk population in order to conduct a randomized trial reduces the required sample size due to the higher event rate, these patients may be less amenable to a positive and protective effect with antioxidant supplementation, particularly when treated for a short period of time, as mentioned above. Some patients who would exert no benefit from supplemental antioxidants may have been included in the studies. This is mostly important for patient being treated with statins, as studies on statins plus antioxidant vitamins have found that there was no additional benefit from the vitamin supplements [187]. On the other hand, instead of treating everyone with antioxidant supplements, as has been done in previous trials, perhaps only patients who are deficient in antioxidants or who have a proven increase in levels of oxidative stress should be treated. This may explain why trials in select populations, such as renal failure or post– cardiac transplantation populations, have been more likely to demonstrate a benefit from antioxidant therapies [48, 188].

CONCLUSION

A healthy endothelium is vital for the preservation and advancement of vascular homeostasis. On the other hand, oxidative stress has a key role in the progression of endothelial dysfunction as well as in atherosclerotic progression. Despite the supportive observational epidemiologic and animal data, most clinical trials have failed to support the putative action of antioxidant treatment in preventing atherosclerotic disease and its complications. Well designed new clinical trials are necessary to establish the exact role of oxidative stress in the pathogenesis of atherosclerosis and the precise impact of antioxidant therapy on primary and secondary prevention of coronary artery disease. Using the best antioxidants at the right dose, for the optimal duration and in the right population will provide the most trustworthy and indisputable results.

CONFLICT OF INTEREST

The authors confirm that this article content has no conflict of interest.

ACKNOWLEDGEMENTS

All authors have contributed significantly to the submitted work, have design the manuscript, have draft and revised the manuscript critically for important intellectual content and approved the submitted manuscript

DISCLOSURE

This is an update Chapter of the earlier review article [by Briasoulis A, Tousoulis D, Antoniades C, Stefanadis C. "The oxidative stress menace to coronary vasculature: any place for antioxidants?" *Curr Pharm Des.* 2009; 15(26): 3078-90], with additional text and references.

REFERENCES

[1] Tousoulis, D.; Psaltopoulou, T.; Androulakis, E.; Papageorgiou, N.; Papaioannou, S.; Oikonomou, E.; Synetos, A.; Stefanadis, C. Oxidative stress and early atherosclerosis: novel antioxidant treatment. *Cardiovasc. Drugs Ther.*, **2015**, *29(*1), 75-88.

[2] Ignarro, L. J. Biosynthesis and metabolism of endothelium-derived nitric oxide. *Annu. Rev. Pharmacol. Toxicol.*, **1990**, *30(*1), 535-560.

[3] Esper, R. J.; Nordaby, R. A.; Vilarino, J. O.; Paragano, A.; Cacharron, J. L.; Machado, R. A. Endothelial dysfunction: a comprehensive appraisal. *Cardiovasc. Diabetol.*, **2006**, *5(*4).

[4] Montezano, A. C.; Touyz, R. M. Oxidative stress, Noxs, and hypertension: experimental evidence and clinical controversies. *Ann Med*, **2012**, 44 Suppl *1(*S2-16).

[5] Vallance, P.; Collier, J.; Moncada, S. Effects of endothelium-derived nitric oxide on peripheral arteriolar tone in man. *Lancet*, **1989**, *2(*8670), 997-1000.

[6] Vanhoutte, P. M. Endothelium and control of vascular function. State of the Art lecture. *Hypertension*, **1989**, *13(*6 Pt 2), 658-667.

[7] Pacher, P.; Beckman, J. S.; Liaudet, L. Nitric oxide and peroxynitrite in health and disease. *Physiol. Rev.*, **2007**, *87(*1), 315-424.

[8] Papageorgiou, N.; Tousoulis, D.; Katsargyris, A.; Charakida, M.; Androulakis, E.; Siasos, G.; Tentolouris, C.; Stefanadis, C. Antioxidant treatment and endothelial dysfunction: is it time for flavonoids? *Recent. Pat. Cardiovasc. Drug. Discov.*, **2013**, *8(*2), 81-92.

[9] Tousoulis, D.; Briasoulis, A.; Papageorgiou, N.; Tsioufis, C.; Tsiamis, E.; Toutouzas, K.; Stefanadis, C. Oxidative stress and endothelial function: therapeutic interventions. *Recent. Pat. Cardiovasc. Drug. Discov.*, **2011**, *6(*2), 103-114.

[10] Griendling, K. K.; FitzGerald, G. A. Oxidative stress and cardiovascular injury: Part II: animal and human studies. *Circulation*, **2003**, *108(*17), 2034-2040.

[11] Mueller, C. F.; Laude, K.; McNally, J. S.; Harrison, D. G. ATVB in focus: redox mechanisms in blood vessels. *Arterioscler. Thromb. Vasc. Biol.*, **2005**, *25(*2), 274-278.

[12] Drummond, G. R.; Sobey, C. G. Endothelial NADPH oxidases: which NOX to target in vascular disease? *Trends Endocrinol. Metab.*, **2014**, *25(*9), 452-463.

[13] Drummond, G. R.; Selemidis, S.; Griendling, K. K.; Sobey, C. G. Combating oxidative stress in vascular disease: NADPH oxidases as therapeutic targets. *Na.t Rev. Drug. Discov.*, **2011**, *10(*6), 453-471.

[14] Lassegue, B.; San Martin, A.; Griendling, K. K. Biochemistry, physiology, and pathophysiology of NADPH oxidases in the cardiovascular system. *Circ. Res.*, **2012**, *110(*10), 1364-1390.

[15] Nishino, T.; Okamoto, K.; Eger, B. T.; Pai, E. F.; Nishino, T. Mammalian xanthine oxidoreductase - mechanism of transition from xanthine dehydrogenase to xanthine oxidase. *FEBS J*, **2008**, *275(*13), 3278-3289.

[16] Luo, S.; Lei, H.; Qin, H.; Xia, Y. Molecular mechanisms of endothelial NO synthase uncoupling. *Curr. Pharm. Des.*, **2014**, *20(*22), 3548-3553.

[17] Montezano, A. C.; Touyz, R. M. Reactive oxygen species and endothelial function--role of nitric oxide synthase uncoupling and Nox family nicotinamide adenine dinucleotide phosphate oxidases. *Basic Clin Pharmacol Toxicol*, **2012**, *110(*1), 87-94.

[18] de Oliveira-Junior, E. B.; Bustamante, J.; Newburger, P. E.; Condino-Neto, A. The human NADPH oxidase: primary and secondary defects impairing the respiratory burst function and the microbicidal ability of phagocytes. *Scand J Immunol*, **2011**, *73(*5), 420-427.

[19] Battelli, M. G.; Bolognesi, A.; Polito, L. Pathophysiology of circulating xanthine oxidoreductase: new emerging roles for a multi-tasking enzyme. *Biochim Biophys Acta*, **2014**, *1842(*9), 1502-1517.

[20] Ohara, Y.; Peterson, T. E.; Harrison, D. G. Hypercholesterolemia increases endothelial superoxide anion production. *J Clin Invest*, **1993**, *91(*6), 2546-2551.

[21] O'Driscoll, J. G.; Green, D. J.; Rankin, J. M.; Taylor, R. R. Nitric oxide-dependent endothelial function is unaffected by allopurinol in hypercholesterolaemic subjects. *Clin Exp Pharmacol Physiol*, **1999**, *26(*10), 779-783.

[22] Li, A. E.; Ito, H.; Rovira, II; Kim, K. S.; Takeda, K.; Yu, Z. Y.; Ferrans, V. J.; Finkel, T. A role for reactive oxygen species in endothelial cell anoikis. *Circ Res*, **1999**, *85(*4), 304-310.

[23] Marui, N.; Offermann, M. K.; Swerlick, R.; Kunsch, C.; Rosen, C. A.; Ahmad, M.; Alexander, R. W.; Medford, R. M. Vascular cell adhesion molecule-1 (VCAM-1) gene transcription and expression are regulated through an antioxidant-sensitive mechanism in human vascular endothelial cells. *J Clin Invest*, **1993**, *92(*4), 1866-1874.

[24] Maulik, N.; Das, D. K. Redox signaling in vascular angiogenesis. *Free Radic Biol Med*, **2002**, *33(*8), 1047-1060.

[25] Antoniades, C.; Tousoulis, D.; Tentolouris, C.; Toutouzas, P.; Stefanadis, C. Oxidative stress, antioxidant vitamins, and atherosclerosis. From basic research to clinical practice. *Herz*, **2003**, *28(*7), 628-638.

[26] Tousoulis, D.; Siasos, G.; Oikonomou, E.; Stougianos, P.; Papageorgiou, N.; Papavassiliou, A. G.; Stefanadis, C. Asymmetric dimethylarginine (ADMA): is really a biomarker for cardiovascular prognosis? *Int J Cardiol*, **2011**, *153(*2), 123-125.

[27] Landmesser, U.; Cai, H.; Dikalov, S.; McCann, L.; Hwang, J.; Jo, H.; Holland, S. M.; Harrison, D. G. Role of p*47(*phox) in vascular oxidative stress and hypertension caused by angiotensin II. *Hypertension*, **2002**, *40(*4), 511-515.

[28] Giacco, F.; Brownlee, M. Oxidative stress and diabetic complications. *Circ Res*, **2010**, *107(*9), 1058-1070.

[29] Funk, S. D.; Yurdagul, A., Jr.; Orr, A. W. Hyperglycemia and endothelial dysfunction in atherosclerosis: lessons from type 1 diabetes. *Int J Vasc Med*, **2012**, *2012(*569654).

[30] Radi, R.; Beckman, J. S.; Bush, K. M.; Freeman, B. A. Peroxynitrite-induced membrane lipid peroxidation: the cytotoxic potential of superoxide and nitric oxide. *Arch Biochem Biophys*, **1991**, *288(*2), 481-487.

[31] Zeiher, A. M.; Schachinger, V.; Minners, J. Long-term cigarette smoking impairs endothelium-dependent coronary arterial vasodilator function. *Circulation*, **1995**, *92(*5), 1094-1100.

[32] Reznick, A. Z.; Cross, C. E.; Hu, M. L.; Suzuki, Y. J.; Khwaja, S.; Safadi, A.; Motchnik, P. A.; Packer, L.; Halliwell, B. Modification of plasma proteins by cigarette smoke as measured by protein carbonyl formation. *Biochem J*, **1992**, 286 (Pt 2)(607-611.

[33] Galle, J.; Bengen, J.; Schollmeyer, P.; Wanner, C. Impairment of endothelium-dependent dilation in rabbit renal arteries by oxidized lipoprotein(a). Role of oxygen-derived radicals. *Circulation*, **1995**, *92(*6), 1582-1589.

[34] Noguchi, N. Novel insights into the molecular mechanisms of the antiatherosclerotic properties of antioxidants: the alternatives to radical scavenging. *Free Radic Biol Med*, **2002**, *33(*11), 1480-1489.

[35] Rimm, E. B.; Stampfer, M. J.; Ascherio, A.; Giovannucci, E.; Colditz, G. A.; Willett, W. C. Vitamin E consumption and the risk of coronary heart disease in men. *N Engl J Med*, **1993**, *328(*20), 1450-1456.

[36] Osganian, S. K.; Stampfer, M. J.; Rimm, E.; Spiegelman, D.; Manson, J. E.; Willett, W. C. Dietary carotenoids and risk of coronary artery disease in women. *Am J Clin Nutr*, **2003**, *77(*6), 1390-1399.

[37] Nyyssonen, K.; Parviainen, M. T.; Salonen, R.; Tuomilehto, J.; Salonen, J. T. Vitamin C deficiency and risk of myocardial infarction: prospective population study of men from eastern Finland. *BMJ*, **1997**, *314(*7081), 634-638.

[38] Kritchevsky, S. B.; Shimakawa, T.; Tell, G. S.; Dennis, B.; Carpenter, M.; Eckfeldt, J. H.; Peacher-Ryan, H.; Heiss, G. Dietary antioxidants and carotid artery wall thickness. The ARIC Study. Atherosclerosis Risk in Communities Study. *Circulation*, **1995**, *92(*8), 2142-2150.

[39] Klipstein-Grobusch, K.; den Breeijen, J. H.; Grobbee, D. E.; Boeing, H.; Hofman, A.; Witteman, J. C. Dietary antioxidants and peripheral arterial disease : the Rotterdam Study. *Am J Epidemiol*, **2001**, *154(*2), 145-149.

[40] Kushi, L. H.; Folsom, A. R.; Prineas, R. J.; Mink, P. J.; Wu, Y.; Bostick, R. M. Dietary antioxidant vitamins and death from coronary heart disease in postmenopausal women. *N Engl J Med*, **1996**, *334(*18), 1156-1162.

[41] Shardell, M. D.; Alley, D. E.; Hicks, G. E.; El-Kamary, S. S.; Miller, R. R.; Semba, R. D.; Ferrucci, L. Low-serum carotenoid concentrations and carotenoid interactions predict mortality in US adults: the Third National Health and Nutrition Examination Survey. *Nutr Res*, **2011**, *31(*3), 178-189.

[42] Simon, J. A.; Hudes, E. S. Serum ascorbic acid and cardiovascular disease prevalence in U.S. adults: the Third National Health and Nutrition Examination Survey (NHANES III). *Ann Epidemiol*, **1999**, *9(*6), 358-365.

[43] Osganian, S. K.; Stampfer, M. J.; Rimm, E.; Spiegelman, D.; Hu, F. B.; Manson, J. E.; Willett, W. C. Vitamin C and risk of coronary heart disease in women. *J Am Coll Cardiol*, **2003**, *42(*2), 246-252.

[44] Leopold, J. A. Antioxidants and coronary artery disease: from pathophysiology to preventive therapy. *Coron Artery Dis*, **2015**, *26(*2), 176-183.

[45] Virtamo, J.; Rapola, J. M.; Ripatti, S.; Heinonen, O. P.; Taylor, P. R.; Albanes, D.; Huttunen, J. K. Effect of vitamin E and beta carotene on the incidence of primary nonfatal myocardial infarction and fatal coronary heart disease. *Arch Intern Med*, **1998**, *158(*6), 668-675.

[46] Rapola, J. M.; Virtamo, J.; Ripatti, S.; Huttunen, J. K.; Albanes, D.; Taylor, P. R.; Heinonen, O. P. Randomised trial of alpha-tocopherol and beta-carotene supplements on incidence of major coronary events in men with previous myocardial infarction. *Lancet*, **1997**, *349(*9067), 1715-1720.

[47] Omenn, G. S.; Goodman, G. E.; Thornquist, M. D.; Balmes, J.; Cullen, M. R.; Glass, A.; Keogh, J. P.; Meyskens, F. L.; Valanis, B.; Williams, J. H.; Barnhart, S.; Hammar, S. Effects of a combination of beta carotene and vitamin A on lung cancer and cardiovascular disease. *N Engl J Med*, **1996**, *334(*18), 1150-1155.

[48] Stephens, N. G.; Parsons, A.; Schofield, P. M.; Kelly, F.; Cheeseman, K.; Mitchinson, M. J. Randomised controlled trial of vitamin E in patients with coronary disease: Cambridge Heart Antioxidant Study (CHAOS). *Lancet*, **1996**, *347(*9004), 781-786.

[49] Yusuf, S.; Dagenais, G.; Pogue, J.; Bosch, J.; Sleight, P. Vitamin E supplementation and cardiovascular events in high-risk patients. The Heart Outcomes Prevention Evaluation Study Investigators. *N Engl J Med*, **2000**, *342(*3), 154-160.

[50] MRC/BHF Heart Protection Study of antioxidant vitamin supplementation in 20,536 high-risk individuals: a randomised placebo-controlled trial. *Lancet*, **2002**, *360(*9326), 23-33.

[51] Lee, I. M.; Cook, N. R.; Gaziano, J. M.; Gordon, D.; Ridker, P. M.; Manson, J. E.; Hennekens, C. H.; Buring, J. E. Vitamin E in the primary prevention of cardiovascular disease and cancer: the Women's Health Study: a randomized controlled trial. *JAMA*, **2005**, *294(*1), 56-65.

[52] Zureik, M.; Galan, P.; Bertrais, S.; Mennen, L.; Czernichow, S.; Blacher, J.; Ducimetiere, P.; Hercberg, S. Effects of long-term daily low-dose supplementation with antioxidant vitamins and minerals on structure and function of large arteries. *Arterioscler Thromb Vasc Biol*, **2004**, *24(*8), 1485-1491.

[53] Cook, N. R.; Albert, C. M.; Gaziano, J. M.; Zaharris, E.; MacFadyen, J.; Danielson, E.; Buring, J. E.; Manson, J. E. A randomized factorial trial of vitamins C and E and beta carotene in the secondary prevention of cardiovascular events in women: results from the Women's Antioxidant Cardiovascular Study. *Arch Intern Med*, **2007**, *167(*15), 1610-1618.

[54] Hennekens, C. H.; Buring, J. E.; Manson, J. E.; Stampfer, M.; Rosner, B.; Cook, N. R.; Belanger, C.; LaMotte, F.; Gaziano, J. M.; Ridker, P. M.; Willett, W.; Peto, R. Lack of effect of long-term supplementation with beta carotene on the incidence of malignant neoplasms and cardiovascular disease. *N Engl J Med*, **1996**, *334(*18), 1145-1149.

[55] Sesso, H. D.; Buring, J. E.; Christen, W. G.; Kurth, T.; Belanger, C.; MacFadyen, J.; Bubes, V.; Manson, J. E.; Glynn, R. J.; Gaziano, J. M. Vitamins E and C in the prevention of cardiovascular disease in men: the Physicians' Health Study II randomized controlled trial. *JAMA*, **2008**, *300(*18), 2123-2133.

[56] Greenberg, E. R.; Baron, J. A.; Karagas, M. R.; Stukel, T. A.; Nierenberg, D. W.; Stevens, M. M.; Mandel, J. S.; Haile, R. W. Mortality associated with low plasma concentration of beta carotene and the effect of oral supplementation. *JAMA*, **1996**, *275(*9), 699-703.

[57] Dietary supplementation with n-3 polyunsaturated fatty acids and vitamin E after myocardial infarction: results of the GISSI-Prevenzione trial. Gruppo Italiano per lo Studio della Sopravvivenza nell'Infarto miocardico. *Lancet*, **1999**, *354(*9177), 447-455.

[58] Boaz, M.; Smetana, S.; Weinstein, T.; Matas, Z.; Gafter, U.; Iaina, A.; Knecht, A.; Weissgarten, Y.; Brunner, D.; Fainaru, M.; Green, M. S. Secondary prevention with antioxidants of cardiovascular disease in endstage renal disease (SPACE): randomised placebo-controlled trial. *Lancet*, **2000**, *356(*9237), 1213-1218.

[59] Lonn, E.; Yusuf, S.; Hoogwerf, B.; Pogue, J.; Yi, Q.; Zinman, B.; Bosch, J.; Dagenais, G.; Mann, J. F.; Gerstein, H. C. Effects of vitamin E on cardiovascular and microvascular outcomes in high-risk patients with diabetes: results of the HOPE study and MICRO-HOPE substudy. *Diabetes Care*, **2002**, *25(*11), 1919-1927.

[60] Salonen, J. T.; Nyyssonen, K.; Salonen, R.; Lakka, H. M.; Kaikkonen, J.; Porkkala-Sarataho, E.; Voutilainen, S.; Lakka, T. A.; Rissanen, T.; Leskinen, L.; Tuomainen, T. P.; Valkonen, V. P.; Ristonmaa, U.; Poulsen, H. E. Antioxidant Supplementation in Atherosclerosis Prevention (ASAP)

study: a randomized trial of the effect of vitamins E and C on 3-year progression of carotid atherosclerosis. *J Intern Med*, **2000**, *248(*5), 377-386.

[61] de Gaetano, G. Low-dose aspirin and vitamin E in people at cardiovascular risk: a randomised trial in general practice. Collaborative Group of the Primary Prevention Project. *Lancet*, **2001**, *357(*9250), 89-95.

[62] Tardif, J. C.; Cote, G.; Lesperance, J.; Bourassa, M.; Lambert, J.; Doucet, S.; Bilodeau, L.; Nattel, S.; de Guise, P. Probucol and multivitamins in the prevention of restenosis after coronary angioplasty. Multivitamins and Probucol Study Group. *N Engl J Med*, **1997**, *337(*6), 365-372.

[63] Lonn, E.; Bosch, J.; Yusuf, S.; Sheridan, P.; Pogue, J.; Arnold, J. M.; Ross, C.; Arnold, A.; Sleight, P.; Probstfield, J.; Dagenais, G. R. Effects of long-term vitamin E supplementation on cardiovascular events and cancer: a randomized controlled trial. *JAMA*, **2005**, *293(*11), 1338-1347.

[64] Ujeyl, A.; Fang, J. C.; Desai, A. S.; Mudge, G. H., Jr.; Givertz, M. M. Early vitamin C and E supplementation and cardiac allograft vasculopathy: 10-year follow-up from a randomized, controlled study. *J Heart Lung Transplant*, **2011**, *30(*10), 1200-1201.

[65] Devaraj, S.; Tang, R.; Adams-Huet, B.; Harris, A.; Seenivasan, T.; de Lemos, J. A.; Jialal, I. Effect of high-dose alpha-tocopherol supplementation on biomarkers of oxidative stress and inflammation and carotid atherosclerosis in patients with coronary artery disease. *Am J Clin Nutr*, **2007**, *86(*5), 1392-1398.

[66] Wang, Z. J.; Hu, W. K.; Liu, Y. Y.; Shi, D. M.; Cheng, W. J.; Guo, Y. H.; Yang, Q.; Zhao, Y. X.; Zhou, Y. J. The effect of intravenous vitamin C infusion on periprocedural myocardial injury for patients undergoing elective percutaneous coronary intervention. *Can J Cardiol*, **2014**, *30(*1), 96-101.

[67] Waters, D. D.; Alderman, E. L.; Hsia, J.; Howard, B. V.; Cobb, F. R.; Rogers, W. J.; Ouyang, P.; Thompson, P.; Tardif, J. C.; Higginson, L.; Bittner, V.; Steffes, M.; Gordon, D. J.; Proschan, M.; Younes, N.; Verter, J. I. Effects of hormone replacement therapy and antioxidant vitamin supplements on coronary atherosclerosis in postmenopausal women: a randomized controlled trial. *JAMA*, **2002**, *288(*19), 2432-2440.

[68] The effect of vitamin E and beta carotene on the incidence of lung cancer and other cancers in male smokers. The Alpha-Tocopherol, Beta Carotene Cancer Prevention Study Group. *N Engl J Med*, **1994**, *330(*15), 1029-1035.

[69] Rapola, J. M.; Virtamo, J.; Haukka, J. K.; Heinonen, O. P.; Albanes, D.; Taylor, P. R.; Huttunen, J. K. Effect of vitamin E and beta carotene on the incidence of angina pectoris. A randomized, double-blind, controlled trial. *JAMA*, **1996**, *275(*9), 693-698.

[70] Redlich, C. A.; Chung, J. S.; Cullen, M. R.; Blaner, W. S.; Van Bennekum, A. M.; Berglund, L. Effect of long-term beta-carotene and vitamin A on serum cholesterol and triglyceride levels among participants in the Carotene and Retinol Efficacy Trial (CARET). *Atherosclerosis*, **1999**, *145(*2), 425-432.

[71] Hodis, H. N.; Mack, W. J.; LaBree, L.; Mahrer, P. R.; Sevanian, A.; Liu, C. R.; Liu, C. H.; Hwang, J.; Selzer, R. H.; Azen, S. P. Alpha-tocopherol supplementation in healthy individuals reduces low-density lipoprotein oxidation but not atherosclerosis: the Vitamin E Atherosclerosis Prevention Study (VEAPS). *Circulation*, **2002**, *106(*12), 1453-1459.

[72] Knekt, P.; Ritz, J.; Pereira, M. A.; O'Reilly, E. J.; Augustsson, K.; Fraser, G. E.; Goldbourt, U.; Heitmann, B. L.; Hallmans, G.; Liu, S.; Pietinen, P.; Spiegelman, D.; Stevens, J.; Virtamo, J.; Willett, W. C.; Rimm, E. B.; Ascherio, A. Antioxidant vitamins and coronary heart disease risk: a pooled analysis of 9 cohorts. *Am J Clin Nutr*, **2004**, *80(*6), 1508-1520.

[73] Ye, Y.; Li, J.; Yuan, Z. Effect of antioxidant vitamin supplementation on cardiovascular outcomes: a meta-analysis of randomized controlled trials. *PLoS One*, **2013**, *8(*2), e56803.

[74] Myung, S. K.; Ju, W.; Cho, B.; Oh, S. W.; Park, S. M.; Koo, B. K.; Park, B. J. Efficacy of vitamin and antioxidant supplements in prevention of cardiovascular disease: systematic review and meta-analysis of randomised controlled trials. *BMJ*, **2013**, *346(*f10).

[75] Lee, B. J.; Tseng, Y. F.; Yen, C. H.; Lin, P. T. Effects of coenzyme Q10 supplementation (300 mg/day) on antioxidation and anti-inflammation in coronary artery disease patients during statins therapy: a randomized, placebo-controlled trial. *Nutr J*, **2013**, *12(*1), 142.

[76] Lee, B. J.; Huang, Y. C.; Chen, S. J.; Lin, P. T. Effects of coenzyme Q10 supplementation on inflammatory markers (high-sensitivity C-reactive protein, interleukin-6, and homocysteine) in patients with coronary artery disease. *Nutrition*, **2012**, *28(*7-8), 767-772.

[77] Tiano, L.; Belardinelli, R.; Carnevali, P.; Principi, F.; Seddaiu, G.; Littarru, G. P. Effect of coenzyme Q10 administration on endothelial function and extracellular superoxide dismutase in patients with ischaemic heart disease: a double-blind, randomized controlled study. *Eur Heart J*, **2007**, *28(*18), 2249-2255.

[78] Tome-Carneiro, J.; Larrosa, M.; Yanez-Gascon, M. J.; Davalos, A.; Gil-Zamorano, J.; Gonzalvez, M.; Garcia-Almagro, F. J.; Ruiz Ros, J. A.; Tomas-Barberan, F. A.; Espin, J. C.; Garcia-Conesa, M. T. One-year supplementation with a grape extract containing resveratrol modulates inflammatory-related microRNAs and cytokines expression in peripheral blood mononuclear cells of type 2 diabetes and hypertensive patients with coronary artery disease. *Pharmacol Res*, **2013**, *72(*69-82.

[79] Tome-Carneiro, J.; Gonzalvez, M.; Larrosa, M.; Yanez-Gascon, M. J.; Garcia-Almagro, F. J.; Ruiz-Ros, J. A.; Tomas-Barberan, F. A.; Garcia-Conesa, M. T.; Espin, J. C. Grape resveratrol increases serum adiponectin and downregulates inflammatory genes in peripheral blood mononuclear cells: a triple-blind, placebo-controlled, one-year clinical trial in patients with stable coronary artery disease. *Cardiovasc Drugs Ther*, **2013**, *27(*1), 37-48.

[80] Magyar, K.; Halmosi, R.; Palfi, A.; Feher, G.; Czopf, L.; Fulop, A.; Battyany, I.; Sumegi, B.; Toth, K.; Szabados, E. Cardioprotection by resveratrol: A human clinical trial in patients with stable coronary artery disease. *Clin Hemorheol Microcirc*, **2012**, *50(*3), 179-187.

[81] Buijsse, B.; Weikert, C.; Drogan, D.; Bergmann, M.; Boeing, H. Chocolate consumption in relation to blood pressure and risk of cardiovascular disease in German adults. *Eur Heart J*, **2010**, *31(*13), 1616-1623.

[82] Zomer, E.; Owen, A.; Magliano, D. J.; Liew, D.; Reid, C. M. The effectiveness and cost effectiveness of dark chocolate consumption as prevention therapy in people at high risk of cardiovascular disease: best case scenario analysis using a Markov model. *BMJ*, **2012**, *344(*e3657.

[83] Inami, S.; Takano, M.; Yamamoto, M.; Murakami, D.; Tajika, K.; Yodogawa, K.; Yokoyama, S.; Ohno, N.; Ohba, T.; Sano, J.; Ibuki, C.; Seino, Y.; Mizuno, K. Tea catechin consumption reduces circulating oxidized low-density lipoprotein. *Int Heart J*, **2007**, *48(*6), 725-732.

[84] Widlansky, M. E.; Hamburg, N. M.; Anter, E.; Holbrook, M.; Kahn, D. F.; Elliott, J. G.; Keaney, J. F., Jr.; Vita, J. A. Acute EGCG supplementation reverses endothelial dysfunction in patients with coronary artery disease. *J Am Coll Nutr*, **2007**, *26(*2), 95-102.

[85] Siani, A.; Pagano, E.; Iacone, R.; Iacoviello, L.; Scopacasa, F.; Strazzullo, P. Blood pressure and metabolic changes during dietary L-arginine supplementation in humans. *Am J Hypertens*, **2000**, *13(*5 Pt 1), 547-551.

[86] Siasos, G.; Tousoulis, D.; Vlachopoulos, C.; Antoniades, C.; Stefanadi, E.; Ioakeimidis, N.; Zisimos, K.; Siasou, Z.; Papavassiliou, A. G.; Stefanadis, C. The impact of oral L-arginine supplementation on acute smoking-induced endothelial injury and arterial performance. *Am J Hypertens*, **2009**, *22(*6), 586-592.

[87] Kelly, B. S.; Alexander, J. W.; Dreyer, D.; Greenberg, N. A.; Erickson, A.; Whiting, J. F.; Ogle, C. K.; Babcock, G. F.; First, M. R. Oral arginine improves blood pressure in renal transplant and hemodialysis patients. *JPEN J Parenter Enteral Nutr*, **2001**, *25(*4), 194-202.

[88] Yin, W. H.; Chen, J. W.; Tsai, C.; Chiang, M. C.; Young, M. S.; Lin, S. J. L-arginine improves endothelial function and reduces LDL oxidation in patients with stable coronary artery disease. *Clin Nutr*, **2005**, *24(*6), 988-997.

[89] Lee, B. J.; Lin, J. S.; Lin, Y. C.; Lin, P. T. Effects of L-carnitine supplementation on oxidative stress and antioxidant enzymes activities in patients with coronary artery disease: a randomized, placebo-controlled trial. *Nutr J*, **2014**, *13(*79.

[90] Schnabel, R.; Lubos, E.; Messow, C. M.; Sinning, C. R.; Zeller, T.; Wild, P. S.; Peetz, D.; Handy, D. E.; Munzel, T.; Loscalzo, J.; Lackner, K. J.; Blankenberg, S. Selenium supplementation improves antioxidant capacity *in vitro* and *in vivo* in patients with coronary artery disease The SElenium Therapy in Coronary Artery disease Patients (SETCAP) Study. *Am Heart J*, **2008**, *156(*6), 1201 e1201-1211.

[91] Mohammadpour, A. H.; Falsoleiman, H.; Shamsara, J.; Allah Abadi, G.; Rasooli, R.; Ramezani, M. Pentoxifylline decreases serum level of adhesion molecules in atherosclerosis patients. *Iran Biomed J*, **2014**, *18(*1), 23-27.

[92] Fernandes, J. L.; de Oliveira, R. T.; Mamoni, R. L.; Coelho, O. R.; Nicolau, J. C.; Blotta, M. H.; Serrano, C. V., Jr. Pentoxifylline reduces pro-inflammatory and increases anti-inflammatory activity

in patients with coronary artery disease--a randomized placebo-controlled study. *Atherosclerosis*, **2008**, *196(*1), 434-442.

[93] Sobenin, I. A.; Pryanishnikov, V. V.; Kunnova, L. M.; Rabinovich, Y. A.; Martirosyan, D. M.; Orekhov, A. N. The effects of time-released garlic powder tablets on multifunctional cardiovascular risk in patients with coronary artery disease. *Lipids Health Dis*, **2010**, *9(*119.

[94] Wu, Y. Z.; Li, S. Q.; Zu, X. G.; Du, J.; Wang, F. F. Ginkgo biloba extract improves coronary artery circulation in patients with coronary artery disease: contribution of plasma nitric oxide and endothelin-1. *Phytother Res*, **2008**, *22(*6), 734-739.

[95] Enseleit, F.; Sudano, I.; Periat, D.; Winnik, S.; Wolfrum, M.; Flammer, A. J.; Frohlich, G. M.; Kaiser, P.; Hirt, A.; Haile, S. R.; Krasniqi, N.; Matter, C. M.; Uhlenhut, K.; Hogger, P.; Neidhart, M.; Luscher, T. F.; Ruschitzka, F.; Noll, G. Effects of Pycnogenol on endothelial function in patients with stable coronary artery disease: a double-blind, randomized, placebo-controlled, cross-over study. *Eur Heart J*, **2012**, *33(*13), 1589-1597.

[96] Overvad, K.; Diamant, B.; Holm, L.; Holmer, G.; Mortensen, S. A.; Stender, S. Coenzyme Q10 in health and disease. *Eur J Clin Nutr*, **1999**, *53(*10), 764-770.

[97] McCarty, M. F. Coenzyme Q versus hypertension: does CoQ decrease endothelial superoxide generation? *Med Hypotheses*, **1999**, *53(*4), 300-304.

[98] Houston, M. C. Nutraceuticals, vitamins, antioxidants, and minerals in the prevention and treatment of hypertension. *Prog Cardiovasc Dis*, **2005**, *47(*6), 396-449.

[99] Lee, B. J.; Lin, Y. C.; Huang, Y. C.; Ko, Y. W.; Hsia, S.; Lin, P. T. The relationship between coenzyme Q10, oxidative stress, and antioxidant enzymes activities and coronary artery disease. *Scientific World Journal*, **2012**, *2012(*792756.

[100] Rosenfeldt, F. L.; Haas, S. J.; Krum, H.; Hadj, A.; Ng, K.; Leong, J. Y.; Watts, G. F. Coenzyme Q10 in the treatment of hypertension: a meta-analysis of the clinical trials. *J Hum Hypertens*, **2007**, *21(*4), 297-306.

[101] Gao, L.; Mao, Q.; Cao, J.; Wang, Y.; Zhou, X.; Fan, L. Effects of coenzyme Q10 on vascular endothelial function in humans: a meta-analysis of randomized controlled trials. *Atherosclerosis*, **2012**, *221(*2), 311-316.

[102] Chu, L. M.; Lassaletta, A. D.; Robich, M. P.; Sellke, F. W. Resveratrol in the prevention and treatment of coronary artery disease. *Curr Atheroscler Rep*, **2011**, *13(*6), 439-446.

[103] Do, G. M.; Kwon, E. Y.; Kim, H. J.; Jeon, S. M.; Ha, T. Y.; Park, T.; Choi, M. S. Long-term effects of resveratrol supplementation on suppression of atherogenic lesion formation and cholesterol synthesis in apo E-deficient mice. *Biochem Biophys Res Commun*, **2008**, *374(*1), 55-59.

[104] Rocha, K. K.; Souza, G. A.; Ebaid, G. X.; Seiva, F. R.; Cataneo, A. C.; Novelli, E. L. Resveratrol toxicity: effects on risk factors for atherosclerosis and hepatic oxidative stress in standard and high-fat diets. *Food Chem Toxicol*, **2009**, *47(*6), 1362-1367.

[105] Robich, M. P.; Osipov, R. M.; Nezafat, R.; Feng, J.; Clements, R. T.; Bianchi, C.; Boodhwani, M.; Coady, M. A.; Laham, R. J.; Sellke, F. W. Resveratrol improves myocardial perfusion in a swine model of hypercholesterolemia and chronic myocardial ischemia. *Circulation*, **2010**, *122(*11 Suppl), S142-149.

[106] Berrougui, H.; Grenier, G.; Loued, S.; Drouin, G.; Khalil, A. A new insight into resveratrol as an atheroprotective compound: inhibition of lipid peroxidation and enhancement of cholesterol efflux. *Atherosclerosis*, **2009**, *207(*2), 420-427.

[107] Li, Y.; Cao, Z.; Zhu, H. Upregulation of endogenous antioxidants and phase 2 enzymes by the red wine polyphenol, resveratrol in cultured aortic smooth muscle cells leads to cytoprotection against oxidative and electrophilic stress. *Pharmacol Res*, **2006**, *53(*1), 6-15.

[108] Fullerton, M. D.; Steinberg, G. R. SIRT1 takes a backseat to AMPK in the regulation of insulin sensitivity by resveratrol. *Diabetes*, **2010**, *59(*3), 551-553.

[109] Olas, B.; Wachowicz, B.; Saluk-Juszczak, J.; Zielinski, T. Effect of resveratrol, a natural polyphenolic compound, on platelet activation induced by endotoxin or thrombin. *Thromb Res*, **2002**, *107(*3-4), 141-145.

[110] Yang, Y.; Wang, X.; Zhang, L.; An, H.; Zao, Z. Inhibitory effects of resveratrol on platelet activation induced by thromboxane a(2) receptor agonist in human platelets. *Am J Chin Med*, **2011**, *39(*1), 145-159.

[111] Lin, K. H.; Hsiao, G.; Shih, C. M.; Chou, D. S.; Sheu, J. R. Mechanisms of resveratrol-induced platelet apoptosis. *Cardiovasc Res*, **2009**, *83(*3), 575-585.
[112] Deng, Y. H.; Alex, D.; Huang, H. Q.; Wang, N.; Yu, N.; Wang, Y. T.; Leung, G. P.; Lee, S. M. Inhibition of TNF-alpha-mediated endothelial cell-monocyte cell adhesion and adhesion molecules expression by the resveratrol derivative, trans-3,5,4'-trimethoxystilbene. *Phytother Res*, **2011**, *25(*3), 451-457.
[113] Rivera, L.; Moron, R.; Zarzuelo, A.; Galisteo, M. Long-term resveratrol administration reduces metabolic disturbances and lowers blood pressure in obese Zucker rats. *Biochem Pharmacol*, **2009**, *77(*6), 1053-1063.
[114] Behbahani, J.; Thandapilly, S. J.; Louis, X. L.; Huang, Y.; Shao, Z.; Kopilas, M. A.; Wojciechowski, P.; Netticadan, T.; Anderson, H. D. Resveratrol and small artery compliance and remodeling in the spontaneously hypertensive rat. *Am J Hypertens*, **2010**, *23(*12), 1273-1278.
[115] Peters, U.; Poole, C.; Arab, L. Does tea affect cardiovascular disease? A meta-analysis. *Am J Epidemiol*, **2001**, *154(*6), 495-503.
[116] Bazzano, L. A.; He, J.; Ogden, L. G.; Loria, C. M.; Vupputuri, S.; Myers, L.; Whelton, P. K. Fruit and vegetable intake and risk of cardiovascular disease in US adults: the first National Health and Nutrition Examination Survey Epidemiologic Follow-up Study. *Am J Clin Nutr*, **2002**, *76(*1), 93-99.
[117] Siasos, G.; Tousoulis, D.; Tsigkou, V.; Kokkou, E.; Oikonomou, E.; Vavuranakis, M.; Basdra, E. K.; Papavassiliou, A. G.; Stefanadis, C. Flavonoids in atherosclerosis: an overview of their mechanisms of action. *Curr Med Chem*, **2013**, *20(*21), 2641-2660.
[118] Aviram, M.; Fuhrman, B. Wine flavonoids protect against LDL oxidation and atherosclerosis. *Ann N Y Acad Sci*, **2002**, *957(*146-161.
[119] Lotito, S. B.; Fraga, C. G. (+)-Catechin prevents human plasma oxidation. *Free Radic Biol Med*, **1998**, *24(*3), 435-441.
[120] Stein, J. H.; Keevil, J. G.; Wiebe, D. A.; Aeschlimann, S.; Folts, J. D. Purple grape juice improves endothelial function and reduces the susceptibility of LDL cholesterol to oxidation in patients with coronary artery disease. *Circulation*, **1999**, *100(*10), 1050-1055.
[121] Lotito, S. B.; Frei, B. Consumption of flavonoid-rich foods and increased plasma antioxidant capacity in humans: cause, consequence, or epiphenomenon? *Free Radic Biol Med*, **2006**, *41(*12), 1727-1746.
[122] Aviram, M.; Dornfeld, L. Pomegranate juice consumption inhibits serum angiotensin converting enzyme activity and reduces systolic blood pressure. *Atherosclerosis*, **2001**, *158(*1), 195-198.
[123] Aviram, M.; Rosenblat, M.; Gaitini, D.; Nitecki, S.; Hoffman, A.; Dornfeld, L.; Volkova, N.; Presser, D.; Attias, J.; Liker, H.; Hayek, T. Pomegranate juice consumption for 3 years by patients with carotid artery stenosis reduces common carotid intima-media thickness, blood pressure and LDL oxidation. *Clin Nutr*, **2004**, *23(*3), 423-433.
[124] Anter, E.; Thomas, S. R.; Schulz, E.; Shapira, O. M.; Vita, J. A.; Keaney, J. F., Jr. Activation of endothelial nitric-oxide synthase by the p38 MAPK in response to black tea polyphenols. *J Biol Chem*, **2004**, *279(*45), 46637-46643.
[125] Diebolt, M.; Bucher, B.; Andriantsitohaina, R. Wine polyphenols decrease blood pressure, improve NO vasodilatation, and induce gene expression. *Hypertension*, **2001**, *38(*2), 159-165.
[126] Siasos, G.; Tousoulis, D.; Kokkou, E.; Oikonomou, E.; Kollia, M. E.; Verveniotis, A.; Gouliopoulos, N.; Zisimos, K.; Plastiras, A.; Maniatis, K.; Stefanadis, C. Favorable effects of concord grape juice on endothelial function and arterial stiffness in healthy smokers. *Am J Hypertens*, **2014**, *27(*1), 38-45.
[127] Taubert, D.; Berkels, R.; Roesen, R.; Klaus, W. Chocolate and blood pressure in elderly individuals with isolated systolic hypertension. *JAMA*, **2003**, *290(*8), 1029-1030.
[128] Grassi, D.; Lippi, C.; Necozione, S.; Desideri, G.; Ferri, C. Short-term administration of dark chocolate is followed by a significant increase in insulin sensitivity and a decrease in blood pressure in healthy persons. *Am J Clin Nutr*, **2005**, *81(*3), 611-614.
[129] Taubert, D.; Roesen, R.; Lehmann, C.; Jung, N.; Schomig, E. Effects of low habitual cocoa intake on blood pressure and bioactive nitric oxide: a randomized controlled trial. *JAMA*, **2007**, *298(*1), 49-60.
[130] Grassi, D.; Desideri, G.; Necozione, S.; Lippi, C.; Casale, R.; Properzi, G.; Blumberg, J. B.; Ferri, C. Blood pressure is reduced and insulin sensitivity increased in glucose-intolerant, hypertensive subjects after 15 days of consuming high-polyphenol dark chocolate. *J Nutr*, **2008**, *138(*9), 1671-1676.

[131] Ried, K.; Frank, O. R.; Stocks, N. P. Dark chocolate or tomato extract for prehypertension: a randomised controlled trial. *BMC Complement Altern Med*, **2009**, *9(*22.

[132] Zilkens, R. R.; Burke, V.; Hodgson, J. M.; Barden, A.; Beilin, L. J.; Puddey, I. B. Red wine and beer elevate blood pressure in normotensive men. *Hypertension*, **2005**, *45(*5), 874-879.

[133] Vlachopoulos, C.; Alexopoulos, N.; Stefanadis, C. Effect of dark chocolate on arterial function in healthy individuals: cocoa instead of ambrosia? *Curr Hypertens Rep*, **2006**, *8(*3), 205-211.

[134] Sudano, I.; Flammer, A. J.; Roas, S.; Enseleit, F.; Ruschitzka, F.; Corti, R.; Noll, G. Cocoa, blood pressure, and vascular function. *Curr Hypertens Rep*, **2012**, *14(*4), 279-284.

[135] Latif, R. Chocolate/cocoa and human health: a review. *Neth J Med*, **2013**, *71(*2), 63-68.

[136] Vita, J. A. Tea consumption and cardiovascular disease: effects on endothelial function. *J Nutr*, **2003**, *133(*10), 3293S-3297S.

[137] Higdon, J. V.; Frei, B. Tea catechins and polyphenols: health effects, metabolism, and antioxidant functions. *Crit Rev Food Sci Nutr*, **2003**, *43(*1), 89-143.

[138] Karim, M.; McCormick, K.; Kappagoda, C. T. Effects of cocoa extracts on endothelium-dependent relaxation. *J Nutr*, **2000**, *130(*8S Suppl), 2105S-2108S.

[139] Lorenz, M.; Wessler, S.; Follmann, E.; Michaelis, W.; Dusterhoft, T.; Baumann, G.; Stangl, K.; Stangl, V. A constituent of green tea, epigallocatechin-3-gallate, activates endothelial nitric oxide synthase by a phosphatidylinositol-3-OH-kinase-, cAMP-dependent protein kinase-, and Akt-dependent pathway and leads to endothelial-dependent vasorelaxation. *J Biol Chem*, **2004**, *279(*7), 6190-6195.

[140] Tiefenbacher, C. P. Tetrahydrobiopterin: a critical cofactor for eNOS and a strategy in the treatment of endothelial dysfunction? *Am J Physiol Heart Circ Physiol*, **2001**, *280(*6), H2484-2488.

[141] Schlaich, M. P.; Parnell, M. M.; Ahlers, B. A.; Finch, S.; Marshall, T.; Zhang, W. Z.; Kaye, D. M. Impaired L-arginine transport and endothelial function in hypertensive and genetically predisposed normotensive subjects. *Circulation*, **2004**, *110(*24), 3680-3686.

[142] Wang, D.; Strandgaard, S.; Iversen, J.; Wilcox, C. S. Asymmetric dimethylarginine, oxidative stress, and vascular nitric oxide synthase in essential hypertension. *Am J Physiol Regul Integr Comp Physiol*, **2009**, *296(*2), R195-200.

[143] Bevers, L. M.; Braam, B.; Post, J. A.; van Zonneveld, A. J.; Rabelink, T. J.; Koomans, H. A.; Verhaar, M. C.; Joles, J. A. Tetrahydrobiopterin, but not L-arginine, decreases NO synthase uncoupling in cells expressing high levels of endothelial NO synthase. *Hypertension*, **2006**, *47(*1), 87-94.

[144] Siasos, G.; Tousoulis, D.; Vlachopoulos, C.; Antoniades, C.; Stefanadi, E.; Ioakeimidis, N.; Andreou, I.; Zisimos, K.; Papavassiliou, A. G.; Stefanadis, C. Short-term treatment with L-arginine prevents the smoking-induced impairment of endothelial function and vascular elastic properties in young individuals. *Int J Cardiol*, **2008**, *126(*3), 394-399.

[145] Chen, P. Y.; Sanders, P. W. L-arginine abrogates salt-sensitive hypertension in Dahl/Rapp rats. *J Clin Invest*, **1991**, *88(*5), 1559-1567.

[146] Martina, V.; Masha, A.; Gigliardi, V. R.; Brocato, L.; Manzato, E.; Berchio, A.; Massarenti, P.; Settanni, F.; Della Casa, L.; Bergamini, S.; Iannone, A. Long-term N-acetylcysteine and L-arginine administration reduces endothelial activation and systolic blood pressure in hypertensive patients with type 2 diabetes. *Diabetes Care*, **2008**, *31(*5), 940-944.

[147] Gokce, N. L-arginine and hypertension. *J Nutr*, **2004**, *134(*10 Suppl), 2807S-2811S; discussion 2818S-2819S.

[148] Loscalzo, J. Adverse effects of supplemental L-arginine in atherosclerosis: consequences of methylation stress in a complex catabolism? *Arterioscler Thromb Vasc Biol*, **2003**, *23(*1), 3-5.

[149] Persky, A. M.; Brazeau, G. A. Clinical pharmacology of the dietary supplement creatine monohydrate. *Pharmacol Rev*, **2001**, *53(*2), 161-176.

[150] Tyagi, N.; Sedoris, K. C.; Steed, M.; Ovechkin, A. V.; Moshal, K. S.; Tyagi, S. C. Mechanisms of homocysteine-induced oxidative stress. *Am J Physiol Heart Circ Physiol*, **2005**, *289(*6), H2649-2656.

[151] Jahangir, E.; Vita, J. A.; Handy, D.; Holbrook, M.; Palmisano, J.; Beal, R.; Loscalzo, J.; Eberhardt, R. T. The effect of L-arginine and creatine on vascular function and homocysteine metabolism. *Vasc Med*, **2009**, *14(*3), 239-248.

[152] Hagen, T. M.; Moreau, R.; Suh, J. H.; Visioli, F. Mitochondrial decay in the aging rat heart: evidence for improvement by dietary supplementation with acetyl-L-carnitine and/or lipoic acid. *Ann N Y Acad Sci*, **2002**, *959(*491-507.

[153] Hagen, T. M.; Ingersoll, R. T.; Wehr, C. M.; Lykkesfeldt, J.; Vinarsky, V.; Bartholomew, J. C.; Song, M. H.; Ames, B. N. Acetyl-L-carnitine fed to old rats partially restores mitochondrial function and ambulatory activity. *Proc Natl Acad Sci U S A*, **1998**, *95(*16), 9562-9566.

[154] Shay, K. P.; Moreau, R. F.; Smith, E. J.; Smith, A. R.; Hagen, T. M. Alpha-lipoic acid as a dietary supplement: molecular mechanisms and therapeutic potential. *Biochim Biophys Acta*, **2009**, *1790(*10), 1149-1160.

[155] Smith, A. R.; Shenvi, S. V.; Widlansky, M.; Suh, J. H.; Hagen, T. M. Lipoic acid as a potential therapy for chronic diseases associated with oxidative stress. *Curr Med Chem*, **2004**, *11(*9), 1135-1146.

[156] Teichert, J.; Kern, J.; Tritschler, H. J.; Ulrich, H.; Preiss, R. Investigations on the pharmacokinetics of alpha-lipoic acid in healthy volunteers. *Int J Clin Pharmacol Ther*, **1998**, *36(*12), 625-628.

[157] Zhang, W. J.; Frei, B. Alpha-lipoic acid inhibits TNF-alpha-induced NF-kappaB activation and adhesion molecule expression in human aortic endothelial cells. *FASEB J*, **2001**, *15(*13), 2423-2432.

[158] McMackin, C. J.; Widlansky, M. E.; Hamburg, N. M.; Huang, A. L.; Weller, S.; Holbrook, M.; Gokce, N.; Hagen, T. M.; Keaney, J. F., Jr.; Vita, J. A. Effect of combined treatment with alpha-Lipoic acid and acetyl-L-carnitine on vascular function and blood pressure in patients with coronary artery disease. *J Clin Hypertens (Greenwich)*, **2007**, *9(*4), 249-255.

[159] Addabbo, F.; Montagnani, M.; Goligorsky, M. S. Mitochondria and reactive oxygen species. *Hypertension*, **2009**, *53(*6), 885-892.

[160] Ruggenenti, P.; Cattaneo, D.; Loriga, G.; Ledda, F.; Motterlini, N.; Gherardi, G.; Orisio, S.; Remuzzi, G. Ameliorating hypertension and insulin resistance in subjects at increased cardiovascular risk: effects of acetyl-L-carnitine therapy. *Hypertension*, **2009**, *54(*3), 567-574.

[161] Rees, K.; Hartley, L.; Day, C.; Flowers, N.; Clarke, A.; Stranges, S. Selenium supplementation for the primary prevention of cardiovascular disease. *Cochrane Database Syst Rev*, **2013**, *1(*CD009671).

[162] Laurat, E.; Poirier, B.; Tupin, E.; Caligiuri, G.; Hansson, G. K.; Bariety, J.; Nicoletti, A. *In vivo* downregulation of T helper cell 1 immune responses reduces atherogenesis in apolipoprotein E-knockout mice. *Circulation*, **2001**, *104(*2), 197-202.

[163] Lutgens, E.; Cleutjens, K. B.; Heeneman, S.; Koteliansky, V. E.; Burkly, L. C.; Daemen, M. J. Both early and delayed anti-CD40L antibody treatment induces a stable plaque phenotype. *Proc Natl Acad Sci U S A*, **2000**, *97(*13), 7464-7469.

[164] Zhao, L.; Cuff, C. A.; Moss, E.; Wille, U.; Cyrus, T.; Klein, E. A.; Pratico, D.; Rader, D. J.; Hunter, C. A.; Pure, E.; Funk, C. D. Selective interleukin-12 synthesis defect in 12/15-lipoxygenase-deficient macrophages associated with reduced atherosclerosis in a mouse model of familial hypercholesterolemia. *J Biol Chem*, **2002**, *277(*38), 35350-35356.

[165] D'Hellencourt, C. L.; Diaw, L.; Cornillet, P.; Guenounou, M. Differential regulation of TNF alpha, IL-1 beta, IL-6, IL-8, TNF beta, and IL-10 by pentoxifylline. *Int J Immunopharmacol*, **1996**, *18(*12), 739-748.

[166] Ramani, M.; Khechai, F.; Ollivier, V.; Ternisien, C.; Bridey, F.; Hakim, J.; de Prost, D. Interleukin-10 and pentoxifylline inhibit C-reactive protein-induced tissue factor gene expression in peripheral human blood monocytes. *FEBS Lett*, **1994**, *356(*1), 86-88.

[167] Ward, A.; Clissold, S. P. Pentoxifylline. A review of its pharmacodynamic and pharmacokinetic properties, and its therapeutic efficacy. *Drugs*, **1987**, *34(*1), 50-97.

[168] Aviello, G.; Abenavoli, L.; Borrelli, F.; Capasso, R.; Izzo, A. A.; Lembo, F.; Romano, B.; Capasso, F. Garlic: empiricism or science? *Nat Prod Commun*, **2009**, *4(*12), 1785-1796.

[169] Gebhardt, R. Multiple inhibitory effects of garlic extracts on cholesterol biosynthesis in hepatocytes. *Lipids*, **1993**, *28(*7), 613-619.

[170] Liu, L.; Yeh, Y. Y. Water-soluble organosulfur compounds of garlic inhibit fatty acid and triglyceride syntheses in cultured rat hepatocytes. *Lipids*, **2001**, *36(*4), 395-400.

[171] Yeh, Y. Y.; Liu, L. Cholesterol-lowering effect of garlic extracts and organosulfur compounds: human and animal studies. *J Nutr*, **2001**, *131(*3s), 989S-993S.

[172] Yeh, Y. Y.; Yeh, S. M. Garlic reduces plasma lipids by inhibiting hepatic cholesterol and triacylglycerol synthesis. *Lipids*, **1994**, *29(*3), 189-193.

[173] Block, E. The chemistry of garlic and onions. *Sci Am*, **1985**, *252(*3), 114-119.
[174] Agarwal, K. C. Therapeutic actions of garlic constituents. *Med Res Rev*, **1996**, *16(*1), 111-124.
[175] Gupta, N.; Porter, T. D. Garlic and garlic-derived compounds inhibit human squalene monooxygenase. *J Nutr*, **2001**, *131(*6), 1662-1667.
[176] Fitzpatrick, D. F.; Bing, B.; Rohdewald, P. Endothelium-dependent vascular effects of Pycnogenol. *J Cardiovasc Pharmacol*, **1998**, *32(*4), 509-515.
[177] Ames, B. N.; Shigenaga, M. K.; Hagen, T. M. Oxidants, antioxidants, and the degenerative diseases of aging. *Proc Natl Acad Sci U S A*, **1993**, *90(*17), 7915-7922.
[178] Jha, P.; Flather, M.; Lonn, E.; Farkouh, M.; Yusuf, S. The antioxidant vitamins and cardiovascular disease. A critical review of epidemiologic and clinical trial data. *Ann Intern Med*, **1995**, *123(*11), 860-872.
[179] Gaziano, J. M.; Manson, J. E.; Branch, L. G.; Colditz, G. A.; Willett, W. C.; Buring, J. E. A prospective study of consumption of carotenoids in fruits and vegetables and decreased cardiovascular mortality in the elderly. *Ann Epidemiol*, **1995**, *5(*4), 255-260.
[180] Losonczy, K. G.; Harris, T. B.; Havlik, R. J. Vitamin E and vitamin C supplement use and risk of all-cause and coronary heart disease mortality in older persons: the Established Populations for Epidemiologic Studies of the Elderly. *Am J Clin Nutr*, **1996**, *64(*2), 190-196.
[181] Tribble, D. L. AHA Science Advisory. Antioxidant consumption and risk of coronary heart disease: emphasison vitamin C, vitamin E, and beta-carotene: A statement for healthcare professionals from the American Heart Association. *Circulation*, **1999**, *99(*4), 591-595.
[182] Estruch, R.; Ros, E.; Salas-Salvado, J.; Covas, M. I.; Corella, D.; Aros, F.; Gomez-Gracia, E.; Ruiz-Gutierrez, V.; Fiol, M.; Lapetra, J.; Lamuela-Raventos, R. M.; Serra-Majem, L.; Pinto, X.; Basora, J.; Munoz, M. A.; Sorli, J. V.; Martinez, J. A.; Martinez-Gonzalez, M. A. Primary prevention of cardiovascular disease with a Mediterranean diet. *N Engl J Med*, **2013**, *368(*14), 1279-1290.
[183] Fito, M.; Guxens, M.; Corella, D.; Saez, G.; Estruch, R.; de la Torre, R.; Frances, F.; Cabezas, C.; Lopez-Sabater Mdel, C.; Marrugat, J.; Garcia-Arellano, A.; Aros, F.; Ruiz-Gutierrez, V.; Ros, E.; Salas-Salvado, J.; Fiol, M.; Sola, R.; Covas, M. I. Effect of a traditional Mediterranean diet on lipoprotein oxidation: a randomized controlled trial. *Arch Intern Med*, **2007**, *167(*11), 1195-1203.
[184] Steinhubl, S. R. Why have antioxidants failed in clinical trials? *Am J Cardiol*, **2008**, *101(*10A), 14D-19D.
[185] Roberts, L. J., 2nd; Oates, J. A.; Linton, M. F.; Fazio, S.; Meador, B. P.; Gross, M. D.; Shyr, Y.; Morrow, J. D. The relationship between dose of vitamin E and suppression of oxidative stress in humans. *Free Radic Biol Med*, **2007**, *43(*10), 1388-1393.
[186] Kaliora, A. C.; Dedoussis, G. V.; Schmidt, H. Dietary antioxidants in preventing atherogenesis. *Atherosclerosis*, **2006**, *187(*1), 1-17.
[187] Arad, Y.; Spadaro, L. A.; Roth, M.; Newstein, D.; Guerci, A. D. Treatment of asymptomatic adults with elevated coronary calcium scores with atorvastatin, vitamin C, and vitamin E: the St. Francis Heart Study randomized clinical trial. *J Am Coll Cardiol*, **2005**, *46(*1), 166-172.
[188] Fang, J. C.; Kinlay, S.; Beltrame, J.; Hikiti, H.; Wainstein, M.; Behrendt, D.; Suh, J.; Frei, B.; Mudge, G. H.; Selwyn, A. P.; Ganz, P. Effect of vitamins C and E on progression of transplant-associated arteriosclerosis: a randomised trial. *Lancet*, **2002**, *359(*9312), 1108-1113.

CHAPTER 3

Atherosclerosis, Mitochondrial Dysfunction and Oxidative Stress: Mitochondria-Targeted Anti-oxidants as Potential Therapy

Irene Escribano-Lopez[1], Noelia Diaz-Morales[1], Susana Rovira-Llopis[1,2,3], Celia Bañuls[1,2], Sandra Lopez-Domenech[1], Raquel Castelló[1,2], Rosa Falcón[1,2], Eva Sola[1,2], Milagros Rocha[1,2,3], Antonio Hernández-Mijares[1,2,4] and Victor M. Víctor[1,2,3,*]

[1]Service of Endocrinology. University Hospital Dr. Peset, Foundation for the Promotion of Health and Biomedical Research in the Valencian Region (FISABIO), Valencia, Spain; [2]Institute of Health Research INCLIVA, University of Valencia, Valencia, Spain; [3]CIBERehd - Department of Pharmacology and Physiology, University of Valencia, Valencia, Spain; [4]Department of Medicine, University of Valencia, Valencia, Spain

Abstract: Chronic and acute overproduction of reactive oxygen species (ROS) under pathophysiologic conditions forms an integral part of the development of cardiovascular diseases (CVD), and in particular atherosclerosis. These ROS are released from different sources, such as xanthine oxidase, lipoxygenase, nicotinamide adenine dinucleotide phosphate oxidase, the uncoupling of nitric oxide synthase and, in particular, mitochondria. Endothelial dysfunction, characterized by a loss of nitric oxide (NO) bioactivity, occurs early on in the development of atherosclerosis, and determines future vascular complications. Although the molecular mechanisms responsible for mitochondria-mediated disease processes are not clear, oxidative stress seems to play an important role. In general, ROS are essential to cell function, but adequate levels of antioxidant defenses are required in order to avoid the harmful effects of excessive ROS production. Mitochondrial oxidative stress damage and dysfunction contribute to a number of cell pathologies that manifest themselves through a range of conditions. This review considers the process of atherosclerosis from a mitochondrial perspective, and assesses strategies for the targeted delivery of antioxidants to mitochondria that are currently under development. We will provide a summary of the following areas: the cellular metabolism of reactive oxygen species (ROS) and its role in pathophysiological processes such as atherosclerosis; currently available antioxidants and possible reasons for their efficacy and inefficacy in ameliorating oxidative stress-mediated diseases; and recent developments in mitochondrially-targeted antioxidants that concentrate on the matrix-facing surface of the inner mitochondrial membrane in order to protect against mitochondrial oxidative damage, and their therapeutic potential as a treatment for atherosclerosis.

***Corresponding Author Victor M. Victor:** University Hospital Doctor Peset, Avda Gaspar Aguilar 90, 46017, Valencia, Spain; Tel: 34961622757; Fax: 34961622492; E-mail: Victor.Victor@uv.es

Atta-ur-Rahman, M. Iqbal Choudhary and Allen B. Reitz (Eds.)

Keywords: Antioxidant, atherosclerosis, autophagy, cardiovascular disease, cholesterol, clinical study, diabetes, endothelium, human, insulin resistance, mitochondria, mitochondrial dynamics, mitoQ, nitric oxide, oxidized-LDL, reactive oxygen species, reticulum stress, superoxide dismutase, targeted-antioxidant, vitamin C, vitamin E.

INTRODUCTION

Cardiovascular diseases (CVD) and related pathologies are the leading cause of death and disability in the developed world [1-4]. An early prognosis and effective therapies depend on a good understanding of the basic pathophysiological mechanisms of CVD. Over the last decade, knowledge regarding mechanisms contributing to CVD in animals has improved remarkably, yet the management of CVD in humans has not improved to a comparable degree. Despite extensive studies of its pathogenesis and steady progress in its treatment, atherosclerotic CVD is an increasingly serious public health concern for developed countries.

There is a extensive number of studies on animal models of atherosclerosis and humans that support the notion that higher levels of oxidative and nitro-oxidative stress from increased production of reactive oxygen species (ROS) and decreased reserves of antioxidants [2, 5, 6] contribute to the pathogenesis of atherosclerosis. ROS can damage cellular lipids, proteins and DNA, thereby disrupting their normal function. Oxidative stress has been implicated in a number of human diseases, and occurs when the balance is disrupted by excessive production of ROS such as superoxide ($O_2^{\cdot-}$), hydrogen peroxide (H_2O_2), hydroxyl radicals ($^{\cdot}HO$), peroxynitrite ($ONOO^-$) and/or inadequate antioxidant defences [7], including superoxide dismutase (SOD), catalase (CAT), vitamins C and E and reduced glutathione (GSH). Both processes can lead to CVD thorough endothelial dysfunction.

The mitochondrial respiratory chain is a major intracellular source of ROS [8], indicating that a specific action of antioxidants on the mitochondrial respiratory chain constitutes an important mechanism of cardiovascular protection. Mitochondrial $O_2^{\cdot-}$generation in endothelial cells is particularly prominent in situations of perturbed metabolism, such as hyperglucemia [9]. In this sense, mitochondria play a significant role in the regulation of vascular cell function and atherogenesis [10].

ROS and reactive nitrogen species (RNS) are important both physiologically and in the pathogenesis of atherosclerosis. Enhanced endogenous or exogenous ROS production is a hallmark of the early stages of the atherogenic process, suggesting a link between ROS and apoptosis [11]. ROS cause non-specific damage to lipids, proteins and DNA, leading to alteration or loss of cellular function. It is, therefore, conceivable that mitochondria are more vulnerable to oxidative damage than other cellular organelles; in fact, mitochondria are exposed to ROS and accumulate oxidative damage more rapidly than the rest of the cell [12].

The recognition of mitochondria as an arbiter of life and death in cells has raised awareness of the need to develop antioxidants and other cytoprotective agents that are target to mitochondria. Physiologically, mitochondria perform a variety of key cellular regulatory processes and use approximately 90% of the O_2 consumed for oxidative phosphorylation.

Mitochondrial oxidative damage and dysfunction contribute to a number of cell pathologies that include CVD. Studies in different models have shown that the deleterious effects of ROS can be counteracted by antioxidants such as α-tocopherol, ubiquinol and N-acetylcysteine, which decrease mitochondrial oxidative damage [13-15]. However, the effectiveness of these compounds is limited as they do not significantly accumulate within mitochondria, [16], and this has fuelled research efforts to develop mitochondria-targeted antioxidants. This is exemplified by studies showing that triphenyl-phosphonium-(TPP)- and amino acid- and peptide-antioxidants protect mitochondria against oxidative stress, as they are cell-permeable and considerably potent in reducing intracellular ROS and preventing cell death [17].

ENDOTHELIAL DYSFUNCTION

A healthy endothelium plays a central role in cardiovascular control. Therefore, endothelial dysfunction (ED), which is characterized by an imbalance between relaxing and contracting factors, procoagulant and anticoagulant substances, and proinflammatory and antiinflammatory mediators, may play a particularly significant role in the pathogenesis of atherosclerosis (Fig. (**1**)).

ED is closely related to different risk factors of atherosclerosis, and to their intensity and duration. A connection between these risk factors and ED is supported by the results of intervention studies that highlight a regression of ED with treatment of risk factors. Given that risk factors are commonly accompanied by a lower bioavailability of nitric oxide (NO), the common denominator by which different risk factors cause ED is most probably increased oxidative stress [18]. ED is an alteration that appears in physiological situations, such as aging, and in different pathologies, including dyslipemia, diabetes mellitus, obesity, atherogenesis and arterial hypertension [19-22]. In these cases, the endothelial cells are modified and there is a subsequent impairment of secretory substances with a minor liberation of vasodilator substances, fundamentally NO, anticoagulants and antiplatelets. In addition, there is an increase of the proinflammatory molecules, which induces a procoagulant, antifibrinolitic, vasoconstrictor and proinflamatory condition induced by the endothelium and known as endothelial dysfunction (Fig. (**2**)).

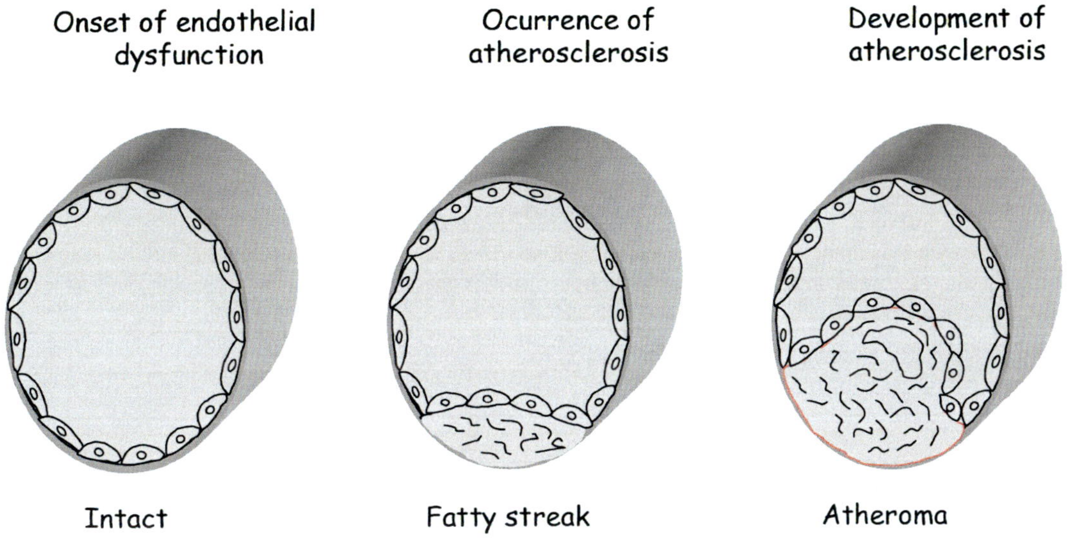

Figure 1: Progression of atherosclerosis in cardiovascular diseases.

In ED a reduction in the synthesis of NO is observed, which has been demonstrated both in the coronary traffic of healthy individuals and in patients with peripheral arterypathology. Diverse factors are implicated in the reduction of NO. Some authors have demonstrated a decrease in tetrahydrobiopterin, a cofactor of its synthesis, whereas others have demonstrated an increase in the levels of asymmetric dimethylarginine (ADMA), a competitive endogenous inhibitor of its synthesis. This increase in the levels of ADMA is a risk factor for coronary events in both healthy subjects and patients with ischemic cardiopathy [23], thereby highlighting an association between endothelial dysfunction and atherosclerosis. Patients with atherosclerosis and ED also exhibit a weaker endothelial response as a result of exposure to exogenous sources of NO.

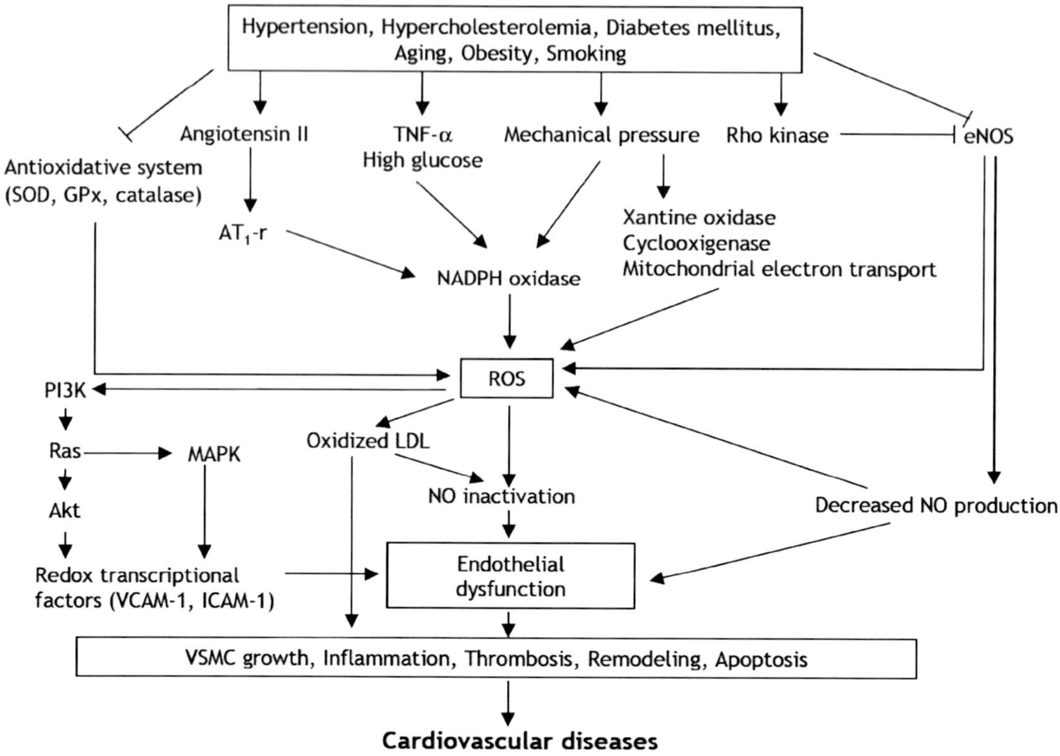

Figure 2: Mechanisms by which ROS induce endothelial dysfunction. NO, nitric oxide; SOD, superoxide dismutase; NADH/NADPH, nicotinamide adenine dinucleotide phosphate; ROS, reactive oxygen species; HSP, heat shock protein; endothelial growth factor; eNOS, endothelial NO synthase; PI3K, phosphatidylinositol-3-kinase; MAPK, mitogen-activated protein kinase; NF-κB, nuclear factor κB; AP-1, plasminogen activator inhibitor-1; VCAM-1, vascular cell adhesion molecule-1; ICAM-1, intracellular adhesion molecule-1; VSMC, vascular smooth muscle cell.

Several researchers have demonstrated that endothelium-dependent vasodilation is impaired in the forearm, coronary, and renal vasculature in patients with CVD [24, 25]. In coronary artery diseases, severe coronary endothelial dysfunction is associated with increased cardiovascular events [26]. Schachinger *et al.* [27] demonstrated a link between coronary ED and subsequent cardiovascular events in patients with coronary artery diseases. Acetylcholine-induced vasodilatation and flow-meditated vasodilatation are also useful for predicting cardiovascular events in such patients [27]. Those clinical studies have shown that endothelial function can be an independent predictor of cardiovascular events. From a clinical perspective, it is important to select an intervention that is effective in improving endothelial function in patients with CVD. Indeed, several interventions have been shown to improve endothelial function and prevent CVD [28], including pharmacological therapy, administration of antihypertensive agents such as angiotensin-converting enzyme inhibitors and AngII type I receptor blockers [29], statins [30], and thiazolidinedione derivatives [31], supplementation therapy consisting of administration of a substrate of NO L-arginine [32], oestrogen replacement [33], administration of a cofactor of NO tetrahydrobiopterine [34], treatment with antioxidant vitamins such as vitamin C [35], and lifestyle modifications such as aerobic exercise [36] and body weight reduction [37] These findings suggest that ED in patients with CVD is reversible.

High levels of ROS damage the endothelium, particularly in arteries of the terminal type. Small vessel disease in diabetes mellitus, which is often associated with mitochondrial disorders, may also result from oxidative stress. The effect of diabetes in atherosclerosis is mediated largely by oxidative stress [38]. ROS may also activate mitogen activated protein kinase (MAPK), which regulates the expression of strain-induced monocyte chemoattractant protein 1 (MCP-1), and thus suggests a potential molecular link between arterial wall strain and atherosclerosis [39].

ATHEROSCLEROSIS

Atherosclerosis is a multifactorial pathology whose molecular etiology involves the interaction of many genes and environmental factors. Increased production of ROS in mitochondria, accumulation of mitochondrial DNA damage and progressive respiratory chain dysfunction are associated with atherosclerosis and cardiomyopathy in human and animal models of oxidative stress. The majority of CVD are a result of complications caused by atherosclerosis. Classic risk factors for the development of atherosclerosis are age, arterial hypertension, diabetes, dyslipemia, a smoking habit, diet, immunological disorder, vascular wall inflammation, and genetic predisposition.

Atherosclerotic vascular disease and its clinical sequelae are the leading causes of morbidity and mortality in the developed world. An elevated level of low-density lipoprotein (LDL) is associated with an increased risk of CVD. Several reviews highlight the role of an activated endothelium and inflammatory cell recruitment in the initiation and progression of early atherosclerosis [40-42]. Many risk factors for atherosclerosis, such as increased levels of modified LDL, a smoking habit, increased ROS and diabetes, have been shown to damage the endothelium, and it has been hypothesized that dysfunction of the endothelium initiates atherosclerotic lesion formation.

Oxidative stress, which occurs in response to an altered metabolic state, and apoptosis and lipid peroxidation are additionally involved in the pathogenesis of atherosclerosis. Oxidative stress seems to be the common denominator for most of the classic risk factors of atherosclerosis [43].

Endothelial cells, smooth muscle cells and macrophages are sources of ROS, while Ox-LDL can damage endothelial cells, thereby inducing the expression of adhesion molecules [44]. ROS limit the bioavailability of NO and induce inflammatory gene expression, cell growth/apoptosis, migration and matrix reorganization, all of which are mechanisms that are central to the initiation and progression of atherosclerosis. NO signalling in the endothelium can be affected by a number of factors during atherosclerosis.

There are numerous reports of a correlation existing between DNA damage and atherosclerosis [45]. MtDNA damage not only correlates with the extent of atherosclerotic lesions in apolipoprotein E (apoE) knockout mice but precedes atherogenesis in young apoE knockout mice [10]. Mitochondrial dysfunction can be a result of manganese superoxide dismutase (SOD2) deficiency, increased mtDNA damage and accelerated atherosclerosis in apoE knockout mice, which is consistent with the notion that increased ROS production and DNA damage in mitochondria constitute events in the initiation of atherosclerosis. Multivariate analysis reveals that DNA-adduct levels are a significant predictor of the stage of atherosclerosis, even after adjustment for age, smoking, obesity and other CVD risk factors [46]. Evidence suggests that there are more DNA repair mechanisms in the atherosclerotic plaques than in control tissues [47]. In accordance with these findings, increased levels of mtDNA damage have been observed in the cardiovascular tissue of CVD patients [10]. However, whether this damage is an effect or initiator of CVD remains unclear. Animal studies have shown that vascular mtDNA damage is more pronounced in animal models of atherosclerosis, and that this damage occurs prior to, or simultaneously with, development of the disease [10].

Oxidized lipids are capable of initiating diverse cellular responses through both receptor-mediated mechanisms and direct post-translational modification of proteins. In fact, incubation of endothelial cells with concentrations of oxLDL induces a mitochondrial complex I activity that seems to depend on the induction of oxidative stress [48]. In fact, with 46 subunits, human mitochondrial complex I is the largest enzyme of the oxidative phosphorylation system. Increased mitochondrial oxidant generation and decreased mitochondrial membrane potential have also been studied in human macrophage cells treated with oxLDL, in which scavengers of peroxides prevented said effects [49]. A standard consequence of increased oxidative stress due to exposure to ox-LDL is the breaking of the nuclear DNA strand, which is cause of p53 upregulation.

High dietary fat intake leads to insulin resistance in skeletal muscle, and this represents a major risk factor for type II diabetes and CVD. Mitochondrial dysfunction and oxidative stress have been implicated in the disease process, and it has been demonstrated that in skeletal muscle of both rodents and humans, a diet high in fat increases the H_2O_2-emitting potential of mitochondria, shifts the cellular redox environment to a more oxidized state, and decreases the redox-buffering capacity in the absence of any change in mitochondrial respiratory function [50]. Furthermore, they show that attenuating mitochondrial H_2O_2 emission, either by treating rats with a mitochondrial-targeted antioxidant or by genetically engineering the overexpression of catalase in mitochondria of muscle in mice, completely preserves insulin sensitivity despite a high-fat diet. High-fat diets reduce the expression of genes involved in free radical scavenging (SOD and glutathione peroxidase (GPX)) and increase the expression of stress response (Hsp 70) and signal transduction genes (Ras, MAPK1) in rats. However, while levels of UCP-2 increase, those of UCP-3 remain unaffected. Antioxidant supplementation of a high-fat diet reduces the magnitude of these differences [51]. Furthermore, it has been described that inflammation and oxidative stress appear to play a key role in the development of increased adiposity in the offspring of Western diet-fed pregnant dams, and restoration of the antioxidant balance during pregnancy in the Western diet-fed dam is associated with decreased adiposity in offspring [52].

LDL Oxidation

ROS and RNS cause oxidation of LDL, and can be consumed by macrophage scavenger receptors [53]. Macrophages that consume excess oxLDL are transformed into foam cells, which accumulate and release growth factors and cytokines. There is now convincing evidence that oxidatively-modified LDL exists in atherosclerotic lesions. In addition to the intrinsic properties of lipoproteins, factors that prolong the life/residence time of LDL may also be related to oxidation. For example, proteoglycans not only "trap" LDL in the extracellular matrix of the intimal space, but also change the structure of apolipoprotein B-100 [54].

Many studies have attempted to determine the oxidants that contribute to oxidative processes in atherosclerotic lesions. In the case of human atherosclerotic lesions, a commonly employed biochemical approach is to first chemically define the "footprint" of products of a particular oxidant with target molecules, and to then analyze vessels and materials derived from them for the presence of these footprints. With the advent of murine models of atherosclerosis, molecular approaches, such as the use of animals deficient in a specific oxidant-forming enzyme, have been increasingly adopted. Together, these studies have provided insight into the origin and nature of the oxidative modifications that take place during atherogenesis. For example, there is now convincing evidence for the participation of myeloperoxidase (MPO) and oxidants derived from this enzyme. In this sense, Brennan *et al.* [55] show that MPO-deficient mice develop larger atherosclerotic lesions and an increase of monocyte number, consistent with larger lesion area.

Vascular cells can oxidize LDL to a form recognized by the LDL receptor. Diet-induced atherosclerosis in primates is associated with excess vascular $O_2^{-}\cdot$ and upregulation of $p22^{phox}$ and $p47^{phox}$, both of which abate with the regression of atherosclerosis that is achieved by lowering cholesterol levels through dietary measures [56]. However, a causal role for NAD(P)H oxidase in atherosclerosis is controversial. No difference in either the apolipoprotein E $^{-/-}$ or diet-induced models of atherosclerosis can be demonstrated in mice lacking $gp91^{phox}$ (Nox2) [57]. In another study, Hsich *et al.* [58] showed that the analysis of wild-type and p47(phox) knockout mice revealed no differences in either basal blood pressure or the rise in blood pressure seen after the pharmacological inhibition of nitric oxide synthase, and they conclude that NADPH oxidase contributes to basal vascular $O_2^{\cdot-}$ production. However, the absence of a functional oxidase does not significantly affect the progression of atherosclerosis in the standard mouse apoE -/- model, nor does it significantly influence basal blood pressure. Other studies in human tissue suggest that Nox isoforms are involved in atherosclerosis. For example, Quesada *et al.* have demonstrated that selective inactivation of NADPH oxidase 2 causes regression of vascularization and the size and stability of atherosclerotic plaques [59]. Vascular production of $O_2^{-}\cdot$ increases with the presence of risk factors for the disease [15, 40, 41], which suggests that NAD(P)H oxidase plays a role in this pathology. Atherosclerotic lesions contain abundant $p22^{phox}$ and $gp91^{phox}$ in the vicinity of macrophages at levels that correlate with the severity of atherosclerosis [60]. Analysis of tissue of patients undergoing bypass surgery revealed that diabetes was characterized by higher expression of $p22^{phox}$, $p47^{phox}$ and $p67^{phox}$ (neutrophil cytosolic factor 2, NCF2) than that seen in non-diabetics [61]. Given that diabetics suffer disproportionately more cardiovascular events than non-diabetics, these data suggest that NADPH oxidase activity plays a role in the clinical expression of atherosclerosis.

In addition to NAD(P)H oxidases, xanthine oxidase has been implicated in endothelial cell $O_2^{-}\cdot$ production in experimental models of hypercholesterolemia. Cholesterol-fed rabbits exhibit an increase in endothelial $O_2^{-}\cdot$ production that is inhibited by oxypurinol [62], a xanthine oxidase inhibitor. This increased $O_2^{-}\cdot$ flux from xanthine oxidase contributes to impaired endothelial NO bioactivity in circumstances of hypercholesterolemia, heavy smoking and coronary artery disease.

It is also possible that eNOS contributes to cellular $O_2^{-}\cdot$ in the setting of vascular disease, as hypercholesterolemia, atherosclerosis, diabetes and hypertension have all been associated with eNOS uncoupling. Moreover, animal models of atherosclerosis that have been treated with tetrahydrobiopterin, a naturally occurring nutrient and essential cofactor of the three aromatic amino acid hydroxylase enzymes, used in the biosynthesis of the neurotransmitters serotonin (5-hydroxytryptamine (5-HT)), melatonin, dopamine, norepinephrine (noradrenaline), epinephrine (adrenaline), and NO, or its precursor exhibit a decrease in vascular O_2^{-} and enhancement of ·NO bioactivity [63]. Thus, considerable evidence exists for eNOS-derived $O_2^{-}\cdot$ production, although the precise mechanism(s) responsible for this phenomenon and how it relates to atherosclerosis is/are still unknown. Furthermore, an important role has been described for endothelial NO in hypoxia by which it regulates O_2 consumption in vessels and tissues, thereby allowing the distribution of O_2 to other cells [64]. As indicated above, the specific role of $O_2^{-}\cdot$ in oxidative modification during atherosclerosis remains unclear, particularly that related to LDL oxidation. Consistent with its relative chemical reactivity, $O_2^{-}\cdot$ is unable to oxidize LDL, in contrast to its protonated form, the hydroperoxyl radical ($HO_2\cdot$) [65]. However, at physiological pH only a fraction of the $O_2^{-}\cdot$ produced is present as $HO_2\cdot$. This suggests that $O_2^{-}\cdot$-generating systems, including NAD(P)H oxidases, are not efficient in oxidizing LDL by themselves [65], in contradiction of the notion that $O_2^{-}\cdot$ is a mediator of LDL oxidation by endothelial cells [66]. However, $O_2^{-}\cdot$ may act as the precursor for chemically more reactive oxidants, such as $ONOO^{-}$ and H_2O_2 and the oxidants derived from them, and these may participate in LDL oxidation in the vessel wall.

Leeuwenburgh *et al.* [67] reported that LDL from aortic atherosclerotic intima exhibited 90-fold higher levels of protein-standardized 3-nitrotyrosine than plasma LDL. Li *et al.*, have described that oxidative modification of lipoproteins and phospholipids are mechanisms that contribute to atherogenesis, and the oxidation of tetrahydrobiopterin may cause eNOS uncoupling and thus potentiation of oxidative stress and reduction of eNOS-derived NO, which is a protective principle in the vasculature [68]. In addition to eNOS, inducible NOS is expressed in atherosclerotic lesions of humans [69] and rabbits [70], providing potential sources for RNS.

In biological samples, $ONOO^-$ and MPO leave similar footprints with regard to protein nitration [71]. Indeed, the MPO/H_2O_2/nitrite system modifies LDL's apolipoprotein B-100 and lipids, probably *via* formation of NO_2, and LDL modified in this way can convert macrophages into foam cells [72]. The involvement of MPO activity in the oxidation reactions that take place during atherogenesis in humans is supported by the detection of footprints of MPO/HOCl, including HOCl-modified proteins, 3-chlorotyrosine, *p*-hydroxyphenylacetaldehyde and *p*-hydroxyphenylacetaldehyde-ethanolamine [73]. The potentially important finding of MPO activity has been confirmed for lesions in humans [74]. In contrast, active MPO appears to be absent in atherosclerotic lesions in LDL receptor $^{-/-}$ and apolipoprotein E $^{-/-}$ mice [55]. This highlights one of the limitations of using animal models to study the pathogenesis of a human disease. Hypochlorite-modified LDL is potentially proatherogenic [75]. *In vitro* studies have shown that HOCl-modified LDL can stimulate macrophage foam cell formation [76], increases leukocyte adherence and migration into blood vessels [77], and augments ROS production by leukocytes [78]. MPO may oxidize LDL *via* reactive amino acid intermediates, such as the tyrosyl radical [79] and *p*-hydroxyphenylacetaldehyde [80]. In addition, MPO-derived HOCl may convert LDL into a high-uptake form of the lipoprotein *via* reaction with apolipoprotein B-100, with little subsequent oxidation of the lipids [81]. There is some evidence that ·OH contributes to oxidative reactions in atherosclerosis, as certain hydroxylated amino acids are known to be increased in human endarterectomy specimens [82] and in the artery walls of monkeys with early diabetic vascular disease [83]. Recently, it has been demonstrated that MPO levels were positively correlated with PMN rolling flux and adhesion in leukocytes from diabetic patients. These results pointed out that type 2 diabetes induces oxidative stress and an increase in MPO levels and leukocyte-endothelium interactions, and that these effects correlate with the development of nephropathy [20].

OXIDATIVE STRESS AND MITOCHONDRIA

Interest in mitochondrial physiology has been renewed due to growing evidence implicating mitochondria in cellular signalling pathways; namely, the modulation of intracellular calcium stores, production of ROS and the interaction of NO with mitochondrial functions such as respiration and biogenesis. Given these important roles, alterations in mitochondrial function are considered to be key to the development of human disease [84-86].

As a consequence of the high-energy requirements of the myocardium, cardiomyocytes contain a significantly higher proportion of mitochondria than other cell types [87]. This and other evidence have given rise to the suggestion that mitochondria from diverse cellular types have distinct morphological and functional roles [88].

Inherited dysfunction of the mitochondrial oxidative phosphorylation system is increasingly recognized as a CVD in humans, and mitochondrial diseases have been related to mitochondrial respiratory-chain pathologies associated with mutations of mitochondrial DNA [89]. Numerous nuclear and mtDNA mutations have been identified as the cause of disorders of individual oxidative phosphorylation enzymes, including mitochondrial ATP synthase in patients [90].

The discovery that acute and chronic stress in the cells leads to structural and functional impairments of mitochondria has redefined the role of mitochondria in disease etiology [91], and is of some relevance to CVD. Mitochondrial dysfunction triggers signalling cascades for cell necrosis and apoptosis, and leads to organ failure and diseases. The list of mitochondria-related conditions is growing rapidly and includes cancer, heart failure, diabetes, obesity, stroke, neurodegenerative diseases and aging, but all are characterised by disturbances of mitochondrial Ca^{2+}, ATP, or ROS metabolism [91]. Some studies have demonstrated that mitochondrial dysfunction contributes to the progression of neurodegenerative diseases, such as Parkinson's, and to stroke [92]. Myocardial ischemia-reperfusion injury also results in mitochondrial Ca^{2+} overload and a subsequent uncontrollable ROS generation and opening of the mitochondrial permeability transition pore [91], which ultimately induces apoptosis. Therefore, compounds that minimize mitochondrial Ca^{2+} overload, decrease mitochondrial ROS accumulation and improve mitochondrial energy production are all potential sources of therapies for the above mentioned diseases.

As mitochondria are the major site of generation of cellular oxidative stress and play a key role in mediating programmed cell death (apoptosis), damage to mtDNA may contribute in a significant way to human ageing, cancer and CVD. Oxidative damage to the mitochondrial membrane may also occur, resulting in membrane depolarisation and the uncoupling of oxidative phosphorylation, and a subsequently altered cellular respiration [93]. Whereas ROS and RNS are capable of targeting a variety of subcellular components, mitochondria are exposed rapidly and constantly to ROS and accumulate oxidative damage much more so than the rest of the cell. In fact, mitochondrial membranes, proteins and mtDNA appear to be particularly sensitive to oxidative and nitrosative damage [94]. Mitochondrial oxidative damage has been implicated in clinical conditions such as ischaemia-reperfusion injury and neurodegenerative diseases. For example, ischaemia-reperfusion leads to the reduction in CXCR4 - C-X-C chemokine receptor type 4 (CXCR4) in cardiomyocytes associated with the dysfunctional energy metabolism, and CXCR4 overexpression can alleviate mitochondrial dysfunction to improve cardiomyocyte survival [95].

ROS and RNS have been implicated in atherogenesis, and increased oxidative stress is a common characteristic of many CVD risk factors, mediating post-translational modifications of mitochondrial proteins [96]. For example, nitration of Tyr34 in SOD2 deactivates it, and increased 3-nitrotyrosine levels of SOD2 *in vivo* have been associated with a decreased specific activity of the enzyme [97].

Mitochondrial damage can alter the capacity of the cell to generate energy, and therefore the redox signalling. ROS generated in the ETC have been highlighted as intermediate messengers of the activation of NF-κB [98]. Consistent with this hypothesis, cells lacking mitochondria also show significant suppression of NF-κB activation [99]. Furthermore, it is possible that mitochondrial damage modifies the effects of signalling factors. Hence, the relative balance between the stimuli of mitochondrial ROS production and the concomitant accumulation of organelle damage can ultimately influence cellular response and function, and the cellular response to increased oxidative stress may, thus, initiate CVD. Consequently, mitochondrial damage would seem to serve as a general, yet direct, index or predictor of mitochondrial dysfunction.

MITOCHONDRIA AS A TARGET FOR ANTIOXIDANT THERAPY

It has been suggested that mitochondria are a pharmacological target, and drugs that modulate mitochondrial function include those that target oxidative stress [100, 101]. This is logical given the crucial role that mitochondria play in energy metabolism and cell death signalling pathways, both of which are related to cancer, diabetes, CVD and aging [102]. These organelles are also a source of ROS, which arise from oxidative phosphorylation. Although the cell has evolved mechanisms to deal with the harmful effects of ROS, such as the enzymes SOD,

CAT and GSH peroxidase, its safeguards do not completely block the formation of ROS, particularly under conditions of elevated or prolonged stress.

Current drug strategies that selectively target mitochondria take advantage of the electrochemical gradient from the outer plasma membrane potential (30 to 60 mV) to the inner mitochondrial matrix (150 to 180mV). This gradient provides a highly driven force for the selective targeting and concentration of large lipophilic cations to the mitochondria [100] (Fig. (**3**)). Two drugs of this type, MitoQ and MitoVitE, have been proposed by Jauslin and colleagues for treatment of Friedreich-s ataxia [103]. Both compounds are based on TPP modifications to coenzyme Q and vitamin E, respectively (Fig. (**2**)). In these studies, MitoVitE was found to be 350 times more potent in human fifroblasts from human patients than Trolox, a water-soluble version of vitamin E, and mitoQ proved 100 times more potent than idebenone, a coenzyme Q derivative.

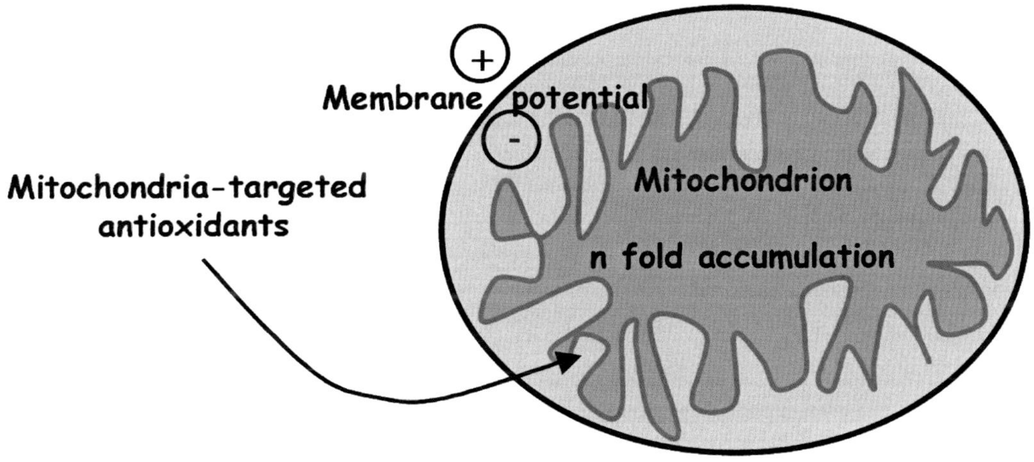

Figure 3: Using lipophilic cations to target compounds to mitochondria, enables reagents to be delivered selectively to mitochondria within cells. Lipophilic cations accumulate within the mitochondria driven by the membrane potential, and can pass through lipid bilayers easily as their positive charge is delocalized over an extended area.

An analog of the SOD-Catalase mimetics was also shown to be effective in a direct test of mammalian mitochondrial oxidative stress in the SOD2 knockout mouse, an animal that lacks mitochondrial SOD [104]. Beneficial effects were observed in the brain and heart and in a number of other tissues [105], and others studies have demonstrated the beneficial effect of antioxidant compounds that limit oxidative damage and restore endothelial function [106].

ANTIOXIDANT STUDIES

Given that LDL oxidation is an essential feature of atherosclerosis, inhibiting LDL oxidation should limit atherosclerosis. This concept has been tested primarily in animals with a number of structurally distinct antioxidant compounds that can inhibit LDL oxidation in *in vitro* assays.

The strongest evidence in support of antioxidant compounds providing protection against atherosclerosis is that reported with probucol [107], a synthetic cholesterol-lowering drug that also expresses antioxidant activity [108].

Early studies with probucol demonstrated inhibition of atherosclerosis in rabbits [109]. Kita *et al.* [110] treated Watanabe heritable hyperlipidemic (WHHL) rabbits with probucol and observed an 87% reduction in lesion area and LDL resistance to oxidation when compared with untreated animals. These findings prompted the authors to conclude that the reduction of atherosclerosis was due to the antioxidant effect of probucol. Carew *et al.* [111] assessed the cholesterol-lowering effect of probucol with lovastatin, an inhibitor of 3-hydroxy-3-methylglutaryl coenzyme A reductase, and similar reductions in total cholesterol were observed with the two drugs; however, the latter's reduction of atherosclerosis was 48% higher, suggesting that it inhibited atherosclerosis due to its antioxidant activity.

Two studies have tested the antiatherosclerotic activity of probucol in humans. The Probucol Quantitative Regression Study (PQRST) found the drug to be ineffective in attenuating lumen loss in the femoral arteries of hyper-cholesterolemic subjects over 3 years, as assessed by quantitative angiography [112]. In contrast, the Fukoaka Atherosclerosis Trial (FAST) observed probucol to significantly decrease atherosclerosis progression in the carotid artery of hypercholesterolemic patients, as assessed by B-mode ultrasound determination of intima-to-media thickness [113].

A number of other antioxidants have also been tested for their ability to inhibit atherosclerosis in animal models of the disease. *N,N'*-diphenyl-phenylenediamine (DPPD) is an aniline compound that attenuates atherosclerosis in the aorta of cholesterol-fed rabbits [114], and similar effects have been reported in a murine model of atherosclerosis [115]. A common feature of all of these compounds is that they are capable of inhibiting LDL oxidation *in vitro*. Thus, a number of lipid-soluble, synthetic antioxidants have been used to demonstrate an association between a reduction in atherosclerosis and *ex vivo* inhibition of LDL oxidation.

LDL contains 6–12 molecules of α-tocopherol per particle, which is its major redox-active constituent, so it is important to discern their role in the process of LDL oxidation, particularly that mediated by radical oxidants. If LDL particles are exposed to a strong oxidant or encounter radicals with a high frequency, α-tocopherol is consumed rapidly and there is little concomitant formation of LOOH [116]. Under these conditions, α-tocopherol ostensibly performs an antioxidant role. Recently, Mathur *et al.*, have described in an interesting review that the conflicting results about the effects of α-tocopherol have led to reconsideration of the importance previously given to α -tocopherol and led to interest in other members of vitamin E family, especially γ-tocopherol, which exerts a much more potent antioxidant, anti-inflammatory, and cardioprotective effect than α - tocopherol [117]. This reconsideration has been backed by solid laboratory and clinical research. They suggest that the absence of γ -tocopherol in traditional preparations may be one reason for the lack of consistent salutary effects of vitamin E preparations in clinical trials.

In sharp contrast, if LDL particles encounter mild oxidants or radicals with low frequency, consumption of α-tocopherol is much slower and is associated with a comparatively large accumulation of LOOH [118]. It is important to consider that depletion of LDL's α-tocopherol is required for accumulation of secondary lipid oxidation products such as aldehydes and F_2-isoprostanes [119], although some transformation of LOOH into hydroxyalkenals takes place in the presence of the vitamin. It is these secondary lipid oxidation products that are largely responsible for the atherogenic properties of *in vitro* oxidized LDL [120]. Thus, under strongly oxidizing conditions, the antioxidant function of α-tocopherol consists in inhibiting an initial step in the overall conversion of native to high uptake LDL.

Antioxidant Status in Atherosclerotic Lesions

Currently available data are restricted primarily to the results of studies of advanced vascular disease and the components released from homogenates of large pieces of aortic tissue. There is a scarcity of systematic studies of antioxidant changes at various disease stages and in focal areas of the intima. In addition, there is a lack of consistency in the available information; for instance, when data of human lesions are compared with those of lesioned material isolated from animal models of the human disease, discrepancies are obvious. Moreover, it is difficult to interpret changes to antioxidant defenses when the material is rarely analyzed also for the presence of oxidized biomolecules. For example, an increase in an antioxidant defense is commonly interpreted as a biological response to increased oxidative stress without establishing whether it is accompanied by a general decrease or increase in oxidative tissue damage, and despite the fact that oxidants do not necessarily affect the expression of antioxidant enzymes [121]. There is evidence that the levels of some important antioxidants in the diseased vascular wall are altered, although information on antioxidant enzyme expression in human atherosclerosis is limited [122].

A) SOD and Catalase

The activity of SOD has been reported to be increased in highly cellular rabbit lesions but decreased in advanced connective tissue-rich human lesions [69]. 't Hoen *et al.* [123] have observed mRNA levels of antioxidant enzymes in the aortic arch of wild-type and apolipoprotein E $^{-/-}$ mice of different ages and fed a regular chow. The mRNA levels of Cu,Zn-SOD (SOD1), Mn-SOD (SOD2) and catalase in the aortic root of apolipoprotein E $^{-/-}$ mice rose prior to lesion formation (weeks 6-12) but dropped at 34 wk, when lesions had developed [122]. A decrease in the activity of SOD [124] may lead to lower bioavailability of ·NO and, hence, dysfunctional endothelium-dependent vasodilation and increased production of $ONOO^-$.

B) Glutathione-Related Antioxidants

Compared with normal internal mammary arteries, selenium-dependent GSH peroxidase and GSH reductase activities are lower in human carotid atherosclerotic plaques, and no measurable changes to GSH transferase activity have been observed [125]. In contrast, selenium-independent GPx activity is increased in the atherosclerotic plaque.

In apolipoprotein E $^{-/-}$ mice, mRNA levels of GSH peroxidases in the aortic arch increase in the period preceding lesion formation in the aortic root, but decrease at 34 week, when lesions have developed, alterations that are similar to those observed for SOD and catalase [123]. In another study with animals fed a cholesterol-enriched diet, the levels of GSH-related enzymes were comparable in control and atherosclerotic arteries and in arteries from cholesterol-fed animals whether they developed macroscopic lesions or not [126].

Heme oxygenase-1 is induced by oxidative stress [127] and expressed in the cells of human atherosclerotic lesions. In this way, increased heme oxygenase-1 activity could represent a local antioxidant response. There is growing evidence that induction of heme oxygenase provides protection against atherosclerosis and related diseases. Indeed, upregulation of heme oxygenase-1 activity by heme, transfection or gene transfer reduces atherosclerosis in mice deficient in LDL receptor, in apolipoprotein E $^{-/-}$ mice and in porcine coronary arteries [128], whereas pharmacological inhibition of heme oxygenase enhances atherosclerosis in LDL receptor –/– rabbits [129]. Furthermore, Yet *et al.* [114] demonstrated in gene deletion experiments that HO-1 plays an essential, protective role in vascular remodeling in the settings of hypercholesterolemia and vein graft stenosis. These results, along with others previous findings in the heart, as well as human studies, indicate that a decrease or absence of HO-1 exacerbates CVD. Induction of heme oxygenase-1 also protects against intimal hyperplasia in the rat aorta following balloon injury [130]. In contrast to that observed in SOD, catalase and GSH peroxidases, the increased expression of heme oxygenase-1 continues throughout the course of the disease's development in apolipoprotein E $^{-/-}$ mice [131]. Whether this apparent protective effect of heme oxygenase depends on the presence of antioxidant activity remains unclear. What is certain is that increased heme oxygenase-1 protects against vascular constriction inducing apoptosis of vascular smooth muscle cells *via* induction of apoptosis [132].

Most of the abovementioned atherosclerosis-associated changes to enzymatic antioxidants are likely to take place within the cells of the vascular wall. While an altered intracellular redox environment may be atherogenic, evidence suggests that changes in extracellular and nonenzymatic antioxidants also play a role in the initiation and/or progression of the disease [133]. In particular, antioxidants associated with LDL and which are required to inhibit lipoprotein lipid peroxidation are thought to be deficient and/or ineffective.

Other studies have demonstrated that cellular GSH levels and GSH/GSSG status can regulate macrophage CD36 expression and cellular oxLDL uptake and demonstrate an important anti-atherogenic function of the GSH-dependent antioxidant system by providing a novel molecular mechanism [134].

Vitamin C. Normal arteries contain approximately one-third the concentration of ascorbate found in normal human plasma [135]. Although it is not clear whether ascorbate originates from extra- or intracellular compartments, it can be assumed that a portion of the ascorbate detected is extracellular. An early study reported that the concentration of ascorbate in diseased human aorta was comparable to that found in the plasma [136]. This result was confirmed by another report in which higher levels of ascorbate were observed in advanced human atherosclerotic plaques than in normal arteries [135]. It may be added that only small amounts of vitamin C were present in the form of dehydroascorbic acid in the atherosclerotic plaque. When considered as a whole, the limited information available to date does not support a gross alteration or deficiency of ascorbate in the arterial vessel during atherosclerosis. Nevertheless, these studies do not rule out the possibility of temporal or local (intra versus extracellular) alterations in ascorbate concentrations during atherosclerosis.

Other studies such as Scioli *et al.*, have demonstrated that the treatment with antioxidant pretreatment restored serum-deprived and TNF-α-induced impaired mitochondrial β -oxidation by reducing flavin adenine dinucleotide level and counteracting increased CAM and Nox4 expression, leukocyte adhesion and inflammatory cytokine secretion [137].

Urate. The concentrations of urate present in advanced human lesions are comparable to those in plasma, and in some lesions are similar to or even exceed the limit of its solubility in an aqueous solution [135]. As in the case of ascorbate, current evidence suggests that vessel wall urate levels do not drop during atherosclerosis. It is still to be clarified whether uric acid provides antioxidant protection *in vivo*, while the role of uric acid in CVD remains unclear, despite a long-standing association between hyperuricemia and atherosclerosis. There are some reports that support a protective role for uric acid [138]. For example, reaction of urate with $ONOO^-$ results in the formation of a compound that relaxes blood vessels, apparently *via* release of·NO [138]. However, there are also reports [139] of a direct association between the concentration of uric acid and atherosclerosis, hypertension and cardiovascular mortality. Indeed, urate is known to stimulate the proliferation of vascular smooth muscle cells *in vitro*, which may be of great relevance in ischemia reperfusion, during which substantial increases in the vessel concentration of uric acid can occur [139]. Moreover, and of potential interest in the context of the suprasaturable concentrations of uric acid present in some human plaque samples [131], urate crystals derived from injured cells have been shown to act as a danger signal in inflammation [140] that may contribute to leukocyte recruitment in atherosclerotic lesions.

Furthermore, it has been described that decreased uric acid excretion is an independent risk factor for diabetic retinopathy but not for lower limb atherosclerosis in patients with type 2 diabetes [141].

Lipid-Soluble Antioxidants

In healthy vessels, the concentrations of lipophilic antioxidants resemble those of lipoprotein lipids present in interstitial fluid [142]. They are considerably lower than their respective plasma levels or the vessel concentration of albumin [143]. This is not surprising, given that lipophilic antioxidants are associated with lipoproteins, whose concentration is low in healthy vessels. In the same way, lipid-soluble antioxidants detected in healthy arteries are normally located within the cells [143]. However, as atherosclerotic lesions develop, lipoproteins including LDL, with their full complement of antioxidants, transfer from the plasma to the vessel wall. Therefore, it is reasonable to assume that, as lesions develop, an increasing proportion of the lipid-soluble antioxidants present is localized extracellularly within lipoproteins and the lipid depositions derived from them.

In the context of atherosclerosis, it should be highlighted that vitamin E also exhibits activity that may not be directly related to its involvement in lipid peroxidation [144]. In short, vitamin E enhances the bioactivity of NO [126], inhibits smooth muscle proliferation [145], and limits platelet aggregation. One mechanism that is common to these effects of vitamin E is the inhibition of protein kinase C stimulation [145]. In atherosclerosis, inhibition of protein kinase C by vitamin E may maintain normal vascular homeostasis and, thus, reduce the clinical incidence of CVD [144].

Coenzyme Q_{10}., (ubiquinone) is present in most eukaryotic cells, primarily in the mitochondria. It is a component of the electron transport chain and participates in aerobic cellular respiration, generating energy in the form of ATP. Little information is available concerning the vessel content of coenzyme Q and the effect that atherosclerosis can have on it. Suarna *et al.* [135] reported similar levels of total coenzyme Q_{10} in homogenates of healthy human arteries and carotid endarterectomy specimens, suggesting that lipoprotein-derived coenzyme Q_{10} in diseased vessels is insignificant in comparison to that from cellular sources. This is consistent with reports that the aortic content of coenzyme Q in apolipoprotein $E^{-/-}$ mice fed an atherogenic diet does not alter as lesions develop [146]. That said, it has also been reported that supplementing these animals with large amounts of coenzyme Q, without and with additional vitamin E, reduces atherosclerosis [147].

Oral coenzyme Q_{10} supplementation increases the concentration of $CoQ_{10}H_2$ in human LDL [148] and in the blood vessels of animal models of CVD [149], and this increase is associated with a decrease in the aortic content of oxidized lipids [149]. In addition to this antioxidant activity, supplemental coenzyme Q_{10} has been reported to reduce the integrin expression of monocytes in humans [150], which points towards anti-inflammatory activity. In addition, there is evidence that supplemental coenzyme Q_{10} improves blood pressure [151] and endothelium-dependent vasodilation in type 2 diabetics [152]. In addition, Coenzyme Q10 promotes macrophage cholesterol efflux by regulation of the activator protein-1/miR-378/ATP-binding cassette transporter G1-signaling pathway [153].

MITOCHONDRIAL-TARGETED ANTIOXIDANTS

A wide range of antioxidants could be targeted to mitochondria by conjugation to the TPP moiety. These include TPP-conjugated derivatives of ubiquinone [154], tocopherol [155], lipoic acid spin traps, and ebselen [156]. Given that lipid peroxidation is an important feature in mitochondrial oxidative damage, and that alkylTPP conjugates are closely associated with the mitochondrial inner membrane, research has tended to focus on antioxidants that are effective against lipid peroxidation. In particular, the targeted version of ubiquinol (MitoQ) is the most extensively evaluated and best understood member of the family [157]. MitoQ is taken up rapidly by isolated mitochondria driven by the mitochondrial membrane potential ($\Delta\psi m$), and within mitochondria, nearly all of the accumulated MitoQ is adsorbed to the matrix surface of the inner membrane. MitoQ is reduced to the active ubiquinol antioxidant by complex II of the respiratory chain. It cannot restore respiration in mitochondria lacking coenzyme Q because the reduced form of MitoQ is not oxidized by complex III. Consequently, all of the effects of MitoQ are probably attributable to the accumulation of the antioxidant ubiquinol form. Importantly, when the ubiquinol form of MitoQ acts as an antioxidant, it is oxidized to the ubiquinone form. This ubiquinone form is then rapidly re-reduced by complex II, which restores its antioxidant efficacy. As MitoQ is generally adsorbed to the mitochondrial inner membrane, and given that its linker chain enables its active ubiquinol antioxidant component to penetrate deeply into the membrane core, it is thought to be an effective antioxidant against lipid peroxidation. MitoQ has also been shown to protect against $ONOO^-$ damage [158]. In addition, it can react with O_2^-, although, as with other ubiquinols, its reactivity with H_2O_2 is negligible. In summary, in isolated mitochondria, MitoQ seems to fulfil most of the requirements of a mitochondria-targeted antioxidant.

Mitochondrial-targeted antioxidants have been used in isolated cells. In fact, toxicity is the first issue to be considered when moving from isolated mitochondria to cells. The extensive accumulation of lipophilic cations within isolated mitochondria can disrupt membrane integrity, respiration and adenosine 5'-triphosphate synthesis as a consequence of the adsorption of the cations to the matrix surface of the inner membrane. Supporting this idea, the more hydrophobic TPP cations disrupt mitochondrial function at lower concentrations, and the degree of disruption correlates with the amount of compound adsorbed to the inner membrane. The nonspecific effects of MitoQ on mitochondria are assessed using the control compound decylTPP, which is similar to MitoQ in hydrophobicity but lacks the antioxidant ubiquinol moiety. The nonspecific mitochondrial disruption of MitoQ and decylTPP occurs at similar concentrations and always limits the amounts of TPP-derived targeted antioxidants that can be used. Therefore, it is essential for compounds to be effective antioxidants at concentrations well below those that disrupt function. When comparing cells to isolated mitochondria, it is important to bear in mind that the targeted antioxidants are accumulated within cells at a concentration 5- to 10-fold greater than that of the extracellular environment due to the plasma membrane potential [159]. For mammalian cells in culture, this concentration is generally in the range of 0.1–0.5 μM, but can vary considerably with cell density, cell type, and incubation conditions. The uptake of MitoQ into cells is far more rapid than that of methylTPP, presumably because it's higher hydrophobicity reduced the activation energy available for passage through the plasma membrane [160]. MitoQ uptake into cells is largely blocked if the mitochondrial membrane potential is blocked with the uncoupler carbonylcyanide-*p*-trifluoromethoxyphenylhydrazone (FCCP). This is consistent with the fact that uptake occurs primarily into the mitochondria rather than other cell compartments. These data are consistent with the rapid equilibration of MitoQ across the plasma membrane and its subsequent accumulation within the mitochondria. One further indication of the membrane potential–dependent mitochondrial concentration of TPP-containing molecules within cells was the FCCP-sensitive protection afforded by MitoQ in a cell model of Friedreich's ataxia, in which FCCP did not affect the potency of the nontargeted compounds decylQ or idebenone [103]. This is fully consistent with the protection afforded by MitoQ against damage in this cell model attributable to its $\Delta\psi$m-dependent uptake into mitochondria. Therefore, upon incubation with cells in culture, MitoQ is accumulated predominantly within the mitochondria, but the amount of MitoQ present in the cell is currently difficult to quantify. MitoQ has been used in a large range of mitochondrial and cell models [157] and has been shown to provide protection against oxidative damage. The interaction of MitoQ with mitochondrial ROS within rotenone-treated fibroblasts has been studied in detail [160] MitoQ did not decrease O_2^- production when measured by dihydroethidium oxidation, but it did prevent lipid peroxidation when measured by the fluorescent probe C11-BODIPY [161]. This finding is consistent with the model for MitoQ action developed based on studies with isolated mitochondria that have demonstrated that MitoQ's main antioxidant action is to prevent lipid peroxidation. Furthermore, MitoQ has been employed in a model of nitrate tolerance in human umbilical endothelial cells, showing protective effects against the deleterious effects of ROS in mitochondria. Nevertheless, it remains to be seen whether this is the primary mechanism by which MitoQ acts as a protective agent in all cell types and all forms of oxidative stress.

Mitochondria-targeted antioxidants have been also used *in vivo* studies. To function as therapy, mitochondria-targeted antioxidants must be delivered to mitochondria within the cells of patients, preferably following oral administration. As TPP cations pass easily through phospholipid bilayers, they should pass from the gut to the bloodstream and from there to most tissues. It has been shown that when simple alkylTPP compounds such as the targeted version of Vitamin E (MitoE) or MitoQ10 are administered intravenously to mice, they are rapidly cleared from the plasma and accumulate in the heart, brain, skeletal muscle, liver and kidney [161]. Importantly, TPP-derived compounds were shown to be made orally bioavailable to mice by feeding them tritiatedmethylTPP, MitoE2 or MitoQ in their drinking water, which led to uptake into the plasma, and from there into the heart, brain, liver, kidney and muscle [161]. The methylTPP was shown to be cleared from all organs at a similar rate by a first-order process with a half-life of about 1.5 days [161]. These studies confirm the distribution of orally administered alkylTPP compounds to all organs due to their facile permeation through biological membranes. The nonspecific toxicity of alkylTPP cations found in mitochondria and cells also occurs *in vivo*, and this is probably the major factor limiting the amounts of the compound that can be administered safely. In crude toxicity assessments [161], methylTPP and MitoE2 showed no toxicity at 300 nmol when administered intravenously (aprox 4–6 mg/kg) but did show toxicity at 500 nmol (aprox 6–10 mg/kg). MitoQ was tolerated marginally better, with no toxicity at 750 nmol (aprox 20 mg/kg) but with toxicity evident at 1000 nmol (aprox 27 mg/kg). When mice were administered 500 μMTPMP, MitoE2 or MitoQ *via* their drinking water, no toxic effects were noted over the course of the experiment, which lasted 43 days, 14 days and 14 days for TPMP, MitoE2, and MitoQ10 respectively [161]. The similarity in the levels of toxicities is consistent with the hypothesis that toxic effects are largely attributable to nonspecific disruption of the mitochondria by the accumulation of large amounts of the lipophilic cation. To summarize, it is possible to administer alkyltriphenylphosphonium compounds to animals orally, as these compounds are taken up into the plasma with reasonable bioavailability and are then rapidly cleared from the plasma when they accumulate in mitochondria within the tissues. Having shown that the long-term administration of mitochondria-targeted antioxidants is viable, the next step is to determine whether the amount of compound accumulated is sufficient to act as an antioxidant *in vivo*. In another study, rats were administered 500 μM MitoQ for 2 weeks *via* their drinking water. Their hearts were then isolated and exposed to ischemia–reperfusion injury in a Langendorff perfusion system, and MitoQ was shown to provide greater protection against the loss of heart function, tissue damage and mitochondrial function than methylTPP or short-chain quinol controls [162]. The most probable reason for the protection observed in these experiments is that lipid peroxidation in the mitochondrial inner membrane was prevented by MitoQ [162]. However, this has yet to be confirmed by showing that MitoQ blocks any increase in markers of oxidative damage in mitochondria. In another work, MitoQ protected against oxidative damage caused by nitroglycerin in the rat aorta in a model of nitrate tolerance [163] Taken as a whole, this evidence suggests that mitochondria-targeted antioxidants may be effective reducing oxidative stress in atherosclerosis and could attenuate lesion/plaque development.

The development of MitoQ as a pharmacological agent is somewhat different from that of most other drugs. Typically in medicinal chemistry, investigations focus on lead compounds that interact with a specific target, such as a receptor binding site. In assessing these compounds, a preliminary screening is often carried out to ensure that drug candidates are soluble, bioavailable and can pass through phospholipid bilayers [164]. However, mitochondria-targeted antioxidants based on TPP lipophilic cations are less constrained by these traditional guidelines, as they have the unusual property of being both relatively water soluble and membrane permeant. Even though the molar mass of MitoQ is relatively large for a pharmaceutical agent and it has a high octanol–PBS partition coefficient, it is readily bioavailable and passes easily through biological membranes. A further unusual feature of TPP-targeted compounds is that they are targeted to an organelle to modify a general, rather than a specific, process (namely oxidative damage). Therefore, if lipophilic cations such as MitoQ prove to be effective drugs, a new approach to medicinal chemistry and drug discovery would be established. MitoQ has been developed as a pharmaceutical agent by Antipodean Pharmaceuticals Inc. (http://www.antipodeanpharma.com/). In the search for a commercially satisfactory stable formulation, the compound was created with the methane sulfonate counter anion and decomposition was inhibited by complexation with β-cyclodextrin. This preparation is easily manufactured as tablets, and has passed conventional animal toxicity tests with no observable adverse effects at 10.6 mg/kg. In Phase 1 trials, MitoQ has shown a positive pharmacokinetic behaviour when administered orally at 80 mg (1 mg/kg), which results in a plasma Cmax = 33.15 ng/mL and Tmax at approx. 1 h. This formulation has good pharmaceutical characteristics and is now in Phase 2 trials for Parkinson's disease and hepatitis C [158].

The cardioprotective effect of Mito-compounds towards doxorubicin has also been shown with MitoTEMPOL in immune-competent spontaneously hypertensive rats and the mechanism for this included increase of the pro-survival autophagy marker LC3-II and decrease of the apoptosis marker caspase-3 in the heart [165]. In this study, doxorubicin-treated rats were implanted with the breast tumor cell line SST-2 and of note, MitoTEMPOL caused significant tumor reduction and appeared to increase the anti-tumor effect of doxorubicin [165] supporting the *in vitro* observed anti-cancer properties of Mito-compounds.

Another mitochondria-targeted antioxidant, MitoE has also been demonstrated utility in a rat model of pneumonia-realted sepsis [166]. This study confirmed the *in vivo* distribution in the heart and in the liver of MitoE 24h after its oral administration and provided evidence that MitoE improved mitochondria-specific antioxidant defense and mitochondrial function (respiration and mitochondrial membrane integrity) in the heart, provided cardiac protection from sepsis and suppressed sepsis-induced inflammation. Importantly, in this study untargeted vitE also exhibited anti-inflammatory and protective effect, however to a lesser extent compared to the mitochondria-targeted molecule.

Stable free radical nitroxides with antioxidant properties (TEMPO derivatives) are another group of mitochondria-targeted antioxidants that has been described. The action of these compounds to metabolize ROS is ascribed primarily to cyclic one- or two electron transfer among three oxidation states: the oxammonium cation, the nitroxide, and the hydroxylamine [167]. Nitroxides undergo a very rapid, one-electron reaction *in vivo* to the corresponding hydroxylamine which has antioxidant activity and can be converted to nitroxides by H_2O_2 or other oxidants such as transition metals. The nitroxide moiety is consided essential for full antioxidant activity, whereas substitution at the 4-position affects potency. One-electron redox cycling of six-member ring nitroxides such as tempol is enhanced by their ability to undergo reversible "boat-and-chair" conformational change-this is not possible with five-member ring nitroxides, which may account for their lesser biological activity [167]. MitoTEMPO which possesses $O_2^{\cdot -}$ and alkyl radical scavenging properties is a combination of the pleiotropic intracellular antioxidant piperidine nitroxide TEMPO (2,2,6,6-tetramethylpiperidin-1-yloxy) and the TPP cation. A related molecule is MitoTEMPOL in which the conjugate molecule is 4-hydroxy-TEMPO (TEMPOL) and is therefore less hydrophobic. TEMPOL is an amphilite nitroxide which concentrates in hydrophobic microdomains of cell membranes.

There are also other types of mitochondrial targeted antioxidants such as unspecific approaches. Although not designed to target mitochondria, several molecules display "mito-protective" activity; i.e. an ability to attenuate mitochondrial injury [168]. They are referred to as "synthetic catalytic scavengers or antioxidants" and were first developed and investigated by Eukarion (hence their code names EUK). These analogues of SOD-catalase mimetic have to be beneficial in various models of oxidative stress. For e.g., salen Mn complexes been shown to be effective in a direct test of mammalian mitochondrial oxidative stress in SOD2 KO mice [169] where EUK-8, EUK-134, or EUK-189 significantly extended the lifespan of these mice [169]. Beneficial effects have been reported not only in the brain but in also in other tissues. Treatment with the salen Mn mimetics has been shown to exert beneficial effects against oxidative stress in animal models of several diseases such PD [170], ALS [171], and AD [169], stroke [172] and age-associated cognitive impairment [173]. Specific animal models of ROS-related injury have revealed the potential of these molecules such as EUK-207 to mitigate radiation damage in the lungs [174] or EUK-189 to prolong survival after receiving a lethal dose of total body radiation [175]. Some of these compounds are orally bioavailable whereas others have been proposed for topic or parenteral administration. The compound EUK-134 has been used as an active ingredient in cosmetics for about 10 years whereas another EUK compound is a clinical candidate for the prevention and treatment of certain neurodegenerative diseases.

Several studies have demonstrated the beneficial effect of antioxidant compounds on the vasculature through protection against oxidative damage and consequent restoration of endothelial function [176]. Novel approaches have been developed in this aspect, and certain compounds have been shown to render protection against these CVD and other related diseases by acting specifically on mitochondria. For e.g., diaxozide, an FDA-approved drug for the management of symptomatic hypoglycemia, has been identified as cardioprotective. Diaxozide is generally believed to inhibit succinate dehydrogenase, a mitochondrial complex II protein [177]. This inhibition, which also takes place in the heart, can occur at concentrations often used to study cardioprotection, and therefore may constitute a mechanism involving partial uncoupling of oxidative phosphorylation and/or modulation of ROS production. However, here is controversial evidence regarding its targets and mechanisms of action. It seems that diazoxide exerts an ambivalent effect on mitochondrial ROS production as shown in a mechanistic study using submitochondrial particles or intact rat heart mitochondria where depending on the metabolic state and $\Delta\psi m$, diazoxide-mediated inhibition of complex II either promoted transient generation of signaling ROS at complex III (during preconditioning) or attenuated the production of deleterious ROS at complex I (during ischaemia and reperfusion) [178]. Whatever the underlying molecular mechanism(s) may be, it is clear that improved mitochondrial function is a principal cardioprotective effect of diazoxide [179].

Other group of antioxidants are cationic plastoquinone derivatives such as SkQs - cationic plastoquinone derivatives containing positively charged phosphonium or rhodamine moieties connected to plastoquinone by decane or pentane linkers. The prototypical example of these molecules is the plastoquinonyl decyltriphenylphosphonium (SkQ1). When electrophoretically entering the cell, assuming the plasma membrane potential of -60 mV, SkQ1 as a monovalent penetrating cation accumulates in the cytosol and the intermembrane space of mitochondria by a factor of 10, whereas further accumulation by factor 1000 occurs when SkQ1 enters, again electrophoretically, the mitochondrial matrix, which is 180 mV more negative than the intermembrane space. The membrane/water distribution coefficient, which is about 10^4 for SkQ1, also has a bearing. Hence, the total magnification of SkQ1 concentration in the inner leaflet of the IMM with respect to the extracellular space is expected to be 10^8. The relative accumulation may even be higher, since the SkQ1 cation forms a complex with cardiolipin anion, which seems to be the prime target of the antioxidant activity of SkQs [180]. In a similar fashion to MitoQ, SkQs are rechargeable antioxidants, since their inactive oxidized forms are reduced (i.e. reactivated), in the case of SkQ by Complex III of ETC [180].

Several recent studies have provided evidence of SkQ1 as a promising candidate for the treatment or prophylaxis of age-related and neurodegenerative disorders particularly AD. The beneficial prophylactic effect of long-term treatment with SkQ1 has been shown in OXYS rats, senescence-accelerated rats that also develop AD-like pathology. SkQ1 reduced the age-related alterations in behavior and spatial memory deficit and slowed down pathological accumulation of AβPP, Aβ, and hyperphosphorylation of tau-protein [181]. Another study revealed that *in vivo* and *in vitro* injection of SkQ1 can compensate for Aβ-induced oxidative damage of long-term synaptic plasticity (LTP) in the rat hippocampus.The neuroprotective activity of the analog of SkQ1 lacking plastoquinone (C12TPP) was found to be much lower than SkQ1, meaning that the plastoquinone part of the SkQ1 molecule is responsible for the LTP rescue [182].

Another important group of mitochondria targeted antioxidants are peptide-based targeting. In fact, Szeto-Schiller (or SS)-peptides [183] and mitochondria-penetrating peptides (MPPs) [184] are peptide-based targeting agents for delivering antioxidants to mitochondria. SS-peptides were developed by Szeto and Schiller and constitute a series of 4, small, cell-permeable antioxidant compounds (SS-19 H-Tyr-D-Arg-Phe-Lys-NH2, SS-02 H-Dmt-D-Arg-Phe-Lys-NH2, SS-31 H-D-Arg-Dmt-Lys-Phe-NH2, and SS-20 H-Phe-D-Arg-Phe-Lys-NH2) with three positive charges in homeostatic pH conditions. SS-peptides display a unique aromatic-cationic sequence motif, which alternates between basic and aromatic residues, enabling their efficient cell up-take as it occurs in a non-saturable manner, independently of the energy status and without the need for peptide transporters. Cell studies performed *in vitro* have shown the rapid up-take of these molecules through the cellular membrane of a wide variety of cell types, including neurons, renal, epithelial, endothelial cells and human embryonic kidney cells. Within cells, SS-peptides accumulate 1000–5000 fold in mitochondria, where they localize at/bind to IMM [185, 186] through a mechanism that is not entirely clear. It is of particular importance that the specific mitochondrial up-take of these compounds does not take place as a result of the presence of $\Delta\Psi_m$ [186-188]. This is clearly an advantage, as mitochondrial membrane polarization is not altered and is not a self-limiting process. Therefore, SS-peptides have adequate characteristics that make them ideal for the treatment of many diseases related to mitochondrial dysfunction and oxidative stress, including CVD and cancer. They are small, readily soluble in water and easily synthetized, and even resistant to peptidase degradation. The presence of a D-amino acid in the first and second positions renders them resistant to aminopeptidase activity, and due to their structure can reach to the mitochondria independently on the tissue and the location [185]. In this sense, SS-O_2 has demonstrated that readily penetrate to a monolayer of intestinal epithelial cells in both basolateral apical and apical-basolateral directions [189]. The same compound has also been shown to cross BBB, a crucial ability of a candidate drug in the development of strategies for neuroprotection.

SS-peptides have been shown to scavenge H_2O_2 and $ONOO^-$ and to inhibit lipid peroxidation; antioxidant actions that are attributed to a tyrosine or dimethyltyrosine residue in their structure, the latter of the two being more effective with respect to ROS scavenging. This moiety is believed to scavenge hydroxyl radicals and $ONOO^-$, and possibly HOCl and peroxyl radicals, but a direct reaction with $O_2\cdot^-$ or H_2O_2 seems unlikely. The specific location of these residues does not appear to be significant, as SS-31 is as effective as SS-02 in scavenging H_2O_2 and inhibiting LDL oxidation [190]. SS-peptides have been reported to be beneficial against the conditions of oxidative stress both in cellular models of disease and in isolated mitochondria, with SS-31 proving to be the most efficient [183, 191, 192].

CONCLUSION

Oxidative stress is linked to the pathogenesis of atherosclerosis and its subsequent cardiovascular events. Despite this, the clarification of the causal relationship between oxidative events and atherosclerosis in general, and the pathophysiological importance of LDL oxidation in particular has been made difficult by the generally poor results achieved with antioxidant strategies for limiting atherosclerosis and its cardiovascular events. It is important to mention that certain HDL-ApoAI mimetic peptides have proven to be effective at reducing atherosclerosis by promoting HDL-associated paraoxonase (PON) antioxidant function and thereby attenuating LDL oxidation [193].

As discussed, the poor outcome of antioxidant trials is comprehensible if we appreciate the importance of oxidative events caused by ROS rather than focusing on classic antioxidants that are effective only against oxidative reactions. In this context, it will be of vital importance to provide future experimental evidence that scavengers of ROS offer protection against disease in suitable animal models of atherosclerosis. Perhaps of equal importance, the relation between atherosclerosis and cellular redox processes, such as those protagonised by ROS and RNS in cell signalling, call for experimental testing with targeted strategies.

The use of the TPP cation or peptide-based targeting to provide localized chemicals that increase the antioxidant defenses of mitochondria has proven to be a viable strategy *in vitro* [194]. It has also been shown that compounds such as MitoQ or SS-31 can be formulated into pharmaceuticals that are successfully delivered orally to humans. Animal experiments have indicated that MitoQ displays antioxidant efficacy in tissues, and therefore the stage is set for testing this and related compounds in human diseases. We trust that work over the next years will indicate in which organs these compounds are effective, whether they can decrease mitochondrial oxidative damage in diseases, and whether this positively affects the outcome for the patient. Finally, there is considerable room for fine-tuning the chemical biology of these compounds to target specific ROS and other mitochondrial targets.

DISCLOSURE

This is an updated chapter of the earlier review article by [Victor *et al.* "Oxidative stress and mitochondrial dysfunction in atherosclerosis: Mitochondria-targeted antioxidants as potential therapy", published in 'Current Medicinal Chemistry', Volume 16, Issue 35, 2009, pp 4654 to 4667], with additional text and references.

ACKNOWLEDGEMENTS

We thank B Normanly for his editorial assistance. This study was financed by grants PI13/1025, PI12/1984, PI13/0073, UGP-14-93, UGP-14-95 and by the European Regional Development Fund (ERDF "A way to build Europe"). V.M.V. and M.R. are recipients of contracts from the Ministry of Health of the Valencian Regional Government and Carlos III Health Institute [CES10/030 and CP10/0360, respectively]. S.R. is the recipient of a postgraduate fellowship from Carlos III Health Institute [FI11/00637]. CB is the recipient of a Sara Borrell contract from Carlos III Health Institute (CD14/00043).

CONFLICT OF INTEREST

The authors confirm that this chapter contents have no conflict of interest.

REFERENCES

[1] Thom, TJ. International mortality from heart disease: rates and trends. *Int. J. Epidemiol.,* **1989**, 18, 20-8.

[2] Victor, V. M.; Rovira-Llopis, S.; Banuls, C.; Diaz-Morales, N.; Lopez-Domenech, S.; Escribano-Lopez, I.; Rios-Navarro, C.; Alvarez, A.; Gomez, M.; Rocha, M.; Hernandez-Mijares, A. Metformin modulates human leukocyte/endothelial cell interactions and proinflammatory cytokines in polycystic ovary syndrome patients. *Atherosclerosis* **2015**, *242*, 167-173.

[3] Apostolova, N.; Rocha, M.; Rovira-Llopis, S.; Banuls, C.; Falcon, R.; Castello, R.; Hernandez-Mijares, A.; Victor, V. M. Mitochondria-targeted antioxidants as a therapeutic strategy for protecting endothelium in cardiovascular diseases. *Curr. Med. Chem.* **2014**, *21*, 2989-3006.

[4] Rocha, M.; Apostolova, N.; Herance, J. R.; Rovira-Llopis, S.; Hernandez-Mijares, A.; Victor, V. M. Perspectives and potential applications of mitochondria-targeted antioxidants in cardiometabolic diseases and type 2 diabetes. *Med. Res. Rev.* **2014**, *34*, 160-189.

[5] Dhalla, N.S.; Temsah, R.M.; Netticadan, T. Role of oxidative stress in cardio-vascular diseases. *J. Hypertens.*, **2000**, 18, 655-73.

[6] Singh, U.; Jialal, I. Oxidative stress and atherosclerosis. *Pathophysiology*, **2006**, 13, 129-42.

[7] Gutteridge, J.M. Lipid peroxidation and antioxidants as biomarkers of tissue damage. *Clin. Chem.*, **1995**, 41, 1819-28.

[8] Green, D.; Kroemer, G. The central executioners of apoptosis: caspases or mitochondria?. *Trends Cell Biol.*, **1998**, 8, 267-71.

[9] Li, J.M.; Shah, A.M. Endothelial cell superoxide generation: regulation and relevance for cardiovascular pathophysiology. *Am. J. Physiol. Regul. Integr. Comp. Physiol.*, **2004**, 287, R1014-30.

[10] Ballinger, S.W.; Patterson, C.; Knight-Lozano, C.A.; Burow, D.L.; Conklin, C.A.; Hu, Z.; Reuf, J.; Horaist, C.; Lebovitz, R.; Hunter, G.C.; McIntyre, K.; Runge, M.S. Mitochondrial integrity and function in atherogenesis. *Circulation*, **2002**, 106, 544-9.

[11] Dimmeler, S.; Zeiher, A.M. Reactive oxygen species and vascular cell apoptosis in response to angiotensin II and pro-atherosclerosic factors. *Regul. Pept.*, **2000**, 90, 19-25.

[12] Kowaltowski, A.J.; Vercesi, A.E. Mitochondrial damage induced by conditions of oxidative stress. *Free Radic. Biol. Med.*, **1999**, 26, 463-71.

[13] Victor, V.M.; Rocha, M.; De la Fuente, M. N-acetylcysteine protects mice from lethal endotoxemia by regulating the redox state of immune cells. *Free Radic. Res.*, **2003**; 37: 919-29.

[14] Victor, V.M.; Rocha, M.; De la Fuente, M. Immune cells: free radicals and antioxidants in sepsis. *Int. Immunopharmacol.*, **2004**, 4, 327-47.

[15] Victor, V.M.; Rocha, M.; Esplugues, J.V.; De la Fuente, M. Role of free radicals in sepsis: antioxidant therapy. *Curr. Pharm. Des.*, **2005**, 11, 3141-58.

[16] Kagan, V.E.; Serbinova, E.A.; Stoyanovsky, D.A.; Khwaja, S.; Packer, L. Assay of ubiquinones and ubiquinols as antioxidants. *Methods Enzymol.*, **1994**, 234, 343-54.

[17] Sheu, S.S., Nauduri, D., Anders, M.W. Targeting antioxidants to mitochondria: a new therapeutic direction. *Biochim. Biophys. Acta*, **2006**, 1762, 256-65.

[18] Banuls, C.; Rocha, M.; Rovira-Llopis, S.; Falcon, R.; Castello, R.; Herance, J. R.; Polo, M.; Blas-Garcia, A.; Hernandez-Mijares, A.; Victor, V. M. The pivotal role of nitric oxide: effects on the nervous and immune systems. *Curr. Pharm. Des.* **2014**, *20*, 4679-4689.

[19] Hernandez-Mijares, A.; Rocha, M.; Rovira-Llopis, S.; Banuls, C.; Bellod, L.; de Pablo, C.; Alvarez, A.; Roldan-Torres, I.; Sola-Izquierdo, E.; Victor, V. M. Human leukocyte/endothelial cell interactions and mitochondrial dysfunction in type 2 diabetic patients and their association with silent myocardial ischemia. *Diabetes Care* **2013**, *36*, 1695-1702.

[20] Rovira-Llopis, S.; Rocha, M.; Falcon, R.; de Pablo, C.; Alvarez, A.; Jover, A.; Hernandez-Mijares, A.; Victor, V. M. Is myeloperoxidase a key component in the ROS-induced vascular damage related to nephropathy in type 2 diabetes? *Antioxid. Redox Signal.* **2013**, *19*, 1452-1458.

[21] Rovira-Llopis, S.; Banuls, C.; Apostolova, N.; Morillas, C.; Hernandez-Mijares, A.; Rocha, M.; Victor, V. M. Is glycemic control modulating endoplasmic reticulum stress in leukocytes of type 2 diabetic patients? *Antioxid. Redox Signal.* **2014**, *21*, 1759-1765.

[22] Rovira-Llopis, S.; Diaz-Morales, N.; Banuls, C.; Blas-Garcia, A.; Polo, M.; Lopez-Domenech, S.; Jover, A.; Rocha, M.; Hernandez-Mijares, A.; Victor, V. M. Is Autophagy Altered in the Leukocytes of Type 2 Diabetic Patients? *Antioxid. Redox Signal.* **2015**.

[23] Valkonen, V.P.; Tuomainen, T.P.; Laaksonen, R. DDAH gene and cardiovascular risk. *Vasc. Med.*, **2005**, 10, S45-8.

[24] Panza, J.A.; Quyyumi, A.A.; Brush, J.E.; Epstein, S.E. Abnormal endothelium-dependent vascular relaxation in patients with essential hypertension. *N. Engl. J. Med.* **1990**, 323, 22-7.

[25] Zeiher, A.M.; Drexler, H.; Saurbier, B.; Just, H. Endothelium-mediated coronary blood flow modulation in humans: Effects of age, atherosclerosis, hypercholesterolemia, and hypertension. *J. Clin. Invest.*, **1993**, 92, 652-62.

[26] Suwaidi, J.A.; Hamasaki, S.; Higano, S.T.; Nishimura, R.A.; Holmes, D.R.; Lerman, A. Long-term follow-up of patients with mild coronary artery disease and endothelial dysfunction. *Circulation,* **2000**, 101, 948-54.

[27] Schachinger, V.; Britten, M.B.; Zeiher, A.M. Prognostic impact of coronary vasodilator dysfunction on adverse long-term outcome of coronary heart disease. *Circulation,* **2000**, 101, 1899-906.

[28] Landmesser, U.; Hornig, B.; Drexler, H. Endothelial function: A critical determinant in atherosclerosis? *Circulation,* **2004**, 109, 27-33.

[29] Ghiadoni, L.; Virdis, A.; Magagna, A.; Taddei, S.; Salvetti, A. Effect of the angiotensin II type 1 receptor blocker candesartan on endothelial function in patients with essential hypertension. *Hypertension,* **2000**, 35, 501-6.

[30] Wolfrum, S.; Jensen, K.S.; Liao, J.K. Endothelium-dependent effects of statins. *Arterioscler. Thromb. Vasc. Biol.,* **2003**, 23, 729-36.

[31] Martens, F.M.; Rabelink, T.J.; Roodt, J.; de Koning, E.J.; Visseren, F.L. TNF-alpha induces endothelial dysfunction in diabetic adults, an effect reversible by the PPAR-gamma agonist pioglitazone. *Eur. Heart J.,* **2006**, 27, 1605-9.

[32] Higashi, Y.; Oshima, T.; Ozono, R.; Watanabe, M.; Matsuura, H.; Kajiyama, G. Effects of L-arginine infusion on renal hemodynamics in patients with mild essential hypertension. *Hypertension,* **1995**, 25, 898-902.

[33] Sanada, M.; Higashi, Y.; Nakagawa, K.; Sasaki, S.; Kodama, I.; Tsuda, M.; Nagai N, Ohama K. Relationship between the angiotensin-converting enzym genotype and forearm vasodilator response to estrogen replacement therapy in postmenopausal. *J. Am. Coll. Cardiol.,* **2001**, 37, 1529-35.

[34] Higashi, Y.; Sasaki, S.; Nakagawa, K.; Fukuda, Y.; Matsuura, H.; Oshima, T.; Chayama, K. Tetrahydrobiopterin enhances forearm vascular response to acetylcholine in both normotensive and hypertensive individuals. *Am. J. Hypertens.,* **2002**, 15, 326-32.

[35] Takeshita, Y.; Katsuki, Y.; Katsuda, Y.; Kai, H.; Saito, Y.; Arima, K.; Sasaki, K.; Imaizumi, T. Vitamin C reversed malfunction of peripheral blood-derived mononuclear cells in smokers through antioxidant properties. *Circ. J.,* **2008**, 72, 654-9.

[36] Goto, C.; Higashi, Y.; Kimura, M.; Noma, K.; Hara, K.; Nakagawa, K.; Kawamura, M.; Chayama, K.; Yoshizumi, M.; Nara, I. Effect of different intensities of exercise on endothelium-dependent vasodilation in humans: role of endothelium-dependent nitric oxide and oxidative stress. *Circulation,* **2003**, 108, 530-5.

[37] Sasaki, S.; Higashi, Y.; Nakagawa, K.; Kimura, M.; Noma, K.; Sasaki, S.; Hara, K.; Matsuura, H.; Goto, C.; Oshima, T.; Chayama, K. A low-calorie diet improves endothelium-dependent vasodilation in obese patients with essential hypertension. *Am. J. Hypertens.,* **2002**, 15, 302-9.

[38] Mehta, J.L.; Rasouli, N.; Sinha, A.K.; Molavi, B. Oxidative stress in diabetes: a mechanistic overview of its effects on atherogenesis and myocardial dysfunction. *Int. J. Biochem. Cell Biol.*, **2006**, 38, 794-803.

[39] Guest, T.M.; Vlastos, G.; Alameddine, F.M.; Taylor, W.R. Mechano-regulation of monocyte chemoattractant protein-1 expression in rat vascular smooth muscle cells. *Antioxid. Redox Signal.*, **2006**, 8, 1461-71.

[40] Victor, V.M.; Rocha, M. Targeting antioxidants to mitochondria: a potential new therapeutic strategy for cardiovascular diseases. *Curr. Pharm. Des.*, **2007**, 13, 845-63.

[41] Rocha, M.; Victor, V.M. Targeting antioxidants to mitochondria and cardiovascular diseases: the effects of mitoquinone. *Med. Sci. Monit.*, **2007**, 13, 132-45.

[42] Ed Rainger, G.; Chimen, M.; Harrison, M. J.; Yates, C. M.; Harrison, P.; Watson, S. P.; Lordkipanidze, M.; Nash, G. B. The role of platelets in the recruitment of leukocytes during vascular disease. *Platelets* **2015**, *26*, 507-520.

[43] Madamanchi, N.R.; Vendrov, A.; Runge, M.S. Oxidative stress and vascular disease. *Arterioscler. Thromb. Vasc. Biol.*, **2005**, 25, 29-38.

[44] Vora, D.K.; Fang, Z.T.; Liva, S.M.; Tyner, T.R.; Parhami, F.; Watson, A.D.; Drake, T.A.; Territo, M.C.; Berliner, J.A. Induction of P-selectin by oxidized lipoproteins. Separate effects on synthesis and surface expression. *Circ. Res.*, **1997**, 80, 810-8.

[45] Andreassi, M.G.; Botto, N. DNA damage as an emerging risk factor in atherosclerosis. *Trends Cardiovasc. Med.*, **2003**, 13, 270-5.

[46] Binkova, B.; Smerhovsky, Z.; Strejc, P.; Boubelik, O.; Stavkova, Z.;, Chvalatova, I., Sram, R.J. DNA-adducts and atherosclerosis: a study of accidental and sudden death males in the Czech Republic. *Mutat. Res.*, **2002**, 501: 115-28.

[47] Martinet, W.; Knaapen, M.W.; DeMeyer, G.R.; Herman, A.G.; Kockx, M.M. Elevated levels of oxidative DNA damage and DNA repair enzymes in human atherosclerotic plaques. *Circulation,* **2002**, 106, 927-32.

[48] Ceaser, E.; Ramachandran, A.; Levonen, A.L.; Darley-Usmar, V.M. Oxidized low-density lipoprotein and 15-deoxy-delta12,14-PGJ2 increase mitochondrial complex I activity in endothelial cells. *Am. J. Physiol. Heart Circ. Physiol.*, **2003**, 285, 2298-308.

[49] Asmis, R.; Begley, J.G. Oxidized LDL promotes peroxide mediated mitochondrial dysfunction and cell death in human macrophages: a caspase-3-independent patway. *Circ. Res.*, **2003**, 92, 20-9.

[50] Anderson, E.J.; Lustig, M.E.; Boyle, K.E.; Woodlief, T.L.; Kane, D.A.; Lin, C.T.; Price. J.W.; Kang, L.; Rabinovitch, P.S.; Szeto, H.H.; Houmard, J.A.; Cortright, R.N.; Wasserman, D.H.; Neufer, P.D.

Mitochondrial H2O2 emission and cellular redox state link excess fat intake to insulin resistance in both rodents and humans. *J. Clin. Invest.*, 2009, 119, 573-581
[51] Sreekumar, R.; Unnikrishnan, J.; Fu, A.; Nygren, K.; Short, K.R.; Schimke, J.; Barazzoni, R.; Nair, K.S. Impact of high-fat diet and antioxidant supplement on mitochondrial functions and gene transcripts in rat muscle. *Am. J. Physiol. Endocrinol. Metab.*, **2002**, 282, 1055-61.
[52] Sen, S.; Simmons, R. A. Maternal antioxidant supplementation prevents adiposity in the offspring of Western diet-fed rats. *Diabetes* **2010**, *59*, 3058-3065.
[53] Rapava, E.A.; Akhvlediani, M.V.; Emukhvari, M.G. Modern view on biochemistry of atherosclerosis. *Georgian Med. News*, **2006**, 133, 124-8.
[54] Camejo, G.; Hurt, E.; Wiklund, O.; Rosengren, B.; Lopez, F.; Bondjers, G. Modifications of low-density lipoprotein induced by arterial proteoglycans and chondroitin-6-sulfate. *Biochim. Biophys. Acta*, **1991**, 1096, 253-61.
[55] Brennan, M.L.; Anderson, M.M.; Shih, D.M.; Qu, X.D.; Wang, X.; Mehta, A.C.; Lim LL, Shi W, Hazen SL, Jacob JS, Crowley JR, Heinecke JW, Lusis AJ. Increased atherosclerosis in myeloperoxidase-deficient mice. *J. Clin. Invest.*, **2001** 107, 419-30.
[56] Hathaway, C.A.; Heistad, D.D.; Piegors, D.J.; Miller, F.J. Regression of atherosclerosis in monkeys reduces vascular superoxide levels. *Circ. Res.*, **2002**, 90, 277-83.
[57] Kirk, E.A.; Dinauer, M.C.; Rosen, H.; Chait, A.; Heinecke, J.W.; LeBoeuf, R.C. Impaired superoxide production due to a deficiency in phagocyte NADPH oxidase fails to inhibit atherosclerosis in mice. *Arterioscler. Thromb. Vasc. Biol.*, **2000**, 20, 1529-35.
[58] Hsich, E.; Segal, B.H.; Pagano, P.J.; Rey, F.E.; Paigen, B.; Deleonardis, J.; Hoyt, R.F.; Holland, S.M.; Finkel, T. Vascular effects following homozygous disruption of p47(phox) : An essential component of NADPH oxidase. *Circulation*. **2000**, 101, 1234-1236
[59] Quesada, I. M.; Lucero, A.; Amaya, C.; Meijles, D. N.; Cifuentes, M. E.; Pagano, P. J.; Castro, C. Selective inactivation of NADPH oxidase 2 causes regression of vascularization and the size and stability of atherosclerotic plaques. *Atherosclerosis* **2015**, *242*, 469-475.
[60] Kalinina, N.; Agrotis, A.; Tararak, E.; Antropova, Y.; Kanellakis, P.; Ilyinskaya, O.; Quinn, M.T.; Smirnov, V.; Bobik, A. Cytochrome *b*558-dependent NAD(P)H oxidase-phox units in smooth muscle and macrophages of atherosclerotic lesions. *Arterioscler. Thromb. Vasc. Biol.*, **2002**, 22, 2037-43.
[61] Guzik, T.J.; Mussa, S.; Gastaldi, D.; Sadowski, J.; Ratnatunga, C.; Pillai, R.; Channon, K.M. Mechanisms of increased vascular superoxide production in human diabetes mellitus: role of NAD(P)H oxidase and endothelial nitric oxide synthase. *Circulation*, **2002**, 105, 1656-62.
[62] Ohara, Y.; Peterson, T.E.; Harrison, D.G. Hypercholesterolemia increases endothelial superoxide anion production. *J. Clin. Invest.*, **1993**, 91, 2546-51.
[63] Laursen, J.B.; Somers, M.; Kurz, S.; McCann, L.; Warnholtz, A.; Freeman, B.A.; Tarpey, M.; Fukay, T.; Harrison, D.G. Endothelial regulation of vasomotion in apoE-deficient mice: implications for interactions between peroxynitrite and tetrahydrobiopterin. *Circulation*, **2001**, 103, 1282-8.
[64] Victor, V.M.; Nuñez, C.; D'Ocon, P.; Taylor, C.T.; Esplugues, J.V.; Moncada, S. Regulation of oxygen distribution in tissues by endothelial nitric oxide. *Circ. Res.*, **2009**, 104, 11**78-83**.
[65] Bedwell, S.; Dean, R.T.; Jessup, W. The action of defined oxygen-centered free radicals on human low-density lipoproteins. *Biochem. J.*, **1989**, 262, 707-12.
[66] Steinbrecher, U.P. Role of superoxide in endothelial-cell modification of low-density lipoproteins. *Biochim. Biophys. Acta*, **1988**, 959, 20-30.
[67] Leeuwenburgh, C.; Hardy, M.M.; Hazen, S.L.; Wagner, P.; Oh-ishi, S.; Steinbrecher, U.P.; Heinecke, J.W. Reactive nitrogen intermediates promote low density lipoprotein oxidation in human atherosclerotic intima. *J. Biol. Chem.*, **1997**, 272, 1433-6.
[68] Li, H.; Horke, S.; Forstermann, U. Vascular oxidative stress, nitric oxide and atherosclerosis. *Atherosclerosis* **2014**, *237*, 208-219.
[69] Buttery, L.D.; Springall, D.R.; Chester, A.H.; Evans, T.J.; Standfield, E.N.; Parums, D.V.; Yacoub, M.H.; Polak, J.M. Inducible nitric oxide synthase is present within human atherosclerotic lesions and promotes the formation and activity of peroxynitrite. *Lab. Invest.*, **1996**, 75, 77-85.
[70] Luoma, J.S.; Stralin, P.; Marklund, S.L.; Hiltunen, T.P.; Sarkioja, T.; Yla-Herttuala S. Expression of extracellular SOD and iNOS in macrophages and smooth muscle cells in human and rabbit atherosclerotic lesions: colocalization with epitopes characteristic of oxidized LDL and peroxynitrite-modified proteins. *Arterioscler. Thromb. Vasc. Biol.*, **1998**, 18: 157-67.

[71] Kettle, A.J.; van Dalen, C.J.; Winterbourn, C.C. Peroxynitrite and myeloperoxidase leave the same footprint in protein nitration. *Redox Report,* **1997**, 3, 257-8.
[72] Podrez, E.A.; Schmitt, D.; Hoff, H.F.; Hazen, S.L. Myeloperoxidase-generated reactive nitrogen species convert LDL into an atherogenic form *in vitro*. *J. Clin. Invest.*, **1999**, 103: 1547-60.
[73] Stocker, R.; Keaney, J.F. Role of oxidative modifications in atherosclerosis. *Physiol. Rev.*, **2004**, 84,1381-478.
[74] Hazell, L.J.; Arnold, L.; Flowers, D.; Waeg, G.; Malle, E.; Stocker, R. Presence of hypochlorite-modified proteins in human atherosclerotic lesions. *J. Clin. Invest.*, **1996**, 97, 1535-44.
[75] Carr, A.C.; McCall, M.R.; Frei, B. Oxidation of LDL by myeloperoxidase and reactive nitrogen species: reaction pathways and antioxidant protection. *Arterioscler. Thromb. Vasc. Biol.*, **2000**, 20, 1716-23.
[76] Hazell, L.J.; Stocker, R. Oxidation of low-density lipoprotein with hypochlorite causes transformation of the lipoprotein into a high-uptake form for macrophages. *Biochem. J.*, **1993**, 290, 165-72.
[77] Liao, L.; Aw, T.Y.; Kvietys, P.R.; Granger, D.N. Oxidized LDLinduced microvascular dysfunction. Dependence on oxidation procedure. *Arterioscler. Thromb. Vasc. Biol.*, **1995**, 15, 2305-11.
[78] Kopprasch, S.; Leonhardt, W.; Pietzsch, J.; Kuhne, H. Hypochlorite-modified low-density lipoprotein stimulates human polymorphonuclear leukocytes for enhanced production of reactive oxygen metabolites, enzyme secretion, and adhesion to endothelial cells. *Atherosclerosis*, **1998**, 136, 315-24.
[79] Savenkova, M.L.; Mueller, D.M.; Heinecke, J.W. Tyrosyl radical generated by myeloperoxidase is a physiological catalyst for the initiation of lipid peroxidation in low density lipoprotein. *J. Biol. Chem.*, **1994**, 269, 20394-400.
[80] Heller, J.I.; Crowley, J.R.; Hazen, S.L.; Salvay, D.M.; Wagner, P.; Pennathur, S.; Heinecke, J.W. *p*-Hydroxyphenylacetaldehyde, an aldehyde generated by myeloperoxidase, modifies phospholipid amino groups of low density lipoprotein in human atherosclerotic intima. *J. Biol. Chem.*, **2000**, 275, 9957-62.
[81] Hazell, L.J.; van den Berg, J.J.; Stocker, R. Oxidation of low-density lipoprotein by hypochlorite causes aggregation that is mediated by modification of lysine residues rather than lipid oxidation. *Biochem. J.*, **1994**, 302, 297-304.
[82] Fu, S.; Davies, M.J.; Stocker, R.; Dean, R.T. Evidence for roles of radicals in protein oxidation in advanced human atherosclerotic plaque. *Biochem J*, **1998**, 333, 519-25.
[83] Pennathur, S.; Wagner, J.D.; Leeuwenburgh, C.; Litwak, K.N.; Heinecke, J.W. A hydroxyl radical-like species oxidizes cynomolgus monkey artery wall proteins in early diabetic vascular disease. *J. Clin. Invest.*, **2001**, 107, 853-60.
[84] White, M.Y.; Edwards, A.V.; Cordwell, J.J.; Van Eyk, J.E. Mitochondria: a mirror into cellular dysfunction in heart disease. *Proteomics Clin. Appl.*, **2008**, 2, 845-61.
[85] Sharma, L.K.; Lu, J.; Bai, Y. Mitochondrial respiratory complex I: structure, function and implication in human diseases. *Curr. Med. Chem.*, **2009**, 16, 1266-77.
[86] Elson, J. L.; Smith, P. M.; Greaves, L. C.; Lightowlers, R. N.; Chrzanowska-Lightowlers, Z. M.; Taylor, R. W.; Vila-Sanjurjo, A. The presence of highly disruptive 16S rRNA mutations in clinical samples indicates a wider role for mutations of the mitochondrial ribosome in human disease. *Mitochondrion* **2015**.
[87] Schlattner, U.; Tokarska-Schlattner, M.; Wallimann, T. Mitochondrial creatine kinase in human health and disease. *Biochim. Biophys. Acta*, **2006**, 1762, 164- 80.
[88] Collins, T.J.; Berridge, M.J.; Lipp, P.; Bootman, MD. Mitochondria are morphologically and functionally heterogeneous within cells. *EMBO J*, **2002**, 21: 1616-27.
[89] DiMauro, S.; Schon, E.A. Mitochondrial respiratory-chain diseases. *N. Engl. J. Med.*, **2003**, 348, 2656-68.
[90] Houstek, J.; Pickova, A.; Vojtiskova, A.; Mracek, T.; Pecina, P.; Jesina, P. Mitochondrial diseases and genetic defects of ATP synthase. *Biochim. Biophys. Acta*, **2006**, 1757, 1400-5.
[91] Brookes, P.S.; Yoon, Y.; Robotham, J.L.; Anders, M.W.; Sheu, S.S. Calcium, ATP, and ROS: a mitochondrial love-hate triangle. *Am. J. Physiol. Cell Physiol.*, **2004**, 287, 817-33.
[92] Stavrovskaya, I.G.;, Kristal, B.S. The powerhouse takes control of the cell: is the mitochondrial permeability transition a viable therapeutic target against neuronal dysfunction and death?. *Free Radic. Biol. Med.*, **2005**, 38: 687-97.

[93] Di Lisa, F.; Bernardi, P. Mitochondrial function and myocardial aging. A critical analysis of the role of permeability transition. *Cardiovasc. Res.*, **2005**, 66: 222-32.
[94] Yakes, F.M.; Van Houten, B. Mitochondrial DNA damage is more extensive and persists longer than nuclear DNA damage in human cells following oxidative stress. *Proc. Natl. Acad. Sci. USA*, **1997**, 94, 514-9.
[95] Cai, W. F.; Kang, K.; Huang, W.; Liang, J. L.; Feng, Y. L.; Liu, G. S.; Chang, D. H.; Wen, Z. L.; Paul, C.; Xu, M.; Millard, R. W.; Wang, Y. CXCR4 attenuates cardiomyocytes mitochondrial dysfunction to resist ischaemia-reperfusion injury. *J. Cell. Mol. Med.* **2015**, *19*, 1825-1835.
[96] Ramachandran, A.; Moellering, D.R.; Ceaser, E.; Shiva, S.; Xu, J.; Darley-Usmar, V. Inhibition of mitochondrial protein synthesis results in increased endothelial cell susceptibility to nitric oxide-induced apoptosis. *Proc. Natl. Acad. Sci. USA*, **2002**, 99, 6643-8.
[97] Knight-Lozano, C.A.; Young, C.G.; Burow, D.L.; Hu, Z.Y.; Uyeminami, D.; Pinkerton, K.E.; Ischiropoulos, H.; Ballinger, SW. Cigarette smoke exposure and hypercholesterolemia increase mitochondrial damage in cardiovascular tissues. *Circulation*, **2002**, 105, 849-54.
[98] Devary, Y.; Rosette, C.; DiDonato, J.A.; Karin, M. NF-kappa B activation by ultraviolet light not dependent on a nuclear signal. *Science*, **1993**, 261, 1442-5.
[99] Schulze-Osthoff, K.; Beyaert, R.; Vandervoorde, V.; Haegeman, G.; Fiers, W. Depletion of the mitochondrial electron transport abrogates the cytotoxic and gene-inductive effects of TNF. *EMBO J.*, **1993**, 12, 3095-104.
[100] Murphy, M.P.; Smith, R.A. Drug delivery to mitochondria: the key to mitochondrial medicine. *Adv. Drug Deliv. Rev.*, **2000**, 41, 235-50.
[101] Bhat, T. A.; Kumar, S.; Chaudhary, A. K.; Yadav, N.; Chandra, D. Restoration of mitochondria function as a target for cancer therapy. *Drug Discov. Today* **2015**, *20*, 635-643.
[102] Wallace, D.C. Mitochondrial diseases in man and mouse. *Science*, **1999**, 283, 1482-8.
[103] Jauslin, M.L.; Meier, T.; Smith, R.A.; Murphy, M.P. Mitochondria-targeted antioxidants protect Friedreich Ataxia fibroblasts from endogenous oxidative stress more effectively than untargeted antioxidants. *FASEB J.*, **2003**, 17, 1972-4.
[104] Melov, S.; Coskun, P.; Patel, M.; Tuinstra, R.; Cottrell, B.; Jun, A.S.; Zastawny, T.H.; Dizdaroglu, M.; Goodman, S.I.; Huang, T.T.; Miziorko, H.; Epstein, C.J.; Wallace, D.C. Mitochondrial disease in superoxide dismutase 2 mutant mice. *Proc. Natl. Acad. Sci. USA*, **1999**, 96, 846-51.
[105] Melov, S.; Doctrow, S.R.; Schneider, J.A.; Haberson, J.; Patel, M.; Coskun, P.E.; Huffman, K.; Wallace, D.C.; Malfroy, B. Lifespan extension and rescue of spongiform encephalopaty in superoxide dismutase 2 nullizygous mice treated with superoxide dismutase-catalase mimetics. *J. Neurosci.*, **2001**, 21, 8348-53.
[106] Schulz, E.; Anter, E.; Keaney, J.F. Oxidative stress, antioxidants, and endothelial function. *Curr. Med. Chem.*, **2004**, 11, 1093-104.
[107] Mashima, R.; Witting, P.K.; Stocker, R. Oxidants and antioxidants in atherosclerosis. *Curr. Opin. Lipidol.*, **2001**, 12: 411-8.
[108] Marshall, F.N. Pharmacology and toxicology of probucol. *Artery*, **1982**, 10, 7-21.
[109] Tawara, K.; Ishihara, M.; Ogawa, H.; Tomikawa, M. Effect of probucol, pantethine and their combinations on serum lipoprotein metabolism and on the incidence of atheromatous lesions in the rabbit. *Jpn. J. Pharmacol.*, **1986**, 41, 211-22.
[110] Kita, T.; Nagano, Y.; Yokode, M.; Ishii, K.; Kume, N.; Ooshima, A.; Yoshida, H.; Kawai, C. Probucol prevents the progression of atherosclerosis in Watanabe heritable hyperlipidemic rabbit, an animal model for familial hypercholesterolemia. *Proc. Natl. Acad. Sci. USA.*, **1987**, 84, 5928-31.
[111] Carew, T.E.; Schwenke, D.C.; Steinberg, D. Antiatherogenic effect of probucol unrelated to its hypocholesterolemic effect: evidence that antioxidants *in vivo* can selectively inhibit low density lipoprotein degradation in macrophage-rich fatty streaks and slow the progression of atherosclerosis in the Watanabe heritable hyperlipidemic rabbit. *Proc. Natl. Acad. Sci. USA.*, **1987**, 84, 7725-9.
[112] Walldius, G.; Erikson, U.; Olsson, A.G.; Bergstrand, L.; Hadell, K.; Johansson, J.; Kaijser, L.; Lassvik C, Mölgaard J, Nilsson S. The effect of probucol on femoral atherosclerosis: the Probucol Quantitative Regression Swedish Trial (PQRST). *Am. J. Cardiol.*, **1994**, 74, 875-83.
[113] Sawayama, Y.; Shimizu, C.; Maeda, N.; Tatsukawa, M.; Kinukawa, N.; Koyanagi, S.; Kashiwagi, S.; Hayashi, J. Effects of probucol and pravastatin on common carotid atherosclerosis in patients with

asymptomatic hypercholesterolemia. Fukuoka Atherosclerosis Trial (FAST). *J. Am. Coll. Cardiol.*, **2002**, 39, 610-6.

[114] Sparrow, C.P.; Doebber, T.W.; Olszewski, J.; Wu, M.S.; Ventre, J.; Stevens, K.A, Chao YS. Low density lipoprotein is protected from oxidation and the progression of atherosclerosis is slowed in cholesterol-fed rabbits by the antioxidant *N,N*-diphenyl-phenylenediamine. *J. Clin. Invest.*, **1992**, 89, 1885-91.

[115] Tangirala, R.K.; Casanada, F.; Miller, E.; Witztum, J.L.; Steinberg, D.; Palinski, W. Effect of the antioxidant *N,N*-diphenyl 1,4-phenylenediamine (DPPD) on atherosclerosis in apoE-deficient mice. *Arterioscler. Thromb. Vasc. Biol.*, **1995**, 15, 1625-30.

[116] Esterbauer, H.; Gebicki, J.; Puhl, H.; Jurgens, G. The role of lipid peroxidation and antioxidants in oxidative modification of LDL. *Free Radic. Biol. Med.*, **1992**, 13, 341-90.

[117] Mathur, P.; Ding, Z.; Saldeen, T.; Mehta, J. L. Tocopherols in the Prevention and Treatment of Atherosclerosis and Related Cardiovascular Disease. *Clin. Cardiol.* **2015**.

[118] Ingold, K.U.; Bowry, V.W.; Stocker, R.; Walling, C. Autoxidation of lipids and antioxidation by alpha-tocopherol and ubiquinol in homogeneous solution and in aqueous dispersions of lipids. The unrecognized consequences of lipid particle size as exemplified by the oxidation of human low density lipoprotein. *Proc. Natl. Acad. Sci. USA*, **1993**, 90, 45-9.

[119] Lynch, S.M.; Morrow, J.D.; Roberts, L.J.; Frei, B. Formation of noncyclooxygenase-derived prostanoids (F2-isoprostanes) in plasma and low-density lipoprotein exposed to oxidative stress *in vitro*. *J. Clin. Invest.*, **1994**, 93, 998-1004.

[120] Podrez, E.A.; Poliakov, E.; Shen, Z.; Zhang, R.; Deng, Y.;, Sun. M.; Finton, P.J.; Shan, L.; Febbraio, M.; Hajjar, D.P.; Silverstein, R.L.; Hoff, H.F.; Salomon, R.G.; Hazen, S.L. A novel family of atherogenic oxidized phospholipids promotes macrophage foam cell formation *via* the scavenger receptor CD36 and is enriched in atherosclerotic lesions. *J. Biol. Chem.*, **2002**, 277, 38517-23.

[121] Stralin, P.; Marklund, S.L. Effects of oxidative stress on expression of extracellular superoxide dismutase, CuZn-superoxide dismutase and Mn-superoxide dismutase in human dermal fibroblasts. *Biochem. J.*, **1994**, 298, 347-52.

[122] Stocker, R. Antioxidant defenses in the vascular wall. In: Oxidative Stress and Vascular Disease, edited by Keaney JF Jr. Boston: Kluwer Academic, **1999**, p. 27-47.

[123] T'Hoen, P.A.; Van der Lans, C.A.; Van Eck, M.; Bijsterbosch, M.K.; Van Berkel, T.J.; Twisk, J. Aorta of ApoE-deficient mice responds to atherogenic stimuli by a prelesional increase and subsequent decrease in the expression of antioxidant enzymes. *Circ. Res.*, **2003**, 93, 262-9.

[124] Jung, O.; Marklund, S.L.; Geiger, H.; Pedrazzini, T.; Busse, R.; Brandes, R.P. Extracellular superoxide dismutase is a major determinant of nitric oxide bioavailability: *in vivo* and ex vivo evidence from ecSOD-deficient mice. *Circ. Res.*, **2003**, 93, 622-9.

[125] Lapenna, D.; de Gioia, S.; Ciofani, G.; Mezzetti, A.; Ucchino, S.; Calafiore, A.M, Napolitano, A.M.; Di Ilio, C.; Cuccurullo, F.. Glutathione-related antioxidant defenses in human atherosclerotic plaques. *Circulation*, **1998**, 97, 1930-4.

[126] Godin, D.V.; Garnett, M.E.; Cheng, K.M.; Nichols, C.R. Sex-related alterations in antioxidant status and susceptibility to atherosclerosis in Japanese quail. *Can. J. Cardiol.*, **1995**, 11, 945-51.

[127] Vile, G.F.; Basu-Modak, S.; Waltner, C.; Tyrrell, R.M. Heme oxygenase 1 mediates an adaptive response to oxidative stress in human skin fibroblasts. *Proc. Natl. Acad. Sci. USA*, **1994**, 91, 2607-10.

[128] Ishikawa, K.; Sugawara, D.; Wang, X.; Suzuki, K.; Itabe, H.; Maruyama, Y.; Lusis, A.J. Heme oxygenase-1 inhibits atherosclerotic lesion formation in ldl-receptor knockout mice. *Circ. Res.*, 2001, 88: 506-12.

[129] Yet, S.F.; Layne, M.D.; Liu, X.; Chen, Y.H.; Ith, B.; Sibinga, N.E.; Perrella, M.A. Absence of heme oxygenase-1 exacerbates atherosclerotic lesion formation and vascular remodeling. *FASEB J.* 2003 17, 1759-61.

[130] Togane, Y.; Morita, T.; Suematsu, M.; Ishimura, Y.; Yamazaki, J.I.; Katayama, S. Protective roles of endogenous carbon monoxide in neointimal development elicited by arterial injury. *Am. J. Physiol. Heart Circ. Physiol.*, **2000**, 278, 623-32.

[131] The Alpha-Tocopherol Beta Carotene Cancer Prevention Study Group. The effect of vitamin E and beta carotene on the incidence of lung cancer and other cancers in male smokers. *N. Engl. J. Med.*, **1994**, 330, 1029-35.

[132] Liu, X.M.; Chapman, G.B.; Wang, H.; Durante, W. Adenovirus mediated heme oxygenase-1 gene expression stimulates apoptosis in vascular smooth muscle cells. *Circulation*, **2002**, 105, 79-84.
[133] Soares, M.P.; Lin, Y.; Anrather, J.; Csizmadia, E.; Takigami, K.; Sato, K.; Grey ST, Colvin RB, Choi AM, Poss KD, Bach FH. Expression of heme-oxygenase-1 can determine cardiac xenograft survival. *Nat. Med.*, **1998**, 4, 1073-7.
[134] Yang, X.; Yao, H.; Chen, Y.; Sun, L.; Li, Y.; Ma, X.; Duan, S.; Li, X.; Xiang, R.; Han, J.; Duan, Y. Inhibition of Glutathione Production Induces Macrophage CD36 Expression and Enhances Cellular-oxidized Low Density Lipoprotein (oxLDL) Uptake. *J. Biol. Chem.* **2015**, *290*, 21788-21799.
[135] Suarna, C.; Dean, R.T.; May, J.; Stocker, R. Human atherosclerotic plaque contains both oxidized lipids and relatively large amounts of α-tocopherol and ascorbate. *Arterioscler. Thromb. Vasc. Biol.*, **1995**, 15, 1616-24.
[136] Willis, G.C.; Fishman, S. Ascorbic acid content of human arterial tissue. *Can. Med. Assoc. J.*, **1955**, 72, 500-3.
[137] Scioli, M. G.; Bielli, A.; Agostinelli, S.; Tarquini, C.; Arcuri, G.; Ferlosio, A.; Costanza, G.; Doldo, E.; Orlandi, A. Antioxidant treatment prevents serum deprivation- and TNF-alpha-induced endothelial dysfunction through the inhibition of NADPH oxidase 4 and the restoration of beta-oxidation. *J. Vasc. Res.* **2014**, *51*, 327-337.
[138] Skinner, K.A.; White, C.R.; Patel, R.; Tan, S.; Barnes, S.; Kirk, M.; Darley-Usmar V, Parks DA. Nitrosation of uric acid by peroxynitrite. Formation of a vasoactive nitric oxide donor. *J. Biol. Chem.*, **1998**, 273, 24491-7.
[139] Rao, G.N.; Corson, M.A.; Berk, B.C. Uric acid stimulates vascular smooth muscle cell proliferation by increasing platelet-derived growth factor A-chain expression. *J. Biol. Chem.*, **1991**, 266, 8604-8.
[140] Shi, Y.; Evans, J.E.;, Rock, K.L. Molecular identification of a danger signal that alerts the immune system to dying cells. *Nature*, **2003**, 425, 516-21.
[141] Li, L. X.; Lu, J. X.; Shuai, H. P.; Xia, H. F.; Zhang, R.; Wang, J. W.; Chen, M. Y.; Li, T. T.; Bao, Y. Q.; Jia, W. P. Decreased urine uric acid excretion is associated with diabetic retinopathy but not with lower limb atherosclerosis in hospitalized patients with type 2 diabetes. *Atherosclerosis* **2015**, *242*, 13-18.
[142] Dabbagh, A.J.; Frei, B. Human suction blister interstitial fluid prevents metal ion-dependent oxidation of low density lipoprotein by macrophages and in cell-free systems. *J. Clin. Invest.*, **1995**, 96, 1958-66.
[143] Smith, E.B. The relationship between plasma and tissue lipids in human atherosclerosis. *Adv. Lipid Res.*, **1974**, 12, 1-49.
[144] Keaney, J.F.; Simon, D.I.; Freedman, J.E. Vitamin E and vascular homeostasis: implications for atherosclerosis. *FASEB J.*, **1999**, 13, 965-75.
[145] Boscoboinik, D.; Szewczyk, A.; Hensey, C.; Azzi, A. Inhibition of cell proliferation by α-tocopherol. Role of protein kinase C. *J. Biol. Chem.*, **1991**, 266, 6188-94.
[146] Letters, J.M.; Witting, P.K.; Christison, J.K.; Westin Eriksson, A.W.; Pettersson, K.; Stocker, R. Time-dependent changes to lipids and antioxidants in plasma and aortae of apoE-knockout mice. *J. Lipid Res.*,**1999**, 40, 1104-12.
[147] Thomas, S.R.; Leichtweis, S.B.; Pettersson, K.; Croft, K.D.; Mori, T.A.; Brown, A.J.; Stocker, R. Dietary co-supplementation with vitamin E and coenzyme Q10 inhibits atherosclerosis in apolipoprotein E gene knockout mice. *Arterioscler. Thromb. Vasc. Biol.*, **2001**, 21, 585-93.
[148] Mohr, D.; Bowry, V.W.; Stocker, R. Dietary supplementation with coenzyme Q10 results in increased levels of ubiquinol-10 within circulating lipoproteins and increased resistance of human low density lipoprotein to the initiation of lipid peroxidation. *Biochim. Biophys. Acta*, **1992**, 1126, 247-54.
[149] Witting, P.K.; Pettersson, K.; Letters, J.; Stocker, R. Anti-atherogenic effect of coenzyme Q10 in apolipoprotein E gene knockout mice. *Free Radic. Biol. Med.*, **2000**, 29, 295-305.
[150] Turunen, M.; Wehlin, L.; Sjoberg, M.; Lundahl, J.; Dallner, G.; Brismar, K.; Sindelar, P.J. beta2-Integrin and lipid modifications indicate a non-antioxidant mechanism for the anti-atherogenic effect of dietary coenzyme Q10. *Biochem. Biophys. Res. Commun.*, **2002**, 296, 255-60.
[151] Hodqson, J.M.; Watts, G.F.; Playford, D.A.; Burke, V.; Croft, K.D. Coenzyme Q10 improves blood pressure and glycaemic control: a controlled trial in subjects with type 2 diabetes. *Eur. J. Clin. Nutr.*, **2002**, 56, 1137-42.

[152] Watts, G.F.; Playford, D.A.; Croft, K.D.; Ward, N.C.; Mori, T.A.; Burke, V. Coenzyme Q10 improves endothelial dysfunction of the brachial artery in type II diabetes mellitus. *Diabetologia*, **2002**, 45, 420-6.

[153] Allen, R. M.; Vickers, K. C. Coenzyme Q10 increases cholesterol efflux and inhibits atherosclerosis through microRNAs. *Arterioscler. Thromb. Vasc. Biol.* **2014**, *34*, 1795-1797.

[154] Asin-Cayuela, J.; Manas, A.R.; James, A.M.; Smith, R.A.; Murphy, MP. Fine-tuning the hydrophobicity of a mitochondria-targeted antioxidant. *FEBS Lett.*, **2004**, 571, 9-16.

[155] Smith, R.A.; Porteous, C.M.; Coulter, C.M.; Murphy, M.P. Selective targeting of an antioxidant to mitochondria. *Eur. J. Biochem.*, **1999**, 263: 709-716.

[156] Brown, S.E.; Ross MF, Sanjuan-Pla A, Manas AR, Smith RA, Murphy MP. Targeting lipoic acid to mitochondria: synthesis and characterization of a triphenylphosphonium-conjugated alpha-lipoyl derivative. *Free Radic. Biol. Med.*, **2007**, 42, 1766-80.

[157] Murphy, M.P.; Smith, R.A. Targeting antioxidants to mitochondria by conjugation to lipophilic cations. *Annu. Rev. Pharmacol. Toxicol.*, **2007**, 47, 629-56.

[158] Smith, R.A.; Adlam, V.J.; Blaikie, F.H.; Manas, A.R.; Porteous, C.M.; James, A.M.; Ross. M.F.; Logan, A.; Cochemé, H.M.; Trnka, J.; Prime, T.A.; Abakumova, I.; Jones, B.A.; Filipovska, A.; Murphy, M.P. Mitochondria-targeted antioxidants in the treatment of disease. *Ann. N. Y. Acad. Sci.* **2008** Dec;1147:105-11

[159] Ross, M.F.; Kelso, G.F.; Blaikie, F.H.; James, A.M.; Cochemé, H.M.; Filipovska, A.; Da Ros, T.; Hurd, T.R.; Smith, R.A.; Murphy, M.P. Lipophilic triphenylphosphonium cations as tools in mitochondrial bioenergetics and free radical biology. *Biochemistry (Mosc.)*, **2005**, 70, 222-30.

[160] Koopman, W.J.; Verkaart, S.; Visch, H.J.; van der Westhuizen, F.H.; Murphy, M.P.; van den Heuvel, L.W.; Smeitink, J.A.; Willems, P.H. Inhibition of complex I of the electron transport chain causes O2-mediated mitochondrial outgrowth. *Am. J. Physiol. Cell. Physiol.*, **2005**, 288, C1440-50.

[161] Smith, R.A.; Porteous, C.M.; Gane, A.M.; Murphy, M.P. Delivery of bioactive molecules to mitochondria *in vivo*. *Proc. Natl. Acad. Sci. USA*, **2003**, 100, 5407-12.

[162] Adlam, V.J.; Harrison, J.C.; Porteous, C.M.; James, A.M.; Smith, R.A.; Murphy, M.P.; Sammut, I.A. Targeting an antioxidant to mitochondria decreases cardiac ischemia-reperfusion injury. *FASEB J.*, ***2005***, 19, 1088-95.

[163] Esplugues, J.V.; Rocha, M.; Nuñez, C.; Bosca, I.; Ibiza, S.; Herance, J.R.; Ortega, A.; Serrador, J.M.; D'Ocon, P.; Victor, V.M. Complex I dysfunction and tolerance to nitroglycerin: an approach based on mitochondrial-targeted antioxidants. *Circ Res*., **2006**, 99, 1067-75.

[164] Lipinski, C.A.; Lombardo F, Dominy BW, Feeney PJ. Experimental and computational approaches to estimate solubility and permeability in drug discovery and development settings. *Adv. Drug Deliv. Rev.*, **2001**, 46, 3-26.

[165] Dickey, J. S.; Gonzalez, Y.; Aryal, B.; Mog, S.; Nakamura, A. J.; Redon, C. E.; Baxa, U.; Rosen, E.; Cheng, G.; Zielonka, J.; Parekh, P.; Mason, K. P.; Joseph, J.; Kalyanaraman, B.; Bonner, W.; Herman, E.; Shacter, E.; Rao, V. A. Mito-tempol and dexrazoxane exhibit cardioprotective and chemotherapeutic effects through specific protein oxidation and autophagy in a syngeneic breast tumor preclinical model. *PLoS One* **2013**, *8*, e70575.

[166] Zang, Q. S.; Sadek, H.; Maass, D. L.; Martinez, B.; Ma, L.; Kilgore, J. A.; Williams, N. S.; Frantz, D. E.; Wigginton, J. G.; Nwariaku, F. E.; Wolf, S. E.; Minei, J. P. Specific inhibition of mitochondrial oxidative stress suppresses inflammation and improves cardiac function in a rat pneumonia-related sepsis model. *Am. J. Physiol. Heart Circ. Physiol.* **2012**, *302*, H1847-59.

[167] Wilcox, C. S.; Pearlman, A. Chemistry and antihypertensive effects of tempol and other nitroxides. *Pharmacol. Rev.* **2008**, *60*, 418-469.

[168] Doctrow, S. R.; Liesa, M.; Melov, S.; Shirihai, O. S.; Tofilon, P. Salen Mn complexes are superoxide dismutase/catalase mimetics that protect the mitochondria. *Curr Inorganic Chem.* **2012**, 2, 325-334.

[169] Melov, S.; Wolf, N.; Strozyk, D.; Doctrow, S. R.; Bush, A. I. Mice transgenic for Alzheimer disease beta-amyloid develop lens cataracts that are rescued by antioxidant treatment. *Free Radic. Biol. Med.* **2005**, *38*, 258-261.

[170] Peng, J.; Stevenson, F. F.; Doctrow, S. R.; Andersen, J. K. Superoxide dismutase/catalase mimetics are neuroprotective against selective paraquat-mediated dopaminergic neuron death in the substantial nigra: implications for Parkinson disease. *J. Biol. Chem.* **2005**, *280*, 29194-29198.

[171] Jung, C.; Rong, Y.; Doctrow, S.; Baudry, M.; Malfroy, B.; Xu, Z. Synthetic superoxide dismutase/catalase mimetics reduce oxidative stress and prolong survival in a mouse amyotrophic lateral sclerosis model. *Neurosci. Lett.* **2001**, *304*, 157-160.

[172] Baker, K.; Marcus, C. B.; Huffman, K.; Kruk, H.; Malfroy, B.; Doctrow, S. R. Synthetic combined superoxide dismutase/catalase mimetics are protective as a delayed treatment in a rat stroke model: a key role for reactive oxygen species in ischemic brain injury. *J. Pharmacol. Exp. Ther.* **1998**, *284*, 215-221.

[173] Liu, R.; Liu, I. Y.; Bi, X.; Thompson, R. F.; Doctrow, S. R.; Malfroy, B.; Baudry, M. Reversal of age-related learning deficits and brain oxidative stress in mice with superoxide dismutase/catalase mimetics. *Proc. Natl. Acad. Sci. U. S. A.* **2003**, *100*, 8526-8531.

[174] Lavoie, S.; Murray, M. M.; Deppen, P.; Knyazeva, M. G.; Berk, M.; Boulat, O.; Bovet, P.; Bush, A. I.; Conus, P.; Copolov, D.; Fornari, E.; Meuli, R.; Solida, A.; Vianin, P.; Cuenod, M.; Buclin, T.; Do, K. Q. Glutathione precursor, N-acetyl-cysteine, improves mismatch negativity in schizophrenia patients. *Neuropsychopharmacology* **2008**, *33*, 2187-2199.

[175] Srinivasan, V.; Doctrow, S.; Singh, V. K.; Whitnall, M. H. Evaluation of EUK-189, a synthetic superoxide dismutase/catalase mimetic as a radiation countermeasure. *Immunopharmacol. Immunotoxicol.* **2008**, *30*, 271-290.

[176] Schulz, E.; Anter, E.; Keaney, J. F.,Jr Oxidative stress, antioxidants, and endothelial function. *Curr. Med. Chem.* **2004**, *11*, 1093-1104.

[177] Coetzee, W. A. Multiplicity of effectors of the cardioprotective agent, diazoxide. *Pharmacol. Ther.* **2013**, *140*, 167-175.

[178] Drose, S.; Hanley, P. J.; Brandt, U. Ambivalent effects of diazoxide on mitochondrial ROS production at respiratory chain complexes I and III. *Biochim. Biophys. Acta* **2009**, *1790*, 558-565.

[179] Colognato, R.; Laurenza, I.; Fontana, I.; Coppede, F.; Siciliano, G.; Coecke, S.; Aruoma, O. I.; Benzi, L.; Migliore, L. Modulation of hydrogen peroxide-induced DNA damage, MAPKs activation and cell death in PC12 by ergothioneine. *Clin. Nutr.* **2006**, *25*, 135-145.

[180] Antonenko, Y. N.; Avetisyan, A. V.; Bakeeva, L. E.; Chernyak, B. V.; Chertkov, V. A.; Domnina, L. V.; Ivanova, O. Y.; Izyumov, D. S.; Khailova, L. S.; Klishin, S. S.; Korshunova, G. A.; Lyamzaev, K. G.; Muntyan, M. S.; Nepryakhina, O. K.; Pashkovskaya, A. A.; Pletjushkina, O. Y.; Pustovidko, A. V.; Roginsky, V. A.; Rokitskaya, T. I.; Ruuge, E. K.; Saprunova, V. B.; Severina, I. I.; Simonyan, R. A.; Skulachev, I. V.; Skulachev, M. V.; Sumbatyan, N. V.; Sviryaeva, I. V.; Tashlitsky, V. N.; Vassiliev, J. M.; Vyssokikh, M. Y.; Yaguzhinsky, L. S.; Zamyatnin, A. A.,Jr; Skulachev, V. P. Mitochondria-targeted plastoquinone derivatives as tools to interrupt execution of the aging program. 1. Cationic plastoquinone derivatives: synthesis and *in vitro* studies. *Biochemistry (Mosc)* **2008**, *73*, 1273-1287.

[181] Stefanova, N. A.; Muraleva, N. A.; Skulachev, V. P.; Kolosova, N. G. Alzheimer's disease-like pathology in senescence-accelerated OXYS rats can be partially retarded with mitochondria-targeted antioxidant SkQ1. *J. Alzheimers Dis.* **2014**, *38*, 681-694.

[182] Kapay, N. A.; Popova, O. V.; Isaev, N. K.; Stelmashook, E. V.; Kondratenko, R. V.; Zorov, D. B.; Skrebitsky, V. G.; Skulachev, V. P. Mitochondria-targeted plastoquinone antioxidant SkQ1 prevents amyloid-beta-induced impairment of long-term potentiation in rat hippocampal slices. *J. Alzheimers Dis.* **2013**, *36*, 377-383.

[183] Zhao, K.; Zhao, G. M.; Wu, D.; Soong, Y.; Birk, A. V.; Schiller, P. W.; Szeto, H. H. Cell-permeable peptide antioxidants targeted to inner mitochondrial membrane inhibit mitochondrial swelling, oxidative cell death, and reperfusion injury. *J. Biol. Chem.* **2004**, *279*, 34682-34690.

[184] Yousif, L. F.; Stewart, K. M.; Horton, K. L.; Kelley, S. O. Mitochondria-penetrating peptides: sequence effects and model cargo transport. *Chembiochem* **2009**, *10*, 2081-2088.

[185] Zhao, K.; Luo, G.; Zhao, G. M.; Schiller, P. W.; Szeto, H. H. Transcellular transport of a highly polar 3+ net charge opioid tetrapeptide. *J. Pharmacol. Exp. Ther.* **2003**, *304*, 425-432.

[186] Szeto, H. H.; Schiller, P. W. Novel therapies targeting inner mitochondrial membrane--from discovery to clinical development. *Pharm. Res.* **2011**, *28*, 2669-2679.

[187] Linker, R. A.; Lee, D. H.; Ryan, S.; van Dam, A. M.; Conrad, R.; Bista, P.; Zeng, W.; Hronowsky, X.; Buko, A.; Chollate, S.; Ellrichmann, G.; Bruck, W.; Dawson, K.; Goelz, S.; Wiese, S.; Scannevin, R. H.; Lukashev, M.; Gold, R. Fumaric acid esters exert neuroprotective effects in neuroinflammation *via* activation of the Nrf2 antioxidant pathway. *Brain* **2011**, *134*, 678-692.

[188] Szeto, H. H. Cell-permeable, mitochondrial-targeted, peptide antioxidants. *AAPS J.* **2006**, *8*, E277-83.
[189] Przyklenk, K. Pharmacologic treatment of the stunned myocardium: the concepts and the challenges. *Coron. Artery Dis.* **2001**, *12*, 363-369.
[190] Reddy, P. H.; Tripathi, R.; Troung, Q.; Tirumala, K.; Reddy, T. P.; Anekonda, V.; Shirendeb, U. P.; Calkins, M. J.; Reddy, A. P.; Mao, P.; Manczak, M. Abnormal mitochondrial dynamics and synaptic degeneration as early events in Alzheimer's disease: implications to mitochondria-targeted antioxidant therapeutics. *Biochim. Biophys. Acta* **2012**, *1822*, 639-649.
[191] Whiteman, M.; Spencer, J. P.; Szeto, H. H.; Armstrong, J. S. Do mitochondriotropic antioxidants prevent chlorinative stress-induced mitochondrial and cellular injury? *Antioxid. Redox Signal.* **2008**, *10*, 641-650.
[192] Manczak, M.; Mao, P.; Calkins, M. J.; Cornea, A.; Reddy, A. P.; Murphy, M. P.; Szeto, H. H.; Park, B.; Reddy, P. H. Mitochondria-targeted antioxidants protect against amyloid-beta toxicity in Alzheimer's disease neurons. *J. Alzheimers Dis.* **2010**, *20 Suppl 2*, S609-31.
[193] White, C.R.; Datta, G.; Zhang, Z.; Gupta, H.; Garber, D.W.; Mishra, V.K.; Palgunachari, M.N.; Handattu, S.P.; Chaddha, M.; Anantharamaiah, G.M. HDL therapy for cardiovascular diseases: the road to HDL mimetics. *Curr Atheroscler. Rep.* **2008** 10, 405-412.
[194] Apostolova, N.; Victor, V.M. Molecular strategies for targeting antioxidants to mitochondria: therapeutic implications. *Antioxid. Redox Signal.* **2015**, *10*, 22, 686-729.

CHAPTER 4

Endothelial Dysfunction in Diabetes: An Update on Mechanisms and Therapeutic Targets

Maria Assunta Potenza, Carmela Nacci, Luca Sgarra, Valentina Leo, Maria Antonietta De Salvia and Monica Montagnani*

Department of Biomedical Sciences and Human Oncology – Pharmacology Section, Medical School – University of Bari "Aldo Moro", Bari, Italy

Abstract: Over recent years, the explosive increase in the worldwide prevalence of type 2 diabetes has transformed this disease into a major public health concern. Cardiovascular complications are the leading causes of morbidity in diabetic patients, whose cardiovascular mortality risk is up to four times higher than in non-diabetic subjects. Morphological and structural changes in large and small vessels are usually preceded by alterations in endothelial function, resulting from the unbalanced production of endothelial-derived vascular mediators. Metabolic disturbances including hyperglycemia, insulin resistance, hyperinsulinemia and dyslipidemia all contribute to endothelial dysfunction *via* both distinct and overlapping mechanisms. In turn, abnormal synthesis and release of endothelial mediators may contribute to exacerbate the impaired function of metabolic tissues, thus creating a vicious circle where these conditions reciprocally reinforce and worsen each other. Increasing understanding on mechanisms underlying endothelial dysfunction in diabetes may serve to identify potential therapeutic targets, and help to develop novel therapeutic approaches for reducing cardiovascular risk rate in diabetic patients. This chapter summarizes the current knowledge on the most recently identified targets and therapies for the treatment of diabetes and its cardiovascular complications.

Keywords: Advanced glycation-end products (AGEs), AMP-activated protein kinase (AMPK), diabetes, dipeptidyl peptidase-4 enzyme (DPP4), endothelial dysfunction, endothelial progenitor cells (EPC), endothelin-1 (ET-1), flood mediated dilation (FMD), glucagon-like peptide 1 (GLP-1), nitric oxide (NO), poly (ADP ribose) polymerase 1 (PARP-1), protein kinase C (PKC), reactive oxygen species (ROS), rho-associated kinase (ROCK), sodium-glucose co-trasporter-2 (SGLT-2), therapeutic strategies, thiazolidine-2-4-diones (TZDs), vascular endothelial growth factor (VEGF).

***Corresponding author Monica Montagnani:** Department of Biomedical Sciences and Human Oncology – Pharmacology Section, Medical School – University of Bari "Aldo Moro", Bari, Italy; Tel: +39 080 5478583; Fax: +39 080 5478449; E-mail: monica.montagnani@uniba.it

Atta-ur-Rahman, M. Iqbal Choudhary and Allen B. Reitz (Eds.)

1. STRUCTURE AND FUNCTION OF ENDOTHELIUM

Once considered a mere physical barrier between circulating blood components and underlying tissues, the endothelium is now acknowledged as an important modulator of vascular function. With the growing understanding of endothelial mediators and their role, it has become increasingly clear that endothelial abnormalities may represent an early sign not only of hemodynamic diseases, but also of metabolic disturbances. Consequently, techniques and tools evaluating endothelial function may help to set a key role for vascular biomarkers as surrogate end-points for the study of diabetes progression, guiding treatments and therapeutic follow-ups [1].

One fundamental aspect of the endothelial function regards the ability of endothelial cells to modulate synthesis and release of mediators with opposing vasodilator/vasoconstrictor, anti-thrombotic/pro-haemostatic, anti-inflammatory/pro-mitogenic properties [2]. The maintenance of adequate blood flow, tissue oxygen supply and nutrient delivery to underneath tissues, as well as the active participation of endothelium to vascular structure remodeling, is ensured by a dynamic cross-talk between independent modulators and activation of specific pathways. The recognition of the functional role of the endothelium, with the subsequent discovery of several endothelial mediators and their respective mechanism of action, spans over three decades and has been extensively described in the previous version of this review [3]. A brief summary of the main endothelial functions and mediators is reported below.

Among other mediators - including endothelial-derived hyperpolarizing factor (EDHF) and prostacyclin (PGI_2) - nitric oxide (NO) is probably the most important molecule synthesized by the endothelial cells [4]. Due to its gaseous nature, NO cannot be stored and is instead rapidly synthesized in response to the activation of the endothelial NO synthase enzyme (eNOS). Under physiological conditions, shear stress is a key activator of eNOS; in addition, the enzyme may be activated by signaling molecules such as bradykinin, adenosine, vascular endothelial growth factor (in response to hypoxia) [5, 6], serotonin (released during platelet aggregation) [7], as well as by circulating metabolic mediators including insulin [8, 9], dehydroepiandrosterone (DHEA) [10], and adiponectin [11-13].

Following eNOS activation, NO diffuses to the vascular smooth muscle cells (VSMC) where it binds to the heme mojety of guanylate cyclase (GC) and triggers the conversion of GTP to cyclic GMP (reviewed in [14]). In addition, NO-mediated vasodilation follows from direct nitrosylation of the K_{ATP} channel, with subsequent hyperpolarization of the VSMC membrane and inhibition of calcium entry [15]. Besides vasodilation, NO inhibits platelet activation, secretion and aggregation [16], prevents leukocyte adhesion, and inhibits VSMC migration and proliferation [17]. Finally, NO participates in maintenance of oxidative/ reductive cell balance by limiting oxidative phosphorylation in mitochondria [18]. As a consequence of its low chemical stability, NO is rapidly converted to less reactive metabolites such as nitrites and nitrates. In the cellular environment, NO may also react with several oxygen free radicals, including superoxide anion produced by NADPH-dependent oxidases, to form highly reactive and harmful compounds known as peroxynitrites. However, under physiological conditions, the levels of superoxide anion are limited by the activity of the superoxide dismutase (SOD) enzymes, which convert superoxide anion to hydrogen peroxide. By controlling the formation of reactive oxygen species (ROS), the red-ox mechanisms ensure optimal bioavailability of NO produced into the cells [19].

Endothelial-mediated activity extends to production of vasoconstrictor factors, whose opposing effects participate not only in regulation of local blood flow, but also in arterial structure and remodeling [20]. Some of these factors, including vasoconstrictor prostanoids [21], angiotensin II (Ang II) [22] and superoxide anion [23] predominantly act in a local context, but their influence may also broaden to systemic effects. Endothelin-1 (ET-1), the most potent vasoactive peptide produced by endothelial cells [24], acts as the natural counterpart of NO in regulation of vascular homeostasis. ET-1 contributes to vascular homeostasis *via* stimulation of ETA and ETB G-protein-coupled receptors [25]. Activated ETA receptors stimulate several distinct pathways, including the mitogen-activated protein kinase (MAPK) cascade [26], involved in the short term regulation of vascular smooth muscle tone as well as in the long term control of cell growth, adhesion and migration in the vasculature and the heart [26]. ETB receptors mediate both the vasoconstrictor responses of ET-1 on muscle cells as well as the release of NO and prostacyclin in endothelium [27]. This last effect may serve as a natural inhibitory feed-back mechanism. Physical factors such as shear stress [28], vasoconstrictor agents including norepinephrine (NE), thrombin, Ang II and vasopressin [29], growth factors [30], cytokines [31], and hormones [32] all are able to increase production and/or release of endothelial ET-1. Dysregulation of ET-1 synthesis, release or activity may represent another important mechanism participating to endothelial dysfunction.

In patients with diabetes, several metabolic alterations concur to alter this complex and highly regulated balance between endothelial mediators. The resulting impairment of vessel dynamic adaptation to specific necessities importantly contributes to progression of vascular and end-organ damage [33, 34].

2. ENDOTHELIAL DYSFUNCTION IN DIABETES

The term endothelial dysfunction refers to a condition in which the endothelium loses its physiological properties and shifts toward a vasoconstrictor, pro-thrombotic and pro-inflammatory state. Although increased production and/or release of endothelial mediators with opposing vascular effects may also participate in altered endothelial function [25, 35], the earliest and most important marker of endothelial dysfunction is represented by a reduction of NO bioactivity [36].

The bioavailability of NO reflects a balance between its production *via* eNOS and its conversion to more stable products such as nitrites and nitrates. A decline in NO bioactivity may result from decreased eNOS protein expression [37], reduced or insufficient amounts of eNOS substrates and/or co-factors [38], impaired eNOS activation or accelerated NO degradation, especially by reactive oxygen species (ROS). Under physiological conditions, key enzymes such as NAD(P)H-dependent oxidases and superoxide dismutases (SOD) carefully regulate superoxide production and intracellular redox balance. When production of superoxide anion exceeds the physiological scavenging ability of endothelial cells, this highly reactive intermediate reacts with NO to form peroxinitrites, strong oxidants that alter structure and function of several cellular components [19, 39].

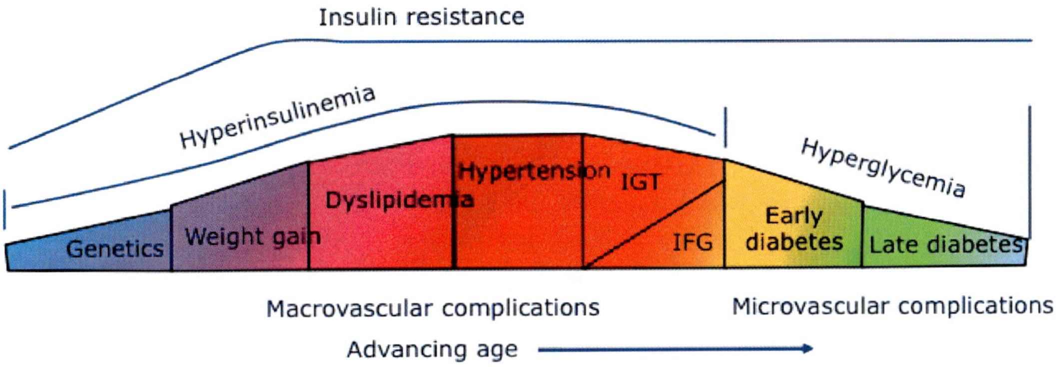

Figure 1: Natural history of diabetes. The interaction of genetic background with environmental factors results in weight gain, dyslipidemia, emodynamic disturbances and glucose intolerance. When pancreatic insulin secretion is no longer sufficient to compensate for insulin resistance, glucose intolerance progresses to chronic hyperglycemia and overt diabetes.

Endothelial dysfunction is considered a major contributing factor in the etiology of diabetic complications such as retinopathy, nephropathy, neuropathy, and impaired wound healing [40]. Moreover, endothelial dysfunction precedes the onset of atherosclerosis, which in diabetic patients is more rapid and more severe than in control population [41]. Diabetes is a progressive metabolic disorder characterized by variable degrees of insulin resistance and pancreatic β-cell dysfunction [42]. The interaction of key genes with environmental factors such as physical inactivity, quality and quantity of nutrients, puberty and ageing, concur to promote adiposity, impair β-cell function, and impair insulin action [43, 44]. When pancreatic insulin secretion is no longer sufficient to compensate for insulin resistance, glucose intolerance progresses to chronic hyperglycemia and overt diabetes [45, 46].

Insulin resistance usually precedes of many years the onset of diabetes [47, 48]. Early in the natural history of this disease, before the emergence of marked hyperglycemia, increased insulin secretion partially compensates for insulin resistance. Over time, chronic exposure to hyperglycemia and dyslipidemia further impair metabolic function *via* glycotoxicity, lipotoxicity, oxidative stress and sustained inflammation [49-51]) (Fig. **1**).

In this complex metabolic derangement, one fundamental notion that helps to explain the tight link between metabolic and vascular impairment is the well-recognized effect of insulin on hemodynamic homeostasis. Besides its essential role on glucose and lipid metabolism, insulin has a number of biological vascular actions including renal reabsorption of sodium, stimulation of sympathetic activity, and stimulated production of both ET-1 and NO from endothelium, that contribute to overall metabolic homeostasis [52] (Fig. **2**).

Physiological actions of insulin in the vasculature are mediated by binding of insulin to its cognate receptor on endothelial cell surface (comprehensively described in [53]). Insulin stimulation of endothelial cells leads to activation of eNOS and production of NO, thereby enhancing vasodilation, blood flow and nutrient delivery, which concur to efficient insulin-mediated glucose uptake on target tissues [54]. Production of NO in response to insulin results from activation of insulin receptor tyrosine kinase (IR), phosphorylation of insulin receptor substrate-1 (IRS-1), subsequent recruitment and activation of phosphatidylinositol (PI) 3-kinase signaling pathway that phosphorylates and activates Akt [8, 9, 55, 56]. Akt then directly phosphorylates eNOS on Ser^{1177} resulting in increased eNOS activity and NO production [5, 9].

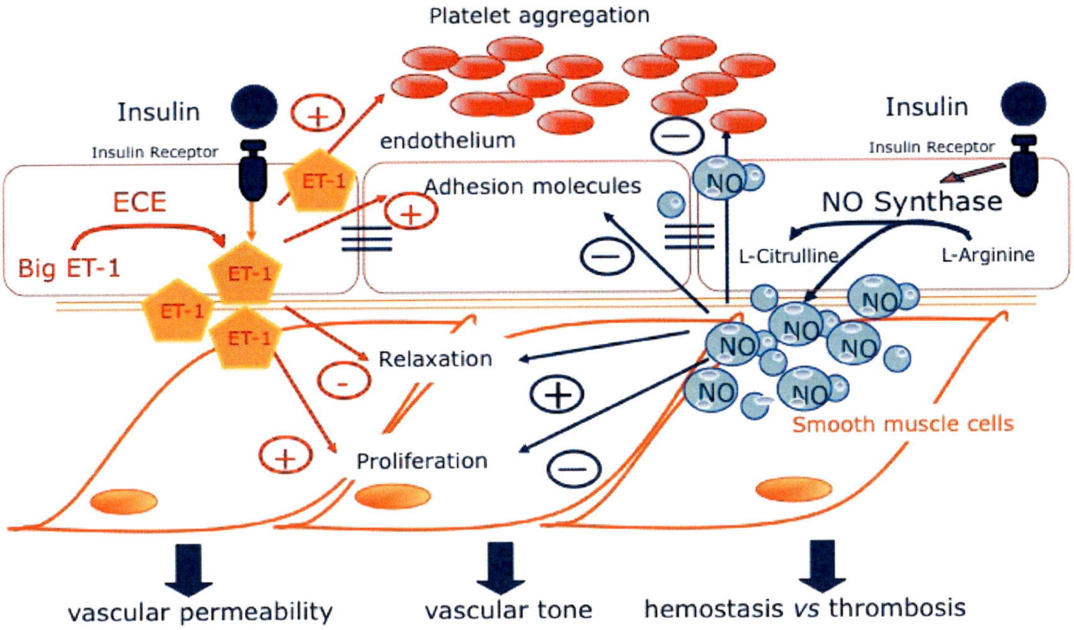

Figure 2: Schematic view of insulin action in endothelium. In endothelium, binding of insulin to its cognate tyrosine kinase receptor results in activation of separate signaling, promoting synthesis and release of mediators with opposing vascular properties: recruitment of IRS/PI 3-kinase/Akt signaling pathway leads to eNOS activation and increased production of NO. Concomitantly, activation of Ras/Raf/MAP kinase signaling results in increased secretion of the vasoconstrictor ET-1 and increased expression of adhesion molecules such as VCAM-1 and E-selectin.

Concomitant activation of the Ras/MAP kinase branch of insulin signaling pathways, on the other hand, is involved in insulin-stimulated secretion of ET-1 from endothelial cells [57]. Indeed, inhibition of MAP-kinase pathways blocks vasoconstrictor effects of insulin in rat skeletal muscle arterioles [58], and vasodilator actions of insulin in humans are potentiated by ET-1 receptor blockade [59]. In addition, MAP kinase-dependent signaling pathways regulate insulin-stimulated expression of vascular cellular adhesion molecule-1 (VCAM-1) and E-selectin [60]. Thus, insulin-stimulated release of endothelial mediators with opposing vascular effects requires activation of distinct intracellular signaling pathways. Interestingly, insulin signaling pathways in vascular endothelium regulating production of NO share striking similarities with metabolic insulin signaling pathways in skeletal muscle and adipose tissue regulating translocation of the glucose carrier GLUT4. To facilitate glucose uptake into the cells, insulin requires activation of the IR, subsequent phosphorylation of IRS-1, binding and activation of PI3-kinase, which then leads to phosphorylation and activation of Akt and PKC-ζ, culminating in the recruitment of GLUT4 transporters to the cell surface. Common insulin signaling pathways in distinct tissues with metabolic or vascular functions may provide a molecular link that tightly couples regulation of glucose metabolism and blood flow under physiological conditions. Impaired sensitivity and increased circulating levels of insulin during diabetes may profoundly affect insulin action on endothelial cell function and participate to the pathophysiology of diabetic vascular complications [61].

Metabolic derangements known to occur in diabetes include, among others, hyperglycemia, excess free fatty acid release, insulin resistance, and compensatory hyperinsulinemia. Each of these abnormalities may impact on endothelial function by disrupting molecular mechanisms regulating NO synthesis or degradation [62, 63]. In addition, a reciprocal interaction between molecular and pathophysiological mechanisms by which these metabolic changes participate to endothelial dysfunction results in a vicious cycle reinforcing their individual contribution to the progression of vascular complications in diabetes [61]. The specific and overall impact of these disturbances on endothelial function has been deeply investigated over time and extensively debated elsewhere [3, 64].

2.1. Endothelial Dysfunction and Metabolic Dysregulation: A Vicious Circle

The recognition that metabolic and endothelial physiology are fully integrated and reciprocally controlled suggests that, beside representing the immediate target of metabolic abnormalities, unbalanced endothelial function may in turn actively contribute to disrupted metabolic homeostasis. This has long been hypothesized for angiotensin II [65], and more recently demonstrated for prostanoids [66] and NO itself [67]. The interference of ET-1 on metabolic function has been deeply investigated: ET-1 infusion results in hyperinsulinemia and insulin resistance *in vivo* [68], and chronically elevated ET-1 levels may contribute to desensitization of metabolic signaling pathways on adipocytes [69]. In addition, ET-1 inhibits adipocyte differentiation [70], reduces lipoprotein lipase activity [71], inhibits insulin-stimulated glucose uptake [72], and stimulates lipolysis [73]. ET-1 seems involved in production and secretion of adiponectin (Ad), a vasodilant and anti-inflammatory adipocytokine whose levels are reduced in obese and diabetic patients [74, 75]. On this respect, we have recently provided evidence that increased circulating levels of ET-1 in obese children directly contribute to reduced levels of adiponectin *via* ET receptor-mediated activation of p42/44 MAPK signaling pathways [76].

Another important aspect on the relationship between endothelial dysfunction and metabolic derangement is the limited ability of endothelium to regenerate overtime. Under healthy conditions, endothelial cell injury resulting from exposure to risk factors is mitigated by endogenous reparative processes mediated by bone marrow-derived endothelial progenitor cells (EPC) [77]. Reduced EPC availability and/or mobilization may disrupt the balance between endothelial damage and endothelial repair and accelerate the onset of endothelial dysfunction [78]. Loss of endothelial integrity under insulin-resistant conditions may be worsened by biochemical abnormalities with the potential to disrupt EPC-based vascular repair (see [79, 80] for review). For example, reduced NO bioavailability, increased production of ROS and downregulation of PI3K/Akt signaling pathways have been related to decreased mobilisation of EPCs from bone marrow in individuals with diabetes [81] and the metabolic syndrome [82]. The migratory defects associated with insulin resistance are, at least partially, attributable to cytoskeletal alterations of EPCs induced by reduced NO bioavailability [83]. Insulin resistance may also affect the ability of EPCs to survive and diminish their capacity for adhesion, endothelial integration, proliferation, and differentiation. For example, the ability of EPCs to survive oxidative stress is impaired in diabetic animals with respect to wild-type controls [84].

Based on these data, EPCs have been proposed as cellular biomarkers of disease and predictors of cardiovascular outcomes [85]. An updated definition of their role and their biological properties in health and disease has been recently published [86].

2.2. Endothelial Function/Dysfunction as a Surrogate Endpoint in Diabetic Cardiovascular Complications

A surrogate endpoint is a biomarker that is intended as a substitute for a clinical endpoint, and it is expected to predict clinical benefit/harm or lack thereof, based on epidemiologic, therapeutic, pathophysiological or other scientific evidence. Changes in surrogate endpoints are detected earlier and at a lower cost than clinical endpoints such as morbidity and mortality. The search for vascular biomarkers that satisfy all the required criteria and may assess diabetic cardiovascular risk has been constantly pursued [1]. The hypothesis that endothelial dysfunction is one fundamental mechanism linking insulin resistance and overt diabetes to cardiovascular risk has been confirmed by a meta-analysis showing that dysfunction of the microcirculation, as assessed by various circulating and vascular biomarkers, is an independent determinant of incident diabetes [87]. On the other hand, the hypothesis of the cardiovascular continuum in diabetes suggests that alterations/dysfunction of macrocirculation take place early in the course of glucometabolic impairment, before the clinical onset of diabetes [88]. One step further, it was recently shown that the presence of carotid plaques is a predictor of incident diabetes independently of the metabolic profile, blood pressure and related drugs [89].

Altogether, these findings support the idea that endothelial dysfunction may possess the essential characteristics to be selected as a surrogate end-point, and be useful to evaluate the progression of diabetes, or even predict the development of this diseases in diabetes-free individuals. In this scenario, increasing attention has been given to non-invasive techniques evaluating endothelial function [90]. Brachial artery flood mediated dilation (FMD) was the first non-invasive technique to assess endothelial function in the peripheral conduit arteries [91] and remains today the most widely applied method [90].

Numerous studies have incorporated FMD measurements because it improves rapidly with physiological and pharmacological interventions (*e.g.* statins, ACE-inhibitors, calcium channel blockers, exercise, weight loss, estrogen, antioxidants, vitamins, dietary constituents). This reinforces the concept that endothelial dysfunction is, under certain circumstances, a reversible process [92, 93]. Nevertheless, whether an improvement in FMD with treatment may also translate into improved clinical outcomes has not been tested in a randomized controlled trial. In non-randomized controlled studies, FMD improvement was associated with a reduction in events in hypertensive postmenopausal women [94], whereas persistent FMD impairment despite optimized therapy was associated with an increase in CV events [95, 96]. At present, FMD does not fulfill all the essential criteria required for a clinical surrogate end-point to predict cardiovascular risk. Nevertheless, FMD remains a safe and useful research tool and should be taken into consideration as an integrative marker in the study of the role of risk factors on the arterial wall [1].

3. CONVENTIONAL TREATMENT OF DIABETIC ENDOTHELIAL DYSFUNCTION

Conventional therapies aim to improve glycaemic control, increase insulin pancreatic secretion, ameliorate insulin sensitivity in peripheral tissue, restore lipid profile and reduce blood pressure levels, thereby preventing several interrelated mechanisms involved in endothelial dysfunction.

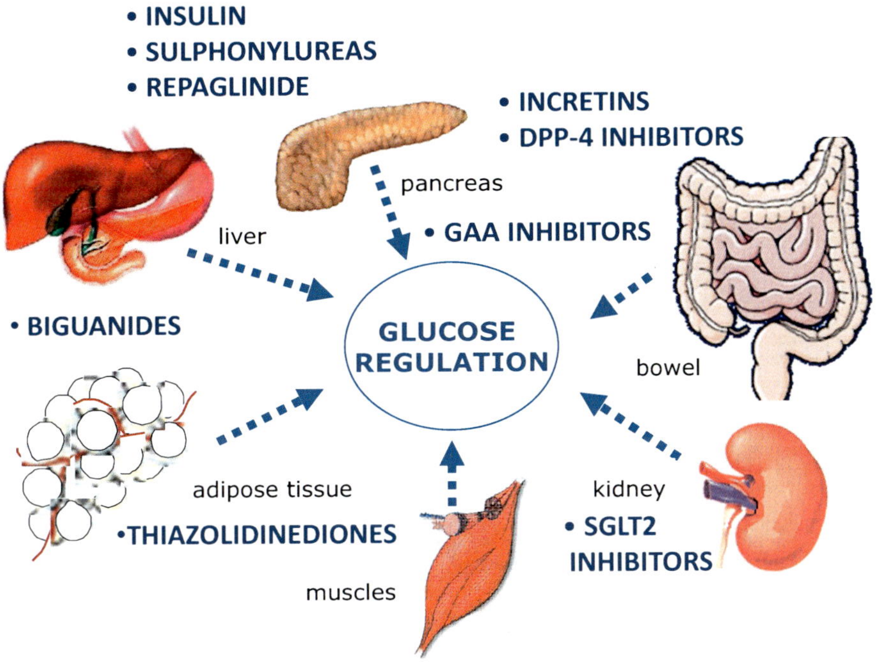

Figure 3: Conventional treatments for glycemic control. Currently available anti-diabetic drugs include the biguanide compound metformin; sulphonylureas derivatives and insulin secretagogues like repaglinide; insulin sensitizers thiazolidinediones; anti-glycaemic agents including inhibitors of alpha-glucosidase; incretins such as glucagon-like peptide 1 (GLP-1); dipeptidyl peptidase-4 enzyme (DPP-4) inhibitors, and selective sodium-glucose co-trasporter-2 (SGLT-2) inhibitors.

Treatment of type 2 diabetes aims to improve glycaemic control by increasing insulin pancreatic secretion and/or ameliorating insulin sensitivity in peripheral tissues. Beside insulin and its analogues (reviewed in [97]), currently available anti-diabetic drugs include the biguanide compound metformin; sulphonylureas derivatives and insulin secretagogues like repaglinide; insulin sensitizers thiazolidinediones; anti-glycaemic agents including inhibitors of alpha-glucosidase; incretins such as glucagon-like peptide 1 (GLP-1) and exenatide; dipeptidyl peptidase-4 enzyme (DPP4) inhibitors; and selective sodium-glucose co-trasporter-2 (SGLT-2) inhibitors [98, 99] (Fig. **3**). Several clinical studies have been conducted to ascertain effects on endothelial function for each of these substances (see [100, 101] for review). Extensive revision of vascular effects for all conventional anti-diabetic drugs is beyond the scope of this review. Among these drugs, we focused on insulin sensitizers thiazolidinediones, whose specific effects on endothelial function and cardiovascular consequences have been intensely debated in recent years, and on the vascular profile of incretins and

DPP4 inhibitors, a novel class of anti-diabetics more recently introduced on the market [102, 103]. In addition, the profile of SGLT-2 inhibitors in terms of endothelial and vascular function is also discussed.

3.1. Thiazolidine-2-4-diones (TZDs)

TZDs are insulin sensitizers acting as synthetic ligands for peroxisome proliferator-activated receptor-γ (PPAR-γ) [104]. TZDs may improve endothelial function, increase forearm blood flow and reduce blood pressure in humans by both direct and indirect mechanisms (reviewed in [105]). First of all, TZDs improve PI3K signaling pathways and increase expression of eNOS in endothelial cells [106]. This may enhance NO production in response to insulin and other mediators that require PI3K-dependent pathways to activate eNOS [10, 11]. TZDs may also increase eNOS activity and NO bioavailability by decreasing ADMA levels in vessels from diabetic patients [107]. Down-regulation of inflammatory NF-kB/STAT/AP-1 signaling pathways by TZDs inhibits the expression of inducible NOS isoform (iNOS) and reduces the amount of oxidative products in monocyte/macrophages [108, 109]. Interestingly, rosiglitazone has been proven to enhance NO availability through a direct, intracellular mechanism that leads to a reduced generation of superoxide anion [110]. The beneficial effects of rosiglitazone in counteracting hyperglycemia-mediated oxidative stress are independent of PPARγ activation, but strictly related to rosiglitazone ability to activate AMPK-dependent signaling pathways in endothelium [111]. Activation of AMPK in response to rosiglitazone correlates with inhibition of the DAG/PKC pathway, subsequent reduction of NADPH oxidase activity, and amelioration of oxidative balance [111].

Anti-inflammatory effects for both pioglitazone and rosiglitazone have been linked to their ability to decrease plasma concentrations of pro-inflammatory mediators such as TNF-α, leptin, PAI-1 and C-reactive protein, increase circulating levels of protective adiponectin, and decrease vascular expression of adhesion molecules [112]. In addition, TZDs may help to maintain plaque stability by reducing expression of matrix metalloproteases-9 (MMP-9) [113, 114], decreasing plasma levels of both FFA and atherogenic LDL cholesterol particles [115], and lowering PAI-1 plasma levels and platelet activity. Decreased platelet aggregation in response to TZDs is likely due to restoration of endothelial NO production, since platelets do not express PPAR receptors. Finally, and possibly as a consequence of their ability to improve endothelial function, TZDs can effectively decrease systolic blood pressure (SBP) in diabetic patients [116].

Several studies consistently reported beneficial effects of TZDs on vascular function, although recent clinical trials designed to assess overall impact of TZDs on cardiovascular risk factors have generated controversial results. The first meta-analysis by Nissen and Wolski [117] reported a 43% increase of myocardial infarction (MI) and a 64% increase of cardiovascular-related death (CVD) in diabetic patients treated with rosiglitazone. Nevertheless, the existence of several limitation observed in Nissen and Wolski meta-analysis prompted a number of other investigations, generating confusing and rather contradictory results [118-125].

In 2010, the unresolved question on cardiovascular risk associated with rosiglitazone prompted the European Medicine Agency (EMA) to recommend withdraw of all licensed rosiglitazone-containing drugs from the European market. Conversely, the FDA decision was to further limit prescription of rosiglitazone and restrict its use in the United States [126, 127]. Thus, up to date, FDA did not remove the drug from the market. Interestingly, after initial worry and in spite of black box warning on increased risk of MI [128], rosiglitazone prescription in USA remains relatively stable [105].

3.2. Glucagon-like Peptide-1 Receptor (GLP-1R) Agonists

Incretins are peptides produced by the gastrointestinal system able to enhance insulin secretion in a glucose-dependent manner [129, 130]. The two main human incretins are glucagon-like peptide (GLP-1) and glucose-dependent insulinotropic polypeptide (GIP) whose secretions are impaired in diabetes [131]. GLP-1-mediated effects may be sustained and improved with two strategies currently available: longer-acting GLP-1 receptor agonists resistant to degradation of the dipeptidyl peptidase 4 (DPP4) enzyme, that can provide supraphysiological stimulation of GLP-1 receptor [132] and inhibitors of DPP4 enzyme, that extends the half-life of endogenous GLP-1 [133].

The receptor for GLP-1 (GLP-1R), originally identified in pancreatic β cells, is expressed on endothelial cells, cardiomyocytes and coronary smooth muscle cells [134]. Activation of GLP-1R improves myocardial function [135], improves endothelial function in high risk cardiac patients [136], and enhances natriuresis [137], with potential positive implications for systolic blood pressure control. Recent studies have confirmed that the beneficial effect of GLP-1 on endothelial cells is mediated by an increased NO production [138]. Moreover, in endothelial cells from human coronary arteries, GLP-1R agonists stimulate proliferation *via* eNOS-, PKA-, and PI3K/Akt-dependent pathways [139]. It has also been shown that GLP-1 exerts a protective effect on atherosclerosis by reducing neointimal formation [140], foam cell formation, and atherosclerotic lesion size [141].

The GLP-1R agonists are peptides and, such as insulin, require subcutaneous injection to avoid degradation by gastrointestinal enzymes. Exenatide, liraglutide and lixisenatide are among the GLP-1R agonists available. Most of the clinical trials, in line with animal studies, have reported a reduction of blood pressure after chronic treatment of GLP-1R agonists [142-145]. GLP-1-associated improvement of several cardiovascular markers suggests that therapies with GLP-1R agonists may have a positive effect on cardiovascular risk factors in patients with diabetes. Nevertheless, their long-term safety profile and the direct clinical benefit to cardiovascular outcomes remain to be determined.

3.3. Dipeptidyl Peptidase 4 (DPP4) Inhibitors

DPP4 is a widely expressed membrane serine exopeptidase involved in degradation of various oligopeptides, including GLP-1[146]. DPP4 is anchored to the cell membrane but, under certain circumstances, may be released in soluble form. Endothelial cells are the main source of soluble DPP4 form, whose expression and enzymatic activity may increase after chronic exposure to high glucose concentrations [147].

Currently approved DPP4 inhibitors include sitagliptin, saxagliptin, linagliptin and vildagliptin. These small molecular-weight substances inhibit more than 90% of DPP4 activity and can be orally administered. Several studies in animal models support the evidence of DPP4 inhibition in improving endothelial function and blood pressure [148-150]. In isolated aorta rings incubated with DPP4-inhibitor, the relaxant effect of DPP4-inhibitor is GLP-1 independent and results from Akt posphorylation and eNOS activation with a rapid increase in NO levels [148]. Saxagliptin treatment has been shown to reduce blood pressure levels in the spontaneously hypertensive rats with a concomitant increase of aortic and glomerular NO release and comparable reductions in peroxynitrite levels [151]. There is a good evidence that DPP4 inhibition mediates protective effect on myocardial infarction, hypertension and atherosclerosis. Pharmacological treatment with sitagliptin has been able to enhance the expression of cardioprotective proteins and improve heart functional recovery after ischemia/reperfusion (I/R) injury [152].

In contrast to the positive effects in animal experiments, the consequences of treatment with DPP4 inhibitors on endothelial functions in humans have not always been consistent. In a double-blind study on patients with diabetes, treatment with vidagliptin for 4 weeks improved forearm blood flow in response to intra-arterially delivered acetylcholine [153]. On the contrary, other studies measuring FMD of the brachial artery have shown diametrically opposite results, suggesting that sitagliptin and alogliptin actually seem to worsen FMD when used to treat diabetic patients [154].

3.4. Sodium Glucose Co-transporter 2 (SGLT-2) Inhibitors

The isoform 2 of the sodium glucose co-transporter (SGLT-2) is located in the proximal convoluted tubule of the kidney and it reabsorbs approximately the 90% of filtered glucose. Inhibition of SGLT-2 represents a novel strategy for achieving glucose control in diabetic patients. It is based on a mechanism of action that targets the kidney to promote urinary glucose excretion and reduce hyperglycemia, and is therefore independent of pancreatic β cell function or the degree of insulin resistance [155, 156].

Among the SGLT-2 inhibitors, canagliflozin, dapagliflozin, and emapgliflozin are currently used. Other gliflozins include ertugliflozin and sotagliflozin and, in Japan, ipragliflozin, tofogliflozin and luseogliflozin. Most of clinical trials of SGLT-2 inhibitors show that patients receiving dapagliflozin, canagliflozin or empagliflozin as add-on therapy exhibit reduction in systolic blood pressure of approximately 3-5 mmHg versus placebo [157, 158]. One possible explanation is that urinary glucose excretion stimulated by inhibition of SGLT-2 causes a diuretic effect responsible for lowering blood pressure levels. However, a recent study on dapagliflozin proposed that SGLT-2 inhibitors may possess an additional diuretic-like capacity to lower blood pressure [159]. The improvements in blood pressure, associated to a moderate decrease in body weight [160] induced by treatment with SGLT-2 inhibitors suggest the potential for reducing the risk of cardiovascular events [161, 162]. Results from large cardiovascular trials underway for dapagliflozin, canagliflozin and empagliflozin will likely provide greater insights into the effects of SGLT-2 inhibition on cardiovascular outcomes. To date, SGLT-2 inhibitors are generally well tolerated, with a favorable safety profile similar to that of placebo. As easily expected, common adverse effects of SGLT-2 inhibitors include genital tract infections and osmotic diuresis.

4. NEW STRATEGIES FOR TREATMENT OF DIABETIC ENDOTHELIAL DYSFUNCTION

With increasing understanding of pathophysiology underlying diabetic endothelial dysfunction, the number of potential therapeutic targets for vascular complications has multiplied as well. In addition to treatments enhancing glycaemic, lipid and blood pressure control, emerging therapies targeting AGEs formation, PKCs activity, oxidative stress or inflammatory conditions may independently improve diabetic endothelial dysfunction. Although for several of these new molecules much remains to be learned before their possible use in clinical practice, they may nevertheless represent alternative approaches to treat or prevent vascular complications of diabetes.

4.1. AGE Inhibitors

Strategies to prevent AGE effects, aside from glycaemic control, include inhibition of AGE formation, blockade of AGE receptor activity, or breakage of the AGE–protein cross-links. In the last decade, numerous studies have been published about compounds with potential AGE-inhibitory effects and their implications for alternative strategies (extensively reviewed in [163-165]). Widespread evaluations were carried out on AGE-inhibitor aminoguanidine, the AGE cross-link breaker alagebrium (ALT-711), and on B vitamins (pyridoxamine, thiamine and its derivative benfotiamine). Aminoguanidine is a small nucleophilic compound that inhibits the formation of AGEs by interaction with and quenching of dicarbonyl compounds [166]. Although aminoguanidine attenuated the formation of diabetes-induced AGEs and the cross-linking of connective tissue proteins in the arterial wall *in vivo* [167], severe side effects such as vasculitis and abnormalities in liver function were reported in large multicentric trials [168, 169]. Owing to safety concerns resulting from its additional effects (*e.g.* pro-oxidant activities [170] and inhibition of NO synthase [171]), it is unlikely that aminoguanidine will be used for the treatment of diabetic complications [172]. Alagebrium chloride (ALT-711), which cleaves AGE and protein cross-links thereby facilitating AGE clearance, improved endothelial function in patients with isolated systolic hypertension [173] and vascular and left ventricular function in atherosclerotic patients with or without diabetes [174, 175], whereas did not show favorable effects in patients with chronic heart failure [176].

A variety of agents with beneficial effects similar to aminoguanidine and a better safety profile have been developed [177]. For example, pyridoxamine and pyridoxamine analogue ALT-946 may improve AGE-related complications through prevention of AGE-dependent oxidative damage [178]. Pyridoxamine treatment reduced the change from baseline creatinine in patients with type 1 or type 2 diabetes mellitus [179]. However, another placebo-controlled randomized clinical trial showed no effects on changes in creatinine [180]. Thus, the clinical evidence on the potential AGE-inhibiting effects of these B6 vitamins is still limited. Thiamine and benfotiamine, a highly bioavailable thiamine derivative, significantly reduced AGE levels in animal experiments [181, 182] and decreased the markers of endothelial dysfunction in diabetic patients [183]. At present, however, the clinical utility of AGE inhibitors remains to be firmly established. More large, specific and well-designed studies are needed to fully elucidate the potential of AGE-inhibiting effects.

4.2. PKC Inhibitors

Administration of PKC-β isoform inhibitors to healthy subjects prevents abnormal endothelium-dependent vasodilation caused by hyperglycemia [184]. Studies on PKC412, the first PKC inhibitor used in treatment of diabetic macular edema [185], have been abandoned due to incidence of hepatotoxic effects. Ruboxistaurin is a new selective PKC-β inhibitor that improved retinal blood flow distribution and decreased diabetic macular edema without significant adverse effects [186, 187]. Furthermore, ruboxistaurin improved the symptoms of diabetic peripheral neuropathy in a randomized, placebo-controlled, double-blind clinical trial [188]. Promising results were obtained in two combined phase III clinical trials about the ability of ruboxistaurin to reduce visual loss of 50% above standard care [189]. However, the experimental event rates were low and statistical significance was not achieved. Therefore, ruboxistaurin is currently pending FDA approval for the treatment of diabetic macular edema. More recently, ruboxistaurin has been proposed as a potential treatment for reducing atherosclerotic plaque in diabetic patients. Since PKC activation promotes endothelial dysfunction by de-regulating IL-18/IL-18BP pathway, leading to increased VCAM-1 expression, monocyte/ macrophage adhesion, and accelerated atherosclerotic plaque formation, inhibition of PKC by ruboxistaurin may represent a potential new mechanism to ameliorate endothelial dysfunction in diabetic patients [190].

4.3. VEGF Inhibitors

VEGF up-regulation is associated with neovascularization in proliferative diabetic retinopathy as well as diabetic macular edema [191]. VEGF exerts its effect on the retina through tyrosine kinase receptors VEGFR-1 and VEGFR-2 [192, 193]. Strategies to inhibit the action of VEGF include neutralizing anti-VEGF antibodies, receptor antagonists, inhibitor of VEGF receptor function, and upstream inhibitors of VEGF regulators, such as PKC (see above). However, since VEGF is vital in processes such as angiogenesis in the myocardium and wound healing [194], systemic therapies against VEGF are not feasible.

Intraocular administration of VEGF inhibitors may avoid their adverse systemic effects. Several drugs have been approved for intraocular treatment of diabetic retinopathy [191, 194] and are currently used in clinical settings: ranibizumab, bevacizumab, pegaptanib, and aflibercept. Ranibizumab is a recombinant antigen-binding Fab fragment of humanized anti-VEGF monoclonal antibody with a high ability to penetrate through the retina. Based on pivotal studies (RESOLVE and RESTORE), since 2011 ranibizumab obtained marketing authorization in several countries for the treatment of neovascular age-related macular degeneration (AMD), and visual impairment in diabetic macular edema (DME) [195, 196]. According to clinical trials, ranibizumab is safe and well tolerated in DME patients. At present, ranibizumab is the only compound approved for the treatment of visual loss due to DME. Extended pharmacovigilance studies, however, are required to confirm the long-term ocular and systemic safety of ranibizumab treatment in patients with DME [197, 198]. Bevacizumab, initially developed for intravenous treatment of metastatic colorectal cancer, has been adapted for off-label use in an intraocular administration [199, 200]. Multiple case reports and studies have suggested that intravitreal administration of bevacizumab is safe and effective in decreasing retinal, disc and iris neovascularization in diabetic patients with proliferative retinopathy and macular edema [201-204]. Some investigators have used bevacizumab as an adjunctive to vitreoctomy surgery in patients with vitreous hemorrhage from proliferative retinopathy [205]. Debates remained in the past years on whether bevacizumab is superior to ranibizumab in terms of potency, efficacy and safety (reviewed in [206]). The smaller size of ranibizumab, which makes it easier for this drug to permeate the retina and inhibit abnormal vessel growth, might suggest a better short-term efficacy of ranibizumab with respect to bevacizumab [207]. On the other hand, bevacizumab has been found more efficacious in the long-term. In part, this may reflect the reduced clearance by virtue of its larger size, with subsequent high accumulation in eye tissues. Results from the Diabetic Retinopathy Clinical Research (DRCR) phase II and Bevacizumab Or Laser Therapy (BOLT) study showed favorable effects of intravitreal bevacizumab administration for the treatment of diabetic ocular neovascularization [208, 209].

Two other anti-VEGF compounds are approved only for neovascular AMD but not for DME: pegaptanib and aflibercept. Pegaptanib is a ribonucleic acid aptamer that blocks specifically the activity of the extracellular VEGF. The effectiveness of pegaptanib was not as ideal as ranibizumab, and visual decline can still be observed in the AMD patients after treatment [210]. The efficacy of pegaptanib was also studied in subjects with DME associated with visual loss [211].

Aflibercept is a recombinant fusion protein with high VEGF affinity, attibuted to the binding sequences from the receptors VEGFR-1 and VEGFR-2 [212]. Recently aflibercept has obtained the regulatory approval from the FDA for the treatment of neovascular AMD administered by intravitreal injection. This treatment in exudative AMD patients was not inferior to ranibizumab. Aflibercept has firmly joined ranibizumab and bevacizumab as an important therapeutic option in the management of neovascular AMD. More recently, aflibercept appears to be contending with ranibizumab as an important therapeutic option in the management of DME [213].

4.4. Antoxidants

Antioxidants include a heterogeneous group of compounds which may potentially improve endothelial dysfunction in diabetes by re-coupling eNOS and mitochondrial function, increasing activity of superoxide scavenging enzymes, or decreasing vascular NAD(P)H oxidase activity [214, 215]. Novel antioxidant therapies aimed to restore production of endothelial-derived nitric oxide and to act as endogenous antioxidant enzymes may represent a more targeted strategy focusing specifically on the mechanisms implicated in diabetes-induced endothelial dysfunction (reviewed in [216].

4.4.1. Vitamin C and Vitamin E

Despite high expectation, administration of traditional antioxidants such as ascorbic acid (vitamin C) or tocopherol (vitamin E) have provided disappointing outcomes in clinical studies. Oral treatments with vitamin C or vitamin E reduced postprandial endothelial dysfunction and increased forearm vasodilation in diabetic patients [217-219], but other studies have not confirmed these effects [205, 220-222]. Surprisingly, an increased risk of cardiovascular events has been associated with high intakes of vitamin C and vitamin E [223-225]. In part, these frustrating results may be explained by the non-selective scavenging properties of vitamin C and E, which probably interfere with physiological important signaling mediated by ROS. Recent experimental studies have ascribed the protective effects of vitamin E to reduced oxidative stress and apoptosis in experimental diabetic cardiomyopathy [226] and to decreased Ox-LDL mediated oxidative stress and vascular muscle cell proliferation in aortic wall [227]. Clinical studies suggest that a combination of vitamin C and insulin [228] or a simultaneous infusion of GLP-1 and vitamin C [229] may help to normalize endothelial dysfunction and reduce both oxidative stress and inflammation in type 1 diabetic patients. On the other hand, clinical trials investigating the antioxidant potential of vitamin C and E in type 2 diabetic complications are still controversial [230-233].

4.4.2. Polyphenols

Polyphenols contained in fruits, vegetables, and beverages such as tea and red wine, are well known for their antioxidant properties [234, 235]. Results from several experimental studies suggest that these compounds may improve endothelial function by multiple mechanisms, but clinical studies addressing their potential benefits are still awaited. Recent studies indicate that resveratrol is able to reverse the effects of hyperglycaemia on mitochondrial function in endothelial cells, exerting protective actions in the early diabetes-associated endothelial dysfunction [236].

4.5. Other Therapeutic Approaches

Novel therapeutic strategies aimed to improve diabetic endothelial dysfunction involve PARP inhibitors, AMPK activators, and Rho-associated coiled-coil protein kinase (ROCK) inhibitors.

4.5.1. PARP Inhibitors

The overactivation of the nuclear enzyme poly (ADP ribose) polymerase 1 (PARP-1) is implicated in acute endothelial dysfunction of diabetic vasculature [237, 238]. Therefore, pharmacological inhibition of PARP-1 appears a potential strategy to approach diabetic vascular complications [239]. Experimental studies indicate a potential beneficial role of PARP-inhibition in diabetic retinopathy [240, 241] and in coronary arteriole dysfunction of db/db mice [242]. The therapeutic effects of PARP-1 inhibitors are currently evaluated in clinical studies as potential candidates in cancer or cardiovascular disease including cardiovascular complications of diabetes [243].

4.5.2. AMPK Activators

AMP-activated protein kinase (AMPK) is a serine/threonine protein kinase which plays a major role in regulating cellular and metabolic homeostasis, insulin sensitivity and mitochondrial function [244]. AMPK, activated in response to a variety of metabolic mediators, is known to regulate endothelial function and eNOS activity [11, 245] and may contribute to ameliorate vascular endothelial function by suppressing diabetes-enhanced degradation of GTP-cyclohydrolase

[246]. Decreased AMPK activity has been found associated to endothelial dysfunction, apoptosis and altered lipid metabolism in aortic endothelium of obese rats [247]. Various natural compounds such as resveratrol, berberine and α-lipoic acid are able to activate AMPK *in vivo* and *in vitro*, resulting in beneficial vascular effects [248]. Numerous pharmacological agents currently used for the treatment of diabetes including biguanides, thiazolidinediones and GLP-1R agonists are able to indirectly activate AMPK [249]. Other pharmacological AMPK activators such as A-769662 and 5-Aminoimidazole-4-carboxamide roboside (AICAR) already tested in animal studies, are unlikely to be used in patients with diabetes or metabolic syndrome due to poor bioavailability and a short half-life [248]. Among endogenous AMPK activators, increasing attention has been pointed out on adiponectin, an adipocytokine recognized for its insulin-sensitizing, anti-inflammatory, antioxidant, antiapoptotic, antiatherogenic, vasodilatory, and antithrombotic activity [250]. AMPK-activated signaling is involved in cardiac and vascular protective effects of adiponectin [11, 251, 252], whose circulating levels are reduced in diabetic patients [253]. Numerous natural products such as fish oil, linoleic acid, green tea extract and pharmacological agents including statins, renin-angiotensin system blockers, PPAR-α agonists and PPAR-γ agonists own their beneficial effects, at least partially, to the ability to restore the endogenous production of adiponectin [250]. Thus, therapeutic strategies aimed to increase adiponectin levels are currently pursued to prevent diabetes-linked cardiovascular complications.

4.5.3. ROCK (Rho-Associated Kinase) Inhibitors

The RhoA/ROCK signaling pathway mediates vascular smooth muscle contraction, downregulates eNOS gene expression and reduces protein kinase B/Akt activation, therefore decreasing eNOS phosphorylation and catalytic activity [254]. Elevated levels of peroxynitrites have been recently associated to an increased RhoA activity, largely responsible for vascular dysfunction in experimental diabetes [255]. On the same line, diabetes-induced endothelial aortic dysfunction is improved in ROCK knockout mice [256]. ROCK inhibitors such as Y-27632 and fasudil have shown promising therapeutic advantages on cardiovascular diseases including atherosclerosis, pulmonary and systemic hypertension and chronic heart failure [257-259], as well as significant beneficial effects on diabetic endothelial dysfunction of retinal [260], coronary [261] and intrarenal arteries [262]. In addition, fasudil has been able to limit TNF-alpha-mediated ICAM-1 expression and eNOS dephosphorylation in diabetic microvasculature [263]. At present, fasudil is under clinical investigation for a potential therapeutical use in diabetic macular edema (NCT01823081) combined with bevacizumab. Ripasudil, a fasudil derivative used to treat glaucoma and ocular hypertension, is in phase II clinical study for the treatment of diabetic retinopathy [264]. Newer ROCK inhibitors with higher specificity among ROCK isoforms are currently under clinical investigation for potential therapeutical use in vascular complications associated to diabetes [265, 266].

CONCLUSION

Experimental and clinical studies have progressively increased our understanding on molecular mechanisms underlying the tight and reciprocal cross-talk between hemodynamic and metabolic regulation. In this chapter, we have described the large body of data associating endothelial dysfunction to the pathogenesis of diabetic vascular complications. Current evidence indicates that interventions aimed at simultaneously improving insulin-stimulated glucose and lipid metabolism represent important clinical options to delay the progression of cardiovascular complications in diabetic patients. Novel drugs targeting intracellular mechanisms underlying metabolic alteration are regarded as promising new strategies to improve vascular dysfunction in diabetes. For most of these compounds, clinical evidence of favorable effects remains to be clearly established. Nevertheless, it is undoubtable that the increasing awareness of the key role played by the endothelium has significantly broaden the spectrum of therapeutic opportunities to treat micro- and macrovascular complications of diabetes.

CONFLICT OF INTEREST

The authors confirm that this chapter contents have no conflict of interest.

ACKNOWLEDGEMENTS

This work was supported in part by a Research Award from MIUR (PRIN 2010YK7Z5K_008) to MM.

DISCLOSURE

This is an updated Chapter of the earlier review article [by M.A. Potenza *et al.*, "Endothelial dysfunction in diabetes: From mechanisms to therapeutic targets", published in 'Current Medicinal Chemistry', Volume 16, Issue 1, 2009, pp 94 to 112], with additional text and references.

REFERENCES

[1] Vlachopoulos C.; Xaplanteris P.; Aboyans V.; Brodmann M.; Cifkova R.; Cosentino F.; De Carlo M.; Gallino A.; Landmesser U.; Laurent S.; Lekakis J.; Mikhailidis D. P.; Naka K. K.; Protogerou A. D.; Rizzoni D.; Schmidt-Trucksass A.; Van Bortel L.; Weber T.; Yamashina A.; Zimlichman R.; Boutouyrie P.; Cockcroft J.; O'Rourke M.; Park J. B.; Schillaci G.; Sillesen H.; Townsend R. R. The role of vascular biomarkers for primary and secondary prevention. A position paper from the European Society of Cardiology Working Group on peripheral circulation: Endorsed by the Association for Research into Arterial Structure and Physiology (ARTERY) Society. *Atherosclerosis,* **2015**, *241*, 507-532.

[2] Furchgott R. F.; Vanhoutte P. M. Endothelium-derived relaxing and contracting factors. *Faseb J.,* **1989**, *3*, 2007-2018.

[3] Potenza M. A.; Addabbo F.; Montagnani M. Vascular actions of insulin with implications for endothelial dysfunction. *Am. J. Physiol. Endocrinol. Metab.,* **2009**, *297*, E568-577.

[4] Furchgott R. F.; Zawadzki J. V. The obligatory role of endothelial cells in the relaxation of arterial smooth muscle by acetylcholine. *Nature,* **1980**, *288*, 373-376.

[5] Dimmeler S.; Fleming I.; Fisslthaler B.; Hermann C.; Busse R.; Zeiher A. M. Activation of nitric oxide synthase in endothelial cells by Akt- dependent phosphorylation. *Nature,* **1999**, *399*, 601-605.

[6] Fulton D.; Gratton J. P.; McCabe T. J.; Fontana J.; Fujio Y.; Walsh K.; Franke T. F.; Papapetropoulos A.; Sessa W. C. Regulation of endothelium-derived nitric oxide production by the protein kinase Akt. *Nature,* **1999**, *399*, 597-601.

[7] Govers R.; Rabelink T. J. Cellular regulation of endothelial nitric oxide synthase. *Am. J. Physiol. Renal. Physiol.,* **2001**, *280*, F193-206.

[8] Zeng G.; Quon M. J. Insulin-stimulated production of nitric oxide is inhibited by wortmannin. Direct measurement in vascular endothelial cells. *J. Clin. Invest.,* **1996**, *98*, 894-898.

[9] Montagnani M.; Chen H.; Barr V. A.; Quon M. J. Insulin-stimulated activation of eNOS is Independent of Ca++ but Requires phosphorylation by Akt at Ser1179. *J. Biol. Chem.,* **2001**, *276*, 30392-30398.

[10] Formoso G.; Chen H.; Kim J. A.; Montagnani M.; Consoli A.; Quon M. J. Dehydroepiandrosterone Mimics Acute Actions of Insulin to Stimulate Production of Both Nitric Oxide and Endothelin 1 via Distinct Phosphatidylinositol 3-Kinase- and Mitogen-Activated Protein Kinase-Dependent Pathways in Vascular Endothelium. *Mol. Endocrinol.,* **2006**, *20*, 1153-1163.

[11] Chen H.; Montagnani M.; Funahashi T.; Shimomura I.; Quon M. J. Adiponectin stimulates production of nitric oxide in vascular endothelial cells. *J. Biol. Chem.,* **2003**, *278*, 45021-45026.

[12] Kadowaki T.; Yamauchi T. Adiponectin and adiponectin receptors. *Endocr. Rev.,* **2005**, *26*, 439-451.

[13] Kadowaki T.; Yamauchi T.; Kubota N.; Hara K.; Ueki K.; Tobe K. Adiponectin and adiponectin receptors in insulin resistance, diabetes, and the metabolic syndrome. *J. Clin. Invest.,* **2006**, *116*, 1784-1792.

[14] Munzel T.; Feil R.; Mulsch A.; Lohmann S. M.; Hofmann F.; Walter U. Physiology and pathophysiology of vascular signaling controlled by guanosine 3',5'-cyclic monophosphate-dependent protein kinase [corrected]. *Circulation,* **2003**, *108*, 2172-2183.

[15] Landry D. W.; Oliver J. A. The pathogenesis of vasodilatory shock. *N. Engl. J. Med.,* **2001**, *345*, 588-595.

[16] Schafer A.; Wiesmann F.; Neubauer S.; Eigenthaler M.; Bauersachs J.; Channon K. M. Rapid regulation of platelet activation in vivo by nitric oxide. *Circulation,* **2004**, *109*, 1819-1822.

[17] Sarkar R.; Meinberg E. G.; Stanley J. C.; Gordon D.; Webb R. C. Nitric oxide reversibly inhibits the migration of cultured vascular smooth muscle cells. *Circ. Res.,* **1996**, *78*, 225-230.

[18] Moncada S.; Erusalimsky J. D. Does nitric oxide modulate mitochondrial energy generation and apoptosis? *Nat. Rev. Mol. Cell. Biol.,* **2002**, *3*, 214-220.

[19] Luo S.; Lei H.; Qin H.; Xia Y. Molecular mechanisms of endothelial NO synthase uncoupling. *Curr. Pharm. Des.,* **2014**, *20*, 3548-3553.

[20] Matsumoto T.; Goulopoulou S.; Taguchi K.; Tostes R. C.; Kobayashi T. Constrictor prostanoids and uridine adenosine tetraphosphate: vascular mediators and therapeutic targets in hypertension and diabetes. *Br. J. Pharmacol.,* **2015**, *172*, 3980-4001.

[21] Davidge S. T. Prostaglandin H synthase and vascular function. *Circ. Res.*, **2001**, *89*, 650-660.
[22] Lever A. F.; Lyall F.; Morton J. J.; Folkow B. Angiotensin II, vascular structure and blood pressure. *Kidney Int. Suppl.*, **1992**, *37*, S51-55.
[23] Kerr S.; Brosnan M. J.; McIntyre M.; Reid J. L.; Dominiczak A. F.; Hamilton C. A. Superoxide anion production is increased in a model of genetic hypertension: role of the endothelium. *Hypertension*, **1999**, *33*, 1353-1358.
[24] Yanagisawa M.; Kurihara H.; Kimura S.; Tomobe Y.; Kobayashi M.; Mitsui Y.; Yazaki Y.; Goto K.; Masaki T. A novel potent vasoconstrictor peptide produced by vascular endothelial cells. *Nature*, **1988**, *332*, 411-415.
[25] Marasciulo F. L.; Montagnani M.; Potenza M. A. Endothelin-1: the yin and yang on vascular function. *Curr. Med. Chem.*, **2006**, *13*, 1655-1665.
[26] Schiffrin E. L. Endothelin and endothelin antagonists in hypertension. *J. Hypertens.*, **1998**, *16*, 1891-1895.
[27] Hirata Y.; Emori T.; Eguchi S.; Kanno K.; Imai T.; Ohta K.; Marumo F. Endothelin receptor subtype B mediates synthesis of nitric oxide by cultured bovine endothelial cells. *J. Clin. Invest.*, **1993**, *91*, 1367-1373.
[28] Malek A.; Izumo S. Physiological fluid shear stress causes downregulation of endothelin-1 mRNA in bovine aortic endothelium. *Am. J. Physiol.*, **1992**, *263*, C389-396.
[29] Barton M.; Shaw S.; d'Uscio L. V.; Moreau P.; Luscher T. F. Angiotensin II increases vascular and renal endothelin-1 and functional endothelin converting enzyme activity in vivo: role of ETA receptors for endothelin regulation. *Biochem. Biophys. Res. Commun.*, **1997**, *238*, 861-865.
[30] Matsuura A.; Yamochi W.; Hirata K.; Kawashima S.; Yokoyama M. Stimulatory interaction between vascular endothelial growth factor and endothelin-1 on each gene expression. *Hypertension*, **1998**, *32*, 89-95.
[31] Corder R.; Carrier M.; Khan N.; Klemm P.; Vane J. R. Cytokine regulation of endothelin-1 release from bovine aortic endothelial cells. *J. Cardiovasc. Pharmacol.*, **1995**, *26* (Suppl 3), S56-58.
[32] Quehenberger P.; Exner M.; Sunder-Plassmann R.; Ruzicka K.; Bieglmayer C.; Endler G.; Muellner C.; Speiser W.; Wagner O. Leptin induces endothelin-1 in endothelial cells in vitro. *Circ. Res.*, **2002**, *90*, 711-718.
[33] Lteif A. A.; Han K.; Mather K. J. Obesity, insulin resistance, and the metabolic syndrome: determinants of endothelial dysfunction in whites and blacks. *Circulation*, **2005**, *112*, 32-38.
[34] Kim J. A.; Koh K. K.; Quon M. J. The union of vascular and metabolic actions of insulin in sickness and in health. *Arterioscler. Thromb. Vasc. Biol.*, **2005**, *25*, 889-891.
[35] Mather K. J.; Lteif A.; Steinberg H. O.; Baron A. D. Interactions between endothelin and nitric oxide in the regulation of vascular tone in obesity and diabetes. *Diabetes*, **2004**, *53*, 2060-2066.
[36] Gimbrone M. A., Jr. Vascular endothelium: an integrator of pathophysiologic stimuli in atherosclerosis. *Am. J. Cardiol.*, **1995**, *75*, 67B-70B.
[37] Wilcox J. N.; Subramanian R. R.; Sundell C. L.; Tracey W. R.; Pollock J. S.; Harrison D. G.; Marsden P. A. Expression of multiple isoforms of nitric oxide synthase in normal and atherosclerotic vessels. *Arterioscler. Thromb. Vasc. Biol.*, **1997**, *17*, 2479-2488.
[38] Landmesser U.; Dikalov S.; Price S. R.; McCann L.; Fukai T.; Holland S. M.; Mitch W. E.; Harrison D. G. Oxidation of tetrahydrobiopterin leads to uncoupling of endothelial cell nitric oxide synthase in hypertension. *J. Clin. Invest.*, **2003**, *111*, 1201-1209.
[39] Addabbo F.; Montagnani M.; Goligorsky M. S. Mitochondria and reactive oxygen species. *Hypertension*, **2009**, *53*, 885-892.
[40] Sheetz M. J.; King G. L. Molecular understanding of hyperglycemia's adverse effects for diabetic complications. *Jama*, **2002**, *288*, 2579-2588.
[41] Boyle P. J. Diabetes mellitus and macrovascular disease: mechanisms and mediators. *Am. J. Med.*, **2007**, *120*, S12-17.
[42] Bajaj M.; Suraamornkul S.; Kashyap S.; Cusi K.; Mandarino L.; DeFronzo R. A. Sustained reduction in plasma free fatty acid concentration improves insulin action without altering plasma adipocytokine levels in subjects with strong family history of type 2 diabetes. *J. Clin. Endocrinol. Metab.*, **2004**, *89*, 4649-4655.
[43] Bajaj M.; Defronzo R. A. Metabolic and molecular basis of insulin resistance. *J. Nucl. Cardiol.*, **2003**, *10*, 311-323.

[44] DeFronzo R. A. Insulin resistance: a multifaceted syndrome responsible for NIDDM, obesity, hypertension, dyslipidaemia and atherosclerosis. *Neth. J. Med.*, **1997**, *50*, 191-197.

[45] Del Prato S.; Matsuda M.; Simonson D. C.; Groop L. C.; Sheehan P.; Leonetti F.; Bonadonna R. C.; DeFronzo R. A. Studies on the mass action effect of glucose in NIDDM and IDDM: evidence for glucose resistance. *Diabetologia,* **1997**, *40*, 687-697.

[46] Bergman R. N.; Finegood D. T.; Kahn S. E. The evolution of beta-cell dysfunction and insulin resistance in type 2 diabetes. *Eur. J. Clin. Invest.*, **2002**, *32* (Suppl 3), 35-45.

[47] DeFronzo R. A. Lilly lecture 1987. The triumvirate: beta-cell, muscle, liver. A collusion responsible for NIDDM. *Diabetes,* **1988**, *37*, 667-687.

[48] Kashyap S. R.; Defronzo R. A. The insulin resistance syndrome: physiological considerations. *Diab. Vasc. Dis. Res.*, **2007**, *4*, 13-19.

[49] Brownlee M. The pathobiology of diabetic complications: a unifying mechanism. *Diabetes,* **2005**, *54*, 1615-1625.

[50] Averill M. M.; Bornfeldt K. E. Lipids versus glucose in inflammation and the pathogenesis of macrovascular disease in diabetes. *Curr. Diab. Rep.*, **2009**, *9*, 18-25.

[51] DeFronzo R. A. Insulin resistance, lipotoxicity, type 2 diabetes and atherosclerosis: the missing links. The Claude Bernard Lecture 2009. *Diabetologia,* **2010**, *53*, 1270-1287.

[52] Montagnani M.; Quon M. J. Insulin action in vascular endothelium: potential mechanisms linking insulin resistance with hypertension. *Diabetes Obes. Metab.*, **2000**, *2*, 285-292.

[53] Muniyappa R.; Montagnani M.; Koh K. K.; Quon M. J. Cardiovascular actions of insulin. *Endocr. Rev.*, **2007**, *28*, 463-491.

[54] Baron A. D.; Steinberg H. O.; Chaker H.; Leaming R.; Johnson A.; Brechtel G. Insulin-mediated skeletal muscle vasodilation contributes to both insulin sensitivity and responsiveness in lean humans. *J. Clin. Invest.*, **1995**, *96*, 786-792.

[55] Zeng G.; Nystrom F. H.; Ravichandran L. V.; Cong L. N.; Kirby M.; Mostowski H.; Quon M. J. Roles for insulin receptor, PI3-kinase, and Akt in insulin-signaling pathways related to production of nitric oxide in human vascular endothelial cells. *Circulation,* **2000**, *101*, 1539-1545.

[56] Montagnani M.; Golovchenko I.; Kim I.; Koh G. Y.; Goalstone M. L.; Mundhekar A. N.; Johansen M.; Kucik D. F.; Quon M. J.; Draznin B. Inhibition of phosphatidylinositol 3-kinase enhances mitogenic actions of insulin in endothelial cells. *J. Biol. Chem.*, **2002**, *277*, 1794-1799.

[57] Potenza M. A.; Marasciulo F. L.; Chieppa D. M.; Brigiani G. S.; Formoso G.; Quon M. J.; Montagnani M. Insulin resistance in spontaneously hypertensive rats is associated with endothelial dysfunction characterized by imbalance between NO and ET-1 production. *Am. J. Physiol. Heart Circ. Physiol.*, **2005**, *289*, H813-822.

[58] Eringa E. C.; Stehouwer C. D.; van Nieuw Amerongen G. P.; Ouwehand L.; Westerhof N.; Sipkema P. Vasoconstrictor effects of insulin in skeletal muscle arterioles are mediated by ERK1/2 activation in endothelium. *Am. J. Physiol. Heart Circ. Physiol.*, **2004**, *287*, H2043-2048.

[59] Cardillo C.; Nambi S. S.; Kilcoyne C. M.; Choucair W. K.; Katz A.; Quon M. J.; Panza J. A. Insulin stimulates both endothelin and nitric oxide activity in the human forearm. *Circulation,* **1999**, *100*, 820-825.

[60] Montagnani M.; Ravichandran L. V.; Chen H.; Esposito D. L.; Quon M. J. Insulin receptor substrate-1 and phosphoinositide-dependent kinase-1 are required for insulin-stimulated production of nitric oxide in endothelial cells. *Mol. Endocrinol.*, **2002**, *16*, 1931-1942.

[61] Kim J. A.; Montagnani M.; Koh K. K.; Quon M. J. Reciprocal relationships between insulin resistance and endothelial dysfunction: molecular and pathophysiological mechanisms. *Circulation,* **2006**, *113*, 1888-1904.

[62] King G. L. The role of hyperglycaemia and hyperinsulinaemia in causing vascular dysfunction in diabetes. *Ann. Med.*, **1996**, *28*, 427-432.

[63] Li H.; Horke S.; Forstermann U. Vascular oxidative stress, nitric oxide and atherosclerosis. *Atherosclerosis,* **2014**, *237*, 208-219.

[64] Potenza M. A.; Gagliardi S.; Nacci C.; Carratu M. R.; Montagnani M. Endothelial dysfunction in diabetes: from mechanisms to therapeutic targets. *Curr. Med. Chem.*, **2009**, *16*, 94-112.

[65] Saint-Marc P.; Kozak L. P.; Ailhaud G.; Darimont C.; Negrel R. Angiotensin II as a trophic factor of white adipose tissue: stimulation of adipose cell formation. *Endocrinology,* **2001**, *142*, 487-492.

[66] Mohite A.; Chillar A.; So S. P.; Cervantes V.; Ruan K. H. Novel mechanism of the vascular protector prostacyclin: regulating microRNA expression. *Biochemistry,* **2011**, *50*, 1691-1699.

[67] Sansbury B. E.; Cummins T. D.; Tang Y.; Hellmann J.; Holden C. R.; Harbeson M. A.; Chen Y.; Patel R. P.; Spite M.; Bhatnagar A.; Hil B. G. Overexpression of endothelial nitric oxide synthase prevents diet-induced obesity and regulates adipocyte phenotype. *Circ. Res.,* **2012**, *111*, 1176-1189.

[68] Wilkes J. J.; Hevener A.; Olefsky J. Chronic endothelin-1 treatment leads to insulin resistance in vivo. *Diabetes,* **2003**, *52*, 1904-1909.

[69] Ishibashi K. I.; Imamura T.; Sharma P. M.; Huang J.; Ugi S.; Olefsky J. M. Chronic endothelin-1 treatment leads to heterologous desensitization of insulin signaling in 3T3-L1 adipocytes. *J. Clin. Invest.,* **2001**, *107*, 1193-1202.

[70] Hauner H.; Petruschke T.; Gries F. A. Endothelin-1 inhibits the adipose differentiation of cultured human adipocyte precursor cells. *Metabolism,* **1994**, *43*, 227-232.

[71] Ishida F.; Saeki K.; Saeki T.; Ishikawa K.; Ihara M.; Kamei T.; Yano M. Suppressive effects of the endothelin receptor (ETA) antagonist BQ-123 on ET-1-induced reduction of lipoprotein lipase activity in 3T3-L1 adipocytes. *Biochem. Pharmacol.,* **1992**, *44*, 1431-1436.

[72] Usui I.; Imamura T.; Babendure J. L.; Satoh H.; Lu J. C.; Hupfeld C. J.; Olefsky J. M. G protein-coupled receptor kinase 2 mediates endothelin-1-induced insulin resistance via the inhibition of both Galphaq/11 and insulin receptor substrate-1 pathways in 3T3-L1 adipocytes. *Mol. Endocrinol.,* **2005**, *19*, 2760-2768.

[73] Juan C. C.; Chang C. L.; Lai Y. H.; Ho L. T. Endothelin-1 induces lipolysis in 3T3-L1 adipocytes. *Am. J. Physiol. Endocrinol. Metab.,* **2005**, *288*, E1146-1152.

[74] Clarke K. J.; Zhong Q.; Schwartz D. D.; Coleman E. S.; Kemppainen R. J.; Judd R. L. Regulation of adiponectin secretion by endothelin-1. *Biochem. Biophys. Res. Commun.,* **2003**, *312*, 945-949.

[75] Bedi D.; Clarke K. J.; Dennis J. C.; Zhong Q.; Brunson B. L.; Morrison E. E.; Judd R. L. Endothelin-1 inhibits adiponectin secretion through a phosphatidylinositol 4,5-bisphosphate/actin-dependent mechanism. *Biochem. Biophys. Res. Commun.,* **2006**, *345*, 332-339.

[76] Nacci C.; Leo V.; De Benedictis L.; Carratu M. R.; Bartolomeo N.; Altomare M.; Giordano P.; Faienza M. F.; Montagnani M. Elevated endothelin-1 (ET-1) levels may contribute to hypoadiponectinemia in childhood obesity. *J. Clin. Endocrinol. Metab.,* **2013**, *98*, E683-693.

[77] Asahara T.; Murohara T.; Sullivan A.; Silver M.; van der Zee R.; Li T.; Witzenbichler B.; Schatteman G.; Isner J. M. Isolation of putative progenitor endothelial cells for angiogenesis. *Science,* **1997**, *275*, 964-967.

[78] Kim K. A.; Shin Y. J.; Kim J. H.; Lee H.; Noh S. Y.; Jang S. H.; Bae O. N. Dysfunction of endothelial progenitor cells under diabetic conditions and its underlying mechanisms. *Arch. Pharm. Res.,* **2012**, *35*, 223-234.

[79] Cubbon R. M.; Rajwani A.; Wheatcroft S. B. The impact of insulin resistance on endothelial function, progenitor cells and repair. *Diab. Vasc. Dis. Res.,* **2007**, *4*, 103-111.

[80] Menegazzo L.; Albiero M.; Avogaro A.; Fadini G. P. Endothelial progenitor cells in diabetes mellitus. *Biofactors,* **2012**, *38*, 194-202.

[81] Tepper O. M.; Galiano R. D.; Capla J. M.; Kalka C.; Gagne P. J.; Jacobowitz G. R.; Levine J. P.; Gurtner G. C. Human endothelial progenitor cells from type II diabetics exhibit impaired proliferation, adhesion, and incorporation into vascular structures. *Circulation,* **2002**, *106*, 2781-2786.

[82] Fadini G. P.; de Kreutzenberg S. V.; Coracina A.; Baesso I.; Agostini C.; Tiengo A.; Avogaro A. Circulating CD34+ cells, metabolic syndrome, and cardiovascular risk. *Eur. Heart. J.,* **2006**, *27*, 2247-2255.

[83] Segal M. S.; Shah R.; Afzal A.; Perrault C. M.; Chang K.; Schuler A.; Beem E.; Shaw L. C.; Li Calzi S.; Harrison J. K.; Tran-Son-Tay R.; Grant M. B. Nitric oxide cytoskeletal-induced alterations reverse the endothelial progenitor cell migratory defect associated with diabetes. *Diabetes,* **2006**, *55*, 102-109.

[84] Awad O.; Jiao C.; Ma N.; Dunnwald M.; Schatteman G. C. Obese diabetic mouse environment differentially affects primitive and monocytic endothelial cell progenitors. *Stem Cells,* **2005**, *23*, 575-583.

[85] Schmidt-Lucke C.; Rossig L.; Fichtlscherer S.; Vasa M.; Britten M.; Kamper U.; Dimmeler S.; Zeiher A. M. Reduced number of circulating endothelial progenitor cells predicts future cardiovascular

events: proof of concept for the clinical importance of endogenous vascular repair. *Circulation,* **2005**, *111*, 2981-2987.

[86] Madonna R.; De Caterina R. Circulating endothelial progenitor cells: Do they live up to their name? *Vascul. Pharmacol.*, **2015**, *67-69*, 2-5.

[87] Muris D. M.; Houben A. J.; Schram M. T.; Stehouwer C. D. Microvascular dysfunction is associated with a higher incidence of type 2 diabetes mellitus: a systematic review and meta-analysis. *Arterioscler. Thromb. Vasc. Biol.*, **2012**, *32*, 3082-3094.

[88] Ryden L.; Grant P. J.; Anker S. D.; Berne C.; Cosentino F.; Danchin N.; Deaton C.; Escaned J.; Hammes H. P.; Huikuri H.; Marre M.; Marx N.; Mellbin L.; Ostergren J.; Patrono C.; Seferovic P.; Uva M. S.; Taskinen M. R.; Tendera M.; Tuomilehto J.; Valensi P.; Zamorano J. L.; Achenbach S.; Baumgartner H.; Bax J. J.; Bueno H.; Dean V.; Erol C.; Fagard R.; Ferrari R.; Hasdai D.; Hoes A. W.; Kirchhof P.; Knuuti J.; Kolh P.; Lancellotti P.; Linhart A.; Nihoyannopoulos P.; Piepoli M. F.; Ponikowski P.; Sirnes P. A.; Tamargo J. L.; Torbicki A.; Wijns W.; Windecker S.; De Backer G.; Ezquerra E. A.; Avogaro A.; Badimon L.; Baranova E.; Betteridge J.; Ceriello A.; Funck-Brentano C.; Gulba D. C.; Kjekshus J. K.; Lev E.; Mueller C.; Neyses L.; Nilsson P. M.; Perk J.; Reiner Z.; Sattar N.; Schachinger V.; Scheen A.; Schirmer H.; Stromberg A.; Sudzhaeva S.; Viigimaa M.; Vlachopoulos C.; Xuereb R. G. ESC Guidelines on diabetes, pre-diabetes, and cardiovascular diseases developed in collaboration with the EASD: the Task Force on diabetes, pre-diabetes, and cardiovascular diseases of the European Society of Cardiology (ESC) and developed in collaboration with the European Association for the Study of Diabetes (EASD). *Eur. Heart J.*, **2013**, *34*, 3035-3087.

[89] Izzo R.; de Simone G.; Trimarco V.; Gerdts E.; Giudice R.; Vaccaro O.; De Luca N.; Trimarco B. Hypertensive target organ damage predicts incident diabetes mellitus. *Eur. Heart J.*, **2013**, *34*, 3419-3426.

[90] Lekakis J.; Abraham P.; Balbarini A.; Blann A.; Boulanger C. M.; Cockcroft J.; Cosentino F.; Deanfield J.; Gallino A.; Ikonomidis I.; Kremastinos D.; Landmesser U.; Protogerou A.; Stefanadis C.; Tousoulis D.; Vassalli G.; Vink H.; Werner N.; Wilkinson I.; Vlachopoulos C. Methods for evaluating endothelial function: a position statement from the European Society of Cardiology Working Group on Peripheral Circulation. *Eur. J. Cardiovasc. Prev. Rehabil.*, **2011**, *18*, 775-789.

[91] Celermajer D. S.; Sorensen K. E.; Gooch V. M.; Spiegelhalter D. J.; Miller O. I.; Sullivan I. D.; Lloyd J. K.; Deanfield J. E. Non-invasive detection of endothelial dysfunction in children and adults at risk of atherosclerosis. *Lancet,* **1992**, *340*, 1111-1115.

[92] Charakida M.; Masi S.; Luscher T. F.; Kastelein J. J.; Deanfield J. E. Assessment of atherosclerosis: the role of flow-mediated dilatation. *Eur. Heart J.*, **2010**, *31*, 2854-2861.

[93] Briet M.; Collin C.; Laurent S.; Tan A.; Azizi M.; Agharazii M.; Jeunemaitre X.; Alhenc-Gelas F.; Boutouyrie P. Endothelial function and chronic exposure to air pollution in normal male subjects. *Hypertension,* **2007**, *50*, 970-976.

[94] Modena M. G.; Bonetti L.; Coppi F.; Bursi F.; Rossi R. Prognostic role of reversible endothelial dysfunction in hypertensive postmenopausal women. *J. Am. Coll. Cardiol.*, **2002**, *40*, 505-510.

[95] Kitta Y.; Obata J. E.; Nakamura T.; Hirano M.; Kodama Y.; Fujioka D.; Saito Y.; Kawabata K.; Sano K.; Kobayashi T.; Yano T.; Nakamura K.; Kugiyama K. Persistent impairment of endothelial vasomotor function has a negative impact on outcome in patients with coronary artery disease. *J. Am. Coll. Cardiol.*, **2009**, *53*, 323-330.

[96] Landmesser U.; von Eckardstein A.; Kastelein J.; Deanfield J.; Luscher T. F. Increasing high-density lipoprotein cholesterol by cholesteryl ester transfer protein-inhibition: a rocky road and lessons learned? The early demise of the dal-HEART programme. *Eur. Heart J.*, **2012**, *33*, 1712-1715.

[97] Cahn A.; Miccoli R.; Dardano A.; Del Prato S. New forms of insulin and insulin therapies for the treatment of type 2 diabetes. *Lancet Diabetes Endocrinol.*, **2015**, *3*(8), 638-652.

[98] Kawalec P.; Mikrut A.; Lopuch S. The safety of dipeptidyl peptidase-4 (DPP-4) inhibitors or sodium-glucose cotransporter 2 (SGLT-2) inhibitors added to metformin background therapy in patients with type 2 diabetes mellitus: a systematic review and meta-analysis. *Diabetes Metab. Res. Rev.*, **2014**, *30*, 269-283.

[99] Inzucchi S. E.; Zinman B.; Wanner C.; Ferrari R.; Fitchett D.; Hantel S.; Espadero R. M.; Woerle H. J.; Broedl U. C.; Johansen O. E. SGLT-2 inhibitors and cardiovascular risk: proposed pathways and review of ongoing outcome trials. *Diab. Vasc. Dis. Res.*, **2015**, *12*, 90-100.

[100] Hamilton S. J.; Chew G. T.; Watts G. F. Therapeutic regulation of endothelial dysfunction in type 2 diabetes mellitus. *Diab. Vasc. Dis. Res.*, **2007**, *4*, 89-102.
[101] Singh T. P.; Vangaveti V. N.; Malabu U. H. Dipeptidyl peptidase-4 inhibitors and their potential role in the management of atherosclerosis-A review. *Diabetes Metab. Syndr.*, **2015**, *9*(4), 223-229.
[102] Knop F. K.; Vilsboll T.; Holst J. J. Incretin-based therapy of type 2 diabetes mellitus. *Curr. Protein Pept. Sci.*, **2009**, *10*, 46-55.
[103] Cernea S.; Raz I. Therapy in the early stage: incretins. *Diabetes Care,* **2011**, *34* (Suppl 2), S264-271.
[104] Hauner H. The mode of action of thiazolidinediones. *Diabetes Metab. Res. Rev.*, **2002**, *18* (Suppl 2), S10-15.
[105] Sgarra L.; Addabbo F.; Potenza M. A.; Montagnani M. Determinants of evolving metabolic and cardiovascular benefit/risk profiles of rosiglitazone therapy during the natural history of diabetes: molecular mechanisms in the context of integrated pathophysiology. *Am. J. Physiol. Endocrinol. Metab.*, **2012**, *302*, E1171-1182.
[106] Kim Y. B.; Ciaraldi T. P.; Kong A.; Kim D.; Chu N.; Mohideen P.; Mudaliar S.; Henry R. R.; Kahn B. B. Troglitazone but not metformin restores insulin-stimulated phosphoinositide 3-kinase activity and increases p110beta protein levels in skeletal muscle of type 2 diabetic subjects. *Diabetes,* **2002**, *51*, 443-448.
[107] Stuhlinger M. C.; Abbasi F.; Chu J. W.; Lamendola C.; McLaughlin T. L.; Cooke J. P.; Reaven G. M.; Tsao P. S. Relationship between insulin resistance and an endogenous nitric oxide synthase inhibitor. *JAMA,* **2002**, *287*, 1420-1426.
[108] Ricote M.; Li A. C.; Willson T. M.; Kelly C. J.; Glass C. K. The peroxisome proliferator-activated receptor-gamma is a negative regulator of macrophage activation. *Nature,* **1998**, *391*, 79-82.
[109] Jiang C.; Ting A. T.; Seed B. PPAR-gamma agonists inhibit production of monocyte inflammatory cytokines. *Nature,* **1998**, *391*, 82-86.
[110] Bagi Z.; Koller A.; Kaley G. PPARgamma activation, by reducing oxidative stress, increases NO bioavailability in coronary arterioles of mice with Type 2 diabetes. *Am. J. Physiol. Heart Circ. Physiol.*, **2004**, *286*, H742-748.
[111] Ceolotto G.; Gallo A.; Papparella I.; Franco L.; Murphy E.; Iori E.; Pagnin E.; Fadini G. P.; Albiero M.; Semplicini A.; Avogaro A. Rosiglitazone reduces glucose-induced oxidative stress mediated by NAD(P)H oxidase via AMPK-dependent mechanism. *Arterioscler. Thromb. Vasc. Biol.*, **2007**, *27*, 2627-2633.
[112] Wakino S.; Law R. E.; Hsueh W. A. Vascular protective effects by activation of nuclear receptor PPARgamma. *J. Diabetes Complications,* **2002**, *16*, 46-49.
[113] Haffner S. M.; Greenberg A. S.; Weston W. M.; Chen H.; Williams K.; Freed M. I. Effect of rosiglitazone treatment on nontraditional markers of cardiovascular disease in patients with type 2 diabetes mellitus. *Circulation,* **2002**, *106*, 679-684.
[114] Sidhu J. S.; Kaposzta Z.; Markus H. S.; Kaski J. C. Effect of rosiglitazone on common carotid intima-media thickness progression in coronary artery disease patients without diabetes mellitus. *Arterioscler. Thromb. Vasc. Biol.*, **2004**, *24*, 930-934.
[115] Bailey C. J. Treating insulin resistance in type 2 diabetes with metformin and thiazolidinediones. *Diabetes Obes. Metab.*, **2005**, *7*, 675-691.
[116] Raji A.; Seely E. W.; Bekins S. A.; Williams G. H.; Simonson D. C. Rosiglitazone improves insulin sensitivity and lowers blood pressure in hypertensive patients. *Diabetes Care,* **2003**, *26*, 172-178.
[117] Nissen S. E.; Wolski K. Effect of rosiglitazone on the risk of myocardial infarction and death from cardiovascular causes. *N. Engl. J. Med.*, **2007**, *356*, 2457-2471.
[118] Psaty B. M.; Furberg C. D. The record on rosiglitazone and the risk of myocardial infarction. *N. Engl. J. Med.*, **2007**, *357*, 67-69.
[119] Diamond G. A.; Bax L.; Kaul S. Uncertain effects of rosiglitazone on the risk for myocardial infarction and cardiovascular death. *Ann. Intern. Med.*, **2007**, *147*, 578-581.
[120] Home P. D.; Pocock S. J.; Beck-Nielsen H.; Gomis R.; Hanefeld M.; Jones N. P.; Komajda M.; McMurray J. J. Rosiglitazone evaluated for cardiovascular outcomes--an interim analysis. *N. Engl. J. Med.*, **2007**, *357*, 28-38.
[121] Singh S.; Loke Y. K.; Furberg C. D. Long-term risk of cardiovascular events with rosiglitazone: a meta-analysis. *JAMA,* **2007**, *298*, 1189-1195.

[122] Home P. D.; Pocock S. J.; Beck-Nielsen H.; Curtis P. S.; Gomis R.; Hanefeld M.; Jones N. P.; Komajda M.; McMurray J. J. Rosiglitazone evaluated for cardiovascular outcomes in oral agent combination therapy for type 2 diabetes (RECORD): a multicentre, randomised, open-label trial. *Lancet,* **2009**, *373*, 2125-2135.

[123] Nissen S. E.; Wolski K. Rosiglitazone revisited: an updated meta-analysis of risk for myocardial infarction and cardiovascular mortality. *Arch. Intern. Med.,* **2010**, *170*, 1191-1201.

[124] Graham D. J.; Ouellet-Hellstrom R.; MaCurdy T. E.; Ali F.; Sholley C.; Worrall C.; Kelman J. A. Risk of acute myocardial infarction, stroke, heart failure, and death in elderly Medicare patients treated with rosiglitazone or pioglitazone. *JAMA,* **2010**, *304*, 411-418.

[125] Kaul S.; Bolger A. F.; Herrington D.; Giugliano R. P.; Eckel R. H. Thiazolidinedione drugs and cardiovascular risks: a science advisory from the American Heart Association and American College Of Cardiology Foundation. *J. Am. Coll .Cardiol.,* **2010**, *55*, 1885-1894.

[126] Rosen C. J. The rosiglitazone story--lessons from an FDA Advisory Committee meeting. *N. Engl. J. Med.,* **2007**, *357*, 844-846.

[127] Rosen C. J. Revisiting the rosiglitazone story--lessons learned. *N. Engl. J. Med.,* **2010**, *363*, 803-806.

[128] Food and Drug Administration. FDA adds boxed warning for heart-related risk to antidiabetes drug Avandia [press release]. Available at: http://www.fda.gov/newsevents/newsroom/pressannouncements/2007/ucm109026.htm Nov 14, **2007**.

[129] Nauck M. A.; Homberger E.; Siegel E. G.; Allen R. C.; Eaton R. P.; Ebert R.; Creutzfeldt W. Incretin effects of increasing glucose loads in man calculated from venous insulin and C-peptide responses. *J. Clin. Endocrinol. Metab.,* **1986**, *63*, 492-498.

[130] Cho Y. M.; Fujita Y.; Kieffer T. J. Glucagon-like peptide-1: glucose homeostasis and beyond. *Annu. Rev. Physiol.,* **2013**, *76*, 535-559.

[131] Nauck M.; Stockmann F.; Ebert R.; Creutzfeldt W. Reduced incretin effect in type 2 (non-insulin-dependent) diabetes. *Diabetologia,* **1986**, *29*, 46-52.

[132] Cho Y. M.; Wideman R. D.; Kieffer T. J. Clinical application of glucagon-like Peptide 1 receptor agonists for the treatment of type 2 diabetes mellitus. *Endocrinol. Metab. (Seoul),* **2013**, *28*, 262-274.

[133] Zhong J.; Gong Q.; Goud A.; Srinivasamaharaj S.; Rajagopalan S. Recent Advances in Dipeptidyl-Peptidase-4 Inhibition Therapy: Lessons from the Bench and Clinical Trials. *J. Diabetes Res.,* **2015**, *2015*, 606031.

[134] Ussher J. R.; Drucker D. J. Cardiovascular biology of the incretin system. *Endocr. Rev.,* **2012**, *33*, 187-215.

[135] Nikolaidis L. A.; Mankad S.; Sokos G. G.; Miske G.; Shah A.; Elahi D.; Shannon R. P. Effects of glucagon-like peptide-1 in patients with acute myocardial infarction and left ventricular dysfunction after successful reperfusion. *Circulation,* **2004**, *109*, 962-965.

[136] Nystrom T.; Gutniak M. K.; Zhang Q.; Zhang F.; Holst J. J.; Ahren B.; Sjoholm A. Effects of glucagon-like peptide-1 on endothelial function in type 2 diabetes patients with stable coronary artery disease. *Am. J. Physiol. Endocrinol. Metab.,* **2004**, *287*, E1209-1215.

[137] Gutzwiller J. P.; Tschopp S.; Bock A.; Zehnder C. E.; Huber A. R.; Kreyenbuehl M.; Gutmann H.; Drewe J.; Henzen C.; Goeke B.; Beglinger C. Glucagon-like peptide 1 induces natriuresis in healthy subjects and in insulin-resistant obese men. *J. Clin. Endocrinol. Metab.,* **2004**, *89*, 3055-3061.

[138] Hattori Y.; Jojima T.; Tomizawa A.; Satoh H.; Hattori S.; Kasai K.; Hayashi T. A glucagon-like peptide-1 (GLP-1) analogue, liraglutide, upregulates nitric oxide production and exerts anti-inflammatory action in endothelial cells. *Diabetologia,* **2010**, *53*, 2256-2263.

[139] Erdogdu O.; Nathanson D.; Sjoholm A.; Nystrom T.; Zhang Q. Exendin-4 stimulates proliferation of human coronary artery endothelial cells through eNOS-, PKA- and PI3K/Akt-dependent pathways and requires GLP-1 receptor. *Mol. Cell. Endocrinol.,* **2010**, *325*, 26-35.

[140] Goto H.; Nomiyama T.; Mita T.; Yasunari E.; Azuma K.; Komiya K.; Arakawa M.; Jin W. L.; Kanazawa A.; Kawamori R.; Fujitani Y.; Hirose T.; Watada H. Exendin-4, a glucagon-like peptide-1 receptor agonist, reduces intimal thickening after vascular injury. *Biochem. Biophys. Res. Commun.,* **2011**, *405*, 79-84.

[141] Nagashima M.; Watanabe T.; Terasaki M.; Tomoyasu M.; Nohtomi K.; Kim-Kaneyama J.; Miyazaki A.; Hirano T. Native incretins prevent the development of atherosclerotic lesions in apolipoprotein E knockout mice. *Diabetologia,* **2011**, *54*, 2649-2659.

[142] Hirata K.; Kume S.; Araki S.; Sakaguchi M.; Chin-Kanasaki M.; Isshiki K.; Sugimoto T.; Nishiyama A.; Koya D.; Haneda M.; Kashiwagi A.; Uzu T. Exendin-4 has an anti-hypertensive effect in salt-sensitive mice model. *Biochem. Biophys. Res. Commun.*, **2009**, *380*, 44-49.

[143] Horton E. S.; Silberman C.; Davis K. L.; Berria R. Weight loss, glycemic control, and changes in cardiovascular biomarkers in patients with type 2 diabetes receiving incretin therapies or insulin in a large cohort database. *Diabetes Care,* **2010**, *33*, 1759-1765.

[144] Buse J. B.; Drucker D. J.; Taylor K. L.; Kim T.; Walsh B.; Hu H.; Wilhelm K.; Trautmann M.; Shen L. Z.; Porter L. E. DURATION-1: exenatide once weekly produces sustained glycemic control and weight loss over 52 weeks. *Diabetes Care,* **2010**, *33*, 1255-1261.

[145] Grimm M.; Han J.; Weaver C.; Griffin P.; Schulteis C. T.; Dong H.; Malloy J. Efficacy, safety, and tolerability of exenatide once weekly in patients with type 2 diabetes mellitus: an integrated analysis of the DURATION trials. *Postgrad. Med.*, **2013**, *125*, 47-57.

[146] Augustyns K.; Bal G.; Thonus G.; Belyaev A.; Zhang X. M.; Bollaert W.; Lambeir A. M.; Durinx C.; Goossens F.; Haemers A. The unique properties of dipeptidyl-peptidase IV (DPP IV / CD26) and the therapeutic potential of DPP IV inhibitors. *Curr. Med. Chem.*, **1999**, *6*, 311-327.

[147] Pala L.; Pezzatini A.; Dicembrini I.; Ciani S.; Gelmini S.; Vannelli B. G.; Cresci B.; Mannucci E.; Rotella C. M. Different modulation of dipeptidyl peptidase-4 activity between microvascular and macrovascular human endothelial cells. *Acta Diabetol.*, **2012**, *49* (Suppl 1), S59-63.

[148] Shah Z.; Pineda C.; Kampfrath T.; Maiseyeu A.; Ying Z.; Racoma I.; Deiuliis J.; Xu X.; Sun Q.; Moffatt-Bruce S.; Villamena F.; Rajagopalan S. Acute DPP-4 inhibition modulates vascular tone through GLP-1 independent pathways. *Vascul. Pharmacol.*, **2011**, *55*, 2-9.

[149] Matsubara J.; Sugiyama S.; Sugamura K.; Nakamura T.; Fujiwara Y.; Akiyama E.; Kurokawa H.; Nozaki T.; Ohba K.; Konishi M.; Maeda H.; Izumiya Y.; Kaikita K.; Sumida H.; Jinnouchi H.; Matsui K.; Kim-Mitsuyama S.; Takeya M.; Ogawa H. A dipeptidyl peptidase-4 inhibitor, des-fluoro-sitagliptin, improves endothelial function and reduces atherosclerotic lesion formation in apolipoprotein E-deficient mice. *J. Am. Coll. Cardiol.*, **2012**, *59*, 265-276.

[150] Aroor A. R.; Sowers J. R.; Bender S. B.; Nistala R.; Garro M.; Mugerfeld I.; Hayden M. R.; Johnson M. S.; Salam M.; Whaley-Connell A.; Demarco V. G. Dipeptidylpeptidase inhibition is associated with improvement in blood pressure and diastolic function in insulin-resistant male Zucker obese rats. *Endocrinology,* **2013**, *154*, 2501-2513.

[151] Mason R. P.; Jacob R. F.; Kubant R.; Ciszewski A.; Corbalan J. J.; Malinski T. Dipeptidyl peptidase-4 inhibition with saxagliptin enhanced nitric oxide release and reduced blood pressure and sICAM-1 levels in hypertensive rats. *J. Cardiovasc. Pharmacol.*, **2012**, *60*, 467-473.

[152] Sauve M.; Ban K.; Momen M. A.; Zhou Y. Q.; Henkelman R. M.; Husain M.; Drucker D. J. Genetic deletion or pharmacological inhibition of dipeptidyl peptidase-4 improves cardiovascular outcomes after myocardial infarction in mice. *Diabetes,* **2010**, *59*, 1063-1073.

[153] van Poppel P. C.; Netea M. G.; Smits P.; Tack C. J. Vildagliptin improves endothelium-dependent vasodilatation in type 2 diabetes. *Diabetes Care,* **2011**, *34*, 2072-2077.

[154] Ayaori M.; Iwakami N.; Uto-Kondo H.; Sato H.; Sasaki M.; Komatsu T.; Iizuka M.; Takiguchi S.; Yakushiji E.; Nakaya K.; Yogo M.; Ogura M.; Takase B.; Murakami T.; Ikewaki K. Dipeptidyl peptidase-4 inhibitors attenuate endothelial function as evaluated by flow-mediated vasodilatation in type 2 diabetic patients. *J. Am. Heart Assoc.*, **2013**, *2*, e003277.

[155] Abdul-Ghani M. A.; DeFronzo R. A. Inhibition of renal glucose reabsorption: a novel strategy for achieving glucose control in type 2 diabetes mellitus. *Endocr. Pract.*, **2008**, *14*, 782-790.

[156] Chao E. C. SGLT-2 Inhibitors: A New Mechanism for Glycemic Control. *Clinical Diabetes,* **2014**, *32*, 4-11.

[157] Zhang M.; Zhang L.; Wu B.; Song H.; An Z.; Li S. Dapagliflozin treatment for type 2 diabetes: a systematic review and meta-analysis of randomized controlled trials. *Diabetes Metab. Res. Rev.*, **2013**, *30*, 204-221.

[158] Kovacs C. S.; Seshiah V.; Merker L.; Christiansen A. V.; Roux F.; Salsali A.; Kim G.; Stella P.; Woerle H. J.; Broedl U. C. Empagliflozin as Add-on Therapy to Pioglitazone With or Without Metformin in Patients With Type 2 Diabetes Mellitus. *Clin. Ther.*, **2015**.

[159] Lambers Heerspink H. J.; de Zeeuw D.; Wie L.; Leslie B.; List J. Dapagliflozin a glucose-regulating drug with diuretic properties in subjects with type 2 diabetes. *Diabetes Obes. Metab.*, **2013**, *15*, 853-862.

[160] Cefalu W. T.; Leiter L. A.; Yoon K. H.; Arias P.; Niskanen L.; Xie J.; Balis D. A.; Canovatchel W.; Meininger G. Efficacy and safety of canagliflozin versus glimepiride in patients with type 2 diabetes inadequately controlled with metformin (CANTATA-SU): 52 week results from a randomised, double-blind, phase 3 non-inferiority trial. *Lancet,* **2013**, *382*, 941-950.

[161] Ptaszynska A.; Hardy E.; Johnsson E.; Parikh S.; List J. Effects of dapagliflozin on cardiovascular risk factors. *Postgrad. Med.,* **2013**, *125*, 181-189.

[162] Dziuba J.; Alperin P.; Racketa J.; Iloeje U.; Goswami D.; Hardy E.; Perlstein I.; Grossman H. L.; Cohen M. Modeling effects of SGLT-2 inhibitor dapagliflozin treatment versus standard diabetes therapy on cardiovascular and microvascular outcomes. *Diabetes Obes. Metab.,* **2014**, *16*, 628-635.

[163] Sourris K. C.; Harcourt B. E.; Forbes J. M. A new perspective on therapeutic inhibition of advanced glycation in diabetic microvascular complications: common downstream endpoints achieved through disparate therapeutic approaches? *Am. J. Nephrol.,* **2009**, *30*, 323-335.

[164] Schalkwijk C. G.; Miyata T. Early- and advanced non-enzymatic glycation in diabetic vascular complications: the search for therapeutics. *Amino Acids,* **2010**, *42*, 1193-1204.

[165] Nagai R.; Murray D. B.; Metz T. O.; Baynes J. W. Chelation: a fundamental mechanism of action of AGE inhibitors, AGE breakers, and other inhibitors of diabetes complications. *Diabetes,* **2012**, *61*, 549-559.

[166] Thornalley P. J. Use of aminoguanidine (Pimagedine) to prevent the formation of advanced glycation endproducts. *Arch. Biochem. Biophys.,* **2003**, *419*, 31-40.

[167] Brownlee M.; Vlassara H.; Kooney A.; Ulrich P.; Cerami A. Aminoguanidine prevents diabetes-induced arterial wall protein cross-linking. *Science,* **1986**, *232*, 1629-1632.

[168] Bolton W. K.; Cattran D. C.; Williams M. E.; Adler S. G.; Appel G. B.; Cartwright K.; Foiles P. G.; Freedman B. I.; Raskin P.; Ratner R. E.; Spinowitz B. S.; Whittier F. C.; Wuerth J. P. Randomized trial of an inhibitor of formation of advanced glycation end products in diabetic nephropathy. *Am. J .Nephrol.,* **2004**, *24*, 32-40.

[169] Freedman B. I.; Wuerth J. P.; Cartwright K.; Bain R. P.; Dippe S.; Hershon K.; Mooradian A. D.; Spinowitz B. S. Design and baseline characteristics for the aminoguanidine Clinical Trial in Overt Type 2 Diabetic Nephropathy (ACTION II). *Control Clin. Trials,* **1999**, *20*, 493-510.

[170] Suji G.; Sivakami S. DNA damage by free radical production by aminoguanidine. *Ann. N. Y. Acad. Sci.,* **2006**, *1067*, 191-199.

[171] Tilton R. G.; Chang K.; Hasan K. S.; Smith S. R.; Petrash J. M.; Misko T. P.; Moore W. M.; Currie M. G.; Corbett J. A.; McDaniel M. L.; et al. Prevention of diabetic vascular dysfunction by guanidines. Inhibition of nitric oxide synthase versus advanced glycation end-product formation. *Diabetes,* **1993**, *42*, 221-232.

[172] Turgut F.; Bolton W. K. Potential new therapeutic agents for diabetic kidney disease. *Am. J. Kidney Dis.,* **2010**, *55*, 928-940.

[173] Zieman S. J.; Melenovsky V.; Clattenburg L.; Corretti M. C.; Capriotti A.; Gerstenblith G.; Kass D. A. Advanced glycation endproduct crosslink breaker (alagebrium) improves endothelial function in patients with isolated systolic hypertension. *J. Hypertens.,* **2007**, *25*, 577-583.

[174] Kass D. A.; Shapiro E. P.; Kawaguchi M.; Capriotti A. R.; Scuteri A.; deGroof R. C.; Lakatta E. G. Improved arterial compliance by a novel advanced glycation end-product crosslink breaker. *Circulation,* **2001**, *104*, 1464-1470.

[175] Little W. C.; Zile M. R.; Kitzman D. W.; Hundley W. G.; O'Brien T. X.; Degroof R. C. The effect of alagebrium chloride (ALT-711), a novel glucose cross-link breaker, in the treatment of elderly patients with diastolic heart failure. *J. Card. Fail.,* **2005**, *11*, 191-195.

[176] Hartog J. W.; Willemsen S.; van Veldhuisen D. J.; Posma J. L.; van Wijk L. M.; Hummel Y. M.; Hillege H. L.; Voors A. A. Effects of alagebrium, an advanced glycation endproduct breaker, on exercise tolerance and cardiac function in patients with chronic heart failure. *Eur. J. Heart Fail.,* **2011**, *13*, 899-908.

[177] Montagnani M. Diabetic cardiomyopathy: how much does it depend on AGE? *Br. J. Pharmacol.,* **2008**, *154*, 725-726.

[178] Voziyan P. A.; Hudson B. G. Pyridoxamine: the many virtues of a maillard reaction inhibitor. *Ann. N. Y. Acad. Sci.,* **2005**, *1043*, 807-816.

[179] Williams M. E.; Bolton W. K.; Khalifah R. G.; Degenhardt T. P.; Schotzinger R. J.; McGill J. B. Effects of pyridoxamine in combined phase 2 studies of patients with type 1 and type 2 diabetes and overt nephropathy. *Am. J. Nephrol.,* **2007**, *27*, 605-614.

[180] Lewis E. J.; Greene T.; Spitalewiz S.; Blumenthal S.; Berl T.; Hunsicker L. G.; Pohl M. A.; Rohde R. D.; Raz I.; Yerushalmy Y.; Yagil Y.; Herskovits T.; Atkins R. C.; Reutens A. T.; Packham D. K.; Lewis J. B. Pyridorin in type 2 diabetic nephropathy. *J. Am. Soc. Nephrol.,* **2011**, *23*, 131-136.

[181] Hammes H. P.; Du X.; Edelstein D.; Taguchi T.; Matsumura T.; Ju Q.; Lin J.; Bierhaus A.; Nawroth P.; Hannak D.; Neumaier M.; Bergfeld R.; Giardino I.; Brownlee M. Benfotiamine blocks three major pathways of hyperglycemic damage and prevents experimental diabetic retinopathy. *Nat. Med.,* **2003**, *9*, 294-299.

[182] Babaei-Jadidi R.; Karachalias N.; Ahmed N.; Battah S.; Thornalley P. J. Prevention of incipient diabetic nephropathy by high-dose thiamine and benfotiamine. *Diabetes,* **2003**, *52*, 2110-2120.

[183] Stirban A.; Negrean M.; Stratmann B.; Gawlowski T.; Horstmann T.; Gotting C.; Kleesiek K.; Mueller-Roesel M.; Koschinsky T.; Uribarri J.; Vlassara H.; Tschoepe D. Benfotiamine prevents macro- and microvascular endothelial dysfunction and oxidative stress following a meal rich in advanced glycation end products in individuals with type 2 diabetes. *Diabetes Care,* **2006**, *29*, 2064-2071.

[184] Beckman J. A.; Goldfine A. B.; Gordon M. B.; Garrett L. A.; Creager M. A. Inhibition of protein kinase Cbeta prevents impaired endothelium-dependent vasodilation caused by hyperglycemia in humans. *Circ. Res.,* **2002**, *90*, 107-111.

[185] Campochiaro P. A. Reduction of diabetic macular edema by oral administration of the kinase inhibitor PKC412. *Invest. Ophthalmol. Vis. Sci.,* **2004**, *45*, 922-931.

[186] Strom C.; Sander B.; Klemp K.; Aiello L. P.; Lund-Andersen H.; Larsen M. Effect of ruboxistaurin on blood-retinal barrier permeability in relation to severity of leakage in diabetic macular edema. *Invest. Ophthalmol. Vis. Sci.,* **2005**, *46*, 3855-3858.

[187] Aiello L. P.; Clermont A.; Arora V.; Davis M. D.; Sheetz M. J.; Bursell S. E. Inhibition of PKC beta by oral administration of ruboxistaurin is well tolerated and ameliorates diabetes-induced retinal hemodynamic abnormalities in patients. *Invest. Ophthalmol. Vis. Sci.,* **2006**, *47*, 86-92.

[188] Vinik A. I.; Bril V.; Kempler P.; Litchy W. J.; Tesfaye S.; Price K. L.; Bastyr E. J., 3rd. Treatment of symptomatic diabetic peripheral neuropathy with the protein kinase C beta-inhibitor ruboxistaurin mesylate during a 1-year, randomized, placebo-controlled, double-blind clinical trial. *Clin. Ther.,* **2005**, *27*, 1164-1180.

[189] Sheetz M. J.; Aiello L. P.; Davis M. D.; Danis R.; Bek T.; Cunha-Vaz J.; Shahri N.; Berg P. H. The effect of the oral PKC beta inhibitor ruboxistaurin on vision loss in two phase 3 studies. *Invest. Ophthalmol. Vis. Sci.,* **2013**, *54*, 1750-1757.

[190] Durpes M. C.; Morin C.; Paquin-Veillet J.; Beland R.; Pare M.; Guimond M. O.; Rekhter M.; King G. L.; Geraldes P. PKC-beta activation inhibits IL-18-binding protein causing endothelial dysfunction and diabetic atherosclerosis. *Cardiovasc. Res.,* **2015**, *106*, 303-313.

[191] Behl T.; Kotwani A. Exploring the various aspects of the pathological role of vascular endothelial growth factor (VEGF) in diabetic retinopathy. *Pharmacol. Res.,* **2015**, *99*, 137-148.

[192] Shen H.; Clauss M.; Ryan J.; Schmidt A. M.; Tijburg P.; Borden L.; Connolly D.; Stern D.; Kao J. Characterization of vascular permeability factor/vascular endothelial growth factor receptors on mononuclear phagocytes. *Blood,* **1993**, *81*, 2767-2773.

[193] Ferrara N. Vascular endothelial growth factor: basic science and clinical progress. *Endocr. Rev.,* **2004**, *25*, 581-611.

[194] van Wijngaarden P.; Coster D. J.; Williams K. A. Inhibitors of ocular neovascularization: promises and potential problems. *JAMA,* **2005**, *293*, 1509-1513.

[195] Massin P.; Bandello F.; Garweg J. G.; Hansen L. L.; Harding S. P.; Larsen M.; Mitchell P.; Sharp D.; Wolf-Schnurrbusch U. E.; Gekkieva M.; Weichselberger A.; Wolf S. Safety and efficacy of ranibizumab in diabetic macular edema (RESOLVE Study): a 12-month, randomized, controlled, double-masked, multicenter phase II study. *Diabetes Care,* **2010**, *33*, 2399-2405.

[196] Mitchell P.; Bandello F.; Schmidt-Erfurth U.; Lang G. E.; Massin P.; Schlingemann R. O.; Sutter F.; Simader C.; Burian G.; Gerstner O.; Weichselberger A. The RESTORE study: ranibizumab monotherapy or combined with laser versus laser monotherapy for diabetic macular edema. *Ophthalmology,* **2011**, *118*, 615-625.

[197] Bandello F.; De Benedetto U.; Knutsson K. A.; Parodi M. B.; Cascavilla M. L.; Iacono P. Ranibizumab in the treatment of patients with visual impairment due to diabetic macular edema. *Clin. Ophthalmol.*, **2011**, *5*, 1303-1308.
[198] Querques G.; Avellis F. O.; Querques L.; Bandello F.; Souied E. H. Age-related macular degeneration. *Clin. Ophthalmol.*, **2011**, *5*, 593-601.
[199] Hurwitz H.; Fehrenbacher L.; Novotny W.; Cartwright T.; Hainsworth J.; Heim W.; Berlin J.; Baron A.; Griffing S.; Holmgren E.; Ferrara N.; Fyfe G.; Rogers B.; Ross R.; Kabbinavar F. Bevacizumab plus irinotecan, fluorouracil, and leucovorin for metastatic colorectal cancer. *N. Engl. J. Med.*, **2004**, *350*, 2335-2342.
[200] Rich R. M.; Rosenfeld P. J.; Puliafito C. A.; Dubovy S. R.; Davis J. L.; Flynn H. W., Jr.; Gonzalez S.; Feuer W. J.; Lin R. C.; Lalwani G. A.; Nguyen J. K.; Kumar G. Short-term safety and efficacy of intravitreal bevacizumab (Avastin) for neovascular age-related macular degeneration. *Retina,* **2006**, *26*, 495-511.
[201] Mason J. O., 3rd; Nixon P. A.; White M. F. Intravitreal injection of bevacizumab (Avastin) as adjunctive treatment of proliferative diabetic retinopathy. *Am. J. Ophthalmol.*, **2006**, *142*, 685-688.
[202] Isaacs T. W.; Barry C. Rapid resolution of severe disc new vessels in proliferative diabetic retinopathy following a single intravitreal injection of bevacizumab (Avastin). *Clin. Experiment. Ophthalmol.*, **2006**, *34*, 802-803.
[203] Avery R. L.; Pearlman J.; Pieramici D. J.; Rabena M. D.; Castellarin A. A.; Nasir M. A.; Giust M. J.; Wendel R.; Patel A. Intravitreal bevacizumab (Avastin) in the treatment of proliferative diabetic retinopathy. *Ophthalmology,* **2006**, *113*, 1695 e1691-1615.
[204] Friedlander S. M.; Welch R. M. Vanishing disc neovascularization following intravitreal bevacizumab (avastin) injection. *Arch. Ophthalmol.*, **2006**, *124*, 1365.
[205] Chen E.; Park C. H. Use of intravitreal bevacizumab as a preoperative adjunct for tractional retinal detachment repair in severe proliferative diabetic retinopathy. *Retina,* **2006**, *26*, 699-700.
[206] Zhang X. Y.; Guo X. F.; Zhang S. D.; He J. N.; Sun C. Y.; Zou Y.; Bi H. S.; Qu Y. Comparison of bevacizumab and ranibizumab in age-related macular degeneration: a systematic review and meta-analysis. *Int. J. Ophthalmol.*, **2014**, *7*, 355-364.
[207] Chang T. S.; Kokame G.; Casey R.; Prenner J.; Feiner L.; Anderson N. Short-term effectiveness of intravitreal bevacizumab versus ranibizumab injections for patients with neovascular age-related macular degeneration. *Retina,* **2009**, *29*, 1235-1241.
[208] Ababneh O. H.; Yousef Y. A.; Gharaibeh A. M.; Abu Ameerh M. A.; Abu-Yaghi N. E.; Al Bdour M. D. Intravitreal bevacizumab in the treatment of diabetic ocular neovascularization. *Retina,* **2013**, *33*, 748-755.
[209] Stefanini F. R.; Arevalo J. F.; Maia M. Bevacizumab for the management of diabetic macular edema. *World J Diabetes,* **2013**, *4*, 19-26.
[210] Gragoudas E. S.; Adamis A. P.; Cunningham E. T., Jr.; Feinsod M.; Guyer D. R. Pegaptanib for neovascular age-related macular degeneration. *N. Engl. J. Med.*, **2004**, *351*, 2805-2816.
[211] Sultan M. B.; Zhou D.; Loftus J.; Dombi T.; Ice K. S. A phase 2/3, multicenter, randomized, double-masked, 2-year trial of pegaptanib sodium for the treatment of diabetic macular edema. *Ophthalmology,* **2011**, *118*, 1107-1118.
[212] Stewart M. W. Aflibercept (VEGF-TRAP): the next anti-VEGF drug. *Inflamm. Allergy Drug Targets,* **2011**, *10*, 497-508.
[213] Palejwala N. V.; Lauer A. K. Aflibercept: an update on recent milestones achieved. *Drugs Today (Barc),* **2015**, *50*, 779-790.
[214] Touyz R. M. Reactive oxygen species, vascular oxidative stress, and redox signaling in hypertension: what is the clinical significance? *Hypertension,* **2004**, *44*, 248-252.
[215] Iacobazzi D.; Mangialardi G.; Gubernator M.; Hofner M.; Wielscher M.; Vierlinger K.; Reni C.; Oikawa A.; Spinetti G.; Vono R.; Sangalli E.; Montagnani M.; Madeddu P. Increased antioxidant defense mechanism in human adventitia-derived progenitor cells is associated with therapeutic benefit in ischemia. *Antioxid. Redox Signal.*, **2014**, *21*, 1591-1604.
[216] Sharma A.; Bernatchez P. N.; de Haan J. B. Targeting endothelial dysfunction in vascular complications associated with diabetes. *Int. J. Vasc. Med.*, **2011**, *2012*, 750126.

[217] Anderson R. A.; Evans L. M.; Ellis G. R.; Khan N.; Morris K.; Jackson S. K.; Rees A.; Lewis M. J.; Frenneaux M. P. Prolonged deterioration of endothelial dysfunction in response to postprandial lipaemia is attenuated by vitamin C in Type 2 diabetes. *Diabet. Med.*, **2006**, *23*, 258-264.
[218] Antoniades C.; Tousoulis D.; Tountas C.; Tentolouris C.; Toutouza M.; Vasiliadou C.; Tsioufis C.; Toutouzas P.; Stefanadis C. Vascular endothelium and inflammatory process, in patients with combined Type 2 diabetes mellitus and coronary atherosclerosis: the effects of vitamin C. *Diabet. Med.*, **2004**, *21*, 552-558.
[219] Paolisso G.; Tagliamonte M. R.; Barbieri M.; Zito G. A.; Gambardella A.; Varricchio G.; Ragno E.; Varricchio M. Chronic vitamin E administration improves brachial reactivity and increases intracellular magnesium concentration in type II diabetic patients. *J. Clin. Endocrinol. Metab.*, **2000**, *85*, 109-115.
[220] Darko D.; Dornhorst A.; Kelly F. J.; Ritter J. M.; Chowienczyk P. J. Lack of effect of oral vitamin C on blood pressure, oxidative stress and endothelial function in Type II diabetes. *Clin. Sci. (Lond)*, **2002**, *103*, 339-344.
[221] Stephens N. G.; Parsons A.; Schofield P. M.; Kelly F.; Cheeseman K.; Mitchinson M. J. Randomised controlled trial of vitamin E in patients with coronary disease: Cambridge Heart Antioxidant Study (CHAOS). *Lancet*, **1996**, *347*, 781-786.
[222] Virtamo J.; Rapola J. M.; Ripatti S.; Heinonen O. P.; Taylor P. R.; Albanes D.; Huttunen J. K. Effect of vitamin E and beta carotene on the incidence of primary nonfatal myocardial infarction and fatal coronary heart disease. *Arch. Intern. Med.*, **1998**, *158*, 668-675.
[223] Lee D. H.; Folsom A. R.; Harnack L.; Halliwell B.; Jacobs D. R., Jr. Does supplemental vitamin C increase cardiovascular disease risk in women with diabetes? *Am. J. Clin. Nutr.*, **2004**, *80*, 1194-1200.
[224] Lonn E.; Bosch J.; Yusuf S.; Sheridan P.; Pogue J.; Arnold J. M.; Ross C.; Arnold A.; Sleight P.; Probstfield J.; Dagenais G. R. Effects of long-term vitamin E supplementation on cardiovascular events and cancer: a randomized controlled trial. *JAMA*, **2005**, *293*, 1338-1347.
[225] Miller E. R., 3rd; Pastor-Barriuso R.; Dalal D.; Riemersma R. A.; Appel L. J.; Guallar E. Meta-analysis: high-dosage vitamin E supplementation may increase all-cause mortality. *Ann. Intern. Med.*, **2005**, *142*, 37-46.
[226] Shirpoor A.; Salami S.; Khadem-Ansari M. H.; Ilkhanizadeh B.; Pakdel F. G.; Khademvatani K. Cardioprotective effect of vitamin E: rescues of diabetes-induced cardiac malfunction, oxidative stress, and apoptosis in rat. *J. Diabetes Complications*, **2009**, *23*, 310-316.
[227] Shirpoor A.; Norouzi L.; Nemati S.; Khadem Ansari M. H. Protective effect of vitamin E against diabetes-induced oxidized LDL and aorta cell wall proliferation in rat. *Iran Biomed. J.*, **2015**, *19*, 117-123.
[228] Ceriello A.; Kumar S.; Piconi L.; Esposito K.; Giugliano D. Simultaneous control of hyperglycemia and oxidative stress normalizes endothelial function in type 1 diabetes. *Diabetes Care*, **2007**, *30*, 649-654.
[229] Ceriello A.; Novials A.; Ortega E.; Canivell S.; La Sala L.; Pujadas G.; Bucciarelli L.; Rondinelli M.; Genovese S. Vitamin C further improves the protective effect of glucagon-like peptide-1 on acute hypoglycemia-induced oxidative stress, inflammation, and endothelial dysfunction in type 1 diabetes. *Diabetes Care*, **2013**, *36*, 4104-4108.
[230] Sheikh-Ali M.; Chehade J. M.; Mooradian A. D. The antioxidant paradox in diabetes mellitus. *Am. J. Ther.*, **2011**, *18*, 266-278.
[231] Bjelakovic G.; Nikolova D.; Gluud C. Antioxidant supplements and mortality. *Curr. Opin. Clin. Nutr. Metab. Care*, **2014**, *17*, 40-44.
[232] Fuchs D.; Sperner-Unterweger B. Can intake of extra antioxidants delay the development and progression of atherosclerosis? *Atherosclerosis*, **2013**, *226*, 43-44.
[233] Sarmento R. A.; Silva F. M.; Sbruzzi G.; Schaan B. D.; Almeida J. C. Antioxidant micronutrients and cardiovascular risk in patients with diabetes: a systematic review. *Arq. Bras. Cardiol.*, **2013**, *101*, 240-248.
[234] Scalbert A.; Williamson G. Dietary intake and bioavailability of polyphenols. *J. Nutr.*, **2000**, *130*, 2073S-2085S.
[235] Maxwell S.; Cruickshank A.; Thorpe G. Red wine and antioxidant activity in serum. *Lancet*, **1994**, *344*, 193-194.

[236] Joshi M. S.; Williams D.; Horlock D.; Samarasinghe T.; Andrews K. L.; Jefferis A. M.; Berger P. J.; Chin-Dusting J. P.; Kaye D. M. Role of mitochondrial dysfunction in hyperglycaemia-induced coronary microvascular dysfunction: Protective role of resveratrol. *Diab. Vasc. Dis. Res.*, **2015**, *12*, 208-216.

[237] Soriano F. G.; Virag L.; Szabo C. Diabetic endothelial dysfunction: role of reactive oxygen and nitrogen species production and poly(ADP-ribose) polymerase activation. *J. Mol. Med.*, **2001**, *79*, 437-448.

[238] Garcia Soriano F.; Virag L.; Jagtap P.; Szabo E.; Mabley J. G.; Liaudet L.; Marton A.; Hoyt D. G.; Murthy K. G.; Salzman A. L.; Southan G. J.; Szabo C. Diabetic endothelial dysfunction: the role of poly(ADP-ribose) polymerase activation. *Nat. Med.*, **2001**, *7*, 108-113.

[239] Pacher P.; Szabo C. Role of poly(ADP-ribose) polymerase 1 (PARP-1) in cardiovascular diseases: the therapeutic potential of PARP inhibitors. *Cardiovasc. Drug Rev.*, **2007**, *25*, 235-260.

[240] Zheng L.; Szabo C.; Kern T. S. Poly(ADP-ribose) polymerase is involved in the development of diabetic retinopathy via regulation of nuclear factor-kappaB. *Diabetes*, **2004**, *53*, 2960-2967.

[241] Mohammad G.; Siddiquei M. M.; Abu El-Asrar A. M. Poly (ADP-ribose) polymerase mediates diabetes-induced retinal neuropathy. *Mediators Inflamm.*, **2013**, *2013*, 510451.

[242] Choi S. K.; Galan M.; Kassan M.; Partyka M.; Trebak M.; Matrougui K. Poly(ADP-ribose) polymerase 1 inhibition improves coronary arteriole function in type 2 diabetes mellitus. *Hypertension*, **2012**, *59*, 1060-1068.

[243] Singh S. S.; Sarma J. A.; Narasu L.; Dayam R.; Xu S.; Neamati N. A review on PARP1 inhibitors: Pharmacophore modeling, virtual and biological screening studies to identify novel PARP1 inhibitors. *Curr. Top Med. Chem.*, **2014**, *14*, 2020-2030.

[244] Gasparrini M.; Giampieri F.; Alvarez Suarez J. M.; Mazzoni L.; Forbes Hernandez T. Y.; Quiles J. L.; Bullon P.; Battino M. AMPK as a New Attractive Therapeutic Target for Disease Prevention: the Role of Dietary Compounds. *Curr. Drug Targets*, **2015**.

[245] Zou M. H.; Wu Y. AMP-activated protein kinase activation as a strategy for protecting vascular endothelial function. *Clin. Exp. Pharmacol. Physiol.*, **2008**, *35*, 535-545.

[246] Wang S.; Xu J.; Song P.; Viollet B.; Zou M. H. In vivo activation of AMP-activated protein kinase attenuates diabetes-enhanced degradation of GTP cyclohydrolase I. *Diabetes*, **2009**, *58*, 1893-1901.

[247] Lee W. J.; Lee I. K.; Kim H. S.; Kim Y. M.; Koh E. H.; Won J. C.; Han S. M.; Kim M. S.; Jo I.; Oh G. T.; Park I. S.; Youn J. H.; Park S. W.; Lee K. U.; Park J. Y. Alpha-lipoic acid prevents endothelial dysfunction in obese rats via activation of AMP-activated protein kinase. *Arterioscler. Thromb. Vasc. Biol.*, **2005**, *25*, 2488-2494.

[248] Fogarty S.; Hardie D. G. Development of protein kinase activators: AMPK as a target in metabolic disorders and cancer. *Biochim. Biophys. Acta*, **2010**, *1804*, 581-591.

[249] Coughlan K. A.; Valentine R. J.; Ruderman N. B.; Saha A. K. AMPK activation: a therapeutic target for type 2 diabetes? *Diabetes Metab. Syndr. Obes.*, **2014**, *7*, 241-253.

[250] Ebrahimi-Mamaeghani M.; Mohammadi S.; Arefhosseini S. R.; Fallah P.; Bazi Z. Adiponectin as a potential biomarker of vascular disease. *Vasc. Health Risk Manag.*, **2015**, *11*, 55-70.

[251] Shibata R.; Sato K.; Pimentel D. R.; Takemura Y.; Kihara S.; Ohashi K.; Funahashi T.; Ouchi N.; Walsh K. Adiponectin protects against myocardial ischemia-reperfusion injury through AMPK- and COX-2-dependent mechanisms. *Nat. Med.*, **2005**, *11*, 1096-1103.

[252] Ouchi N.; Kobayashi H.; Kihara S.; Kumada M.; Sato K.; Inoue T.; Funahashi T.; Walsh K. Adiponectin stimulates angiogenesis by promoting cross-talk between AMP-activated protein kinase and Akt signaling in endothelial cells. *J. Biol. Chem.*, **2004**, *279*, 1304-1309.

[253] Shimabukuro M.; Higa N.; Asahi T.; Oshiro Y.; Takasu N.; Tagawa T.; Ueda S.; Shimomura I.; Funahashi T.; Matsuzawa Y. Hypoadiponectinemia is closely linked to endothelial dysfunction in man. *J. Clin. Endocrinol. Metab.*, **2003**, *88*, 3236-3240.

[254] Ming X. F.; Viswambharan H.; Barandier C.; Ruffieux J.; Kaibuchi K.; Rusconi S.; Yang Z. Rho GTPase/Rho kinase negatively regulates endothelial nitric oxide synthase phosphorylation through the inhibition of protein kinase B/Akt in human endothelial cells. *Mol. Cell. Biol.*, **2002**, *22*, 8467-8477.

[255] Ali T. K.; Al-Gayyar M. M.; Matragoon S.; Pillai B. A.; Abdelsaid M. A.; Nussbaum J. J.; El-Remessy A. B. Diabetes-induced peroxynitrite impairs the balance of pro-nerve growth factor and nerve growth factor, and causes neurovascular injury. *Diabetologia*, **2011**, *54*, 657-668.

[256] Yao L.; Chandra S.; Toque H. A.; Bhatta A.; Rojas M.; Caldwell R. B.; Caldwell R. W. Prevention of diabetes-induced arginase activation and vascular dysfunction by Rho kinase (ROCK) knockout. *Cardiovasc. Res.*, **2013**, *97*, 509-519.

[257] Mallat Z.; Gojova A.; Sauzeau V.; Brun V.; Silvestre J. S.; Esposito B.; Merval R.; Groux H.; Loirand G.; Tedgui A. Rho-associated protein kinase contributes to early atherosclerotic lesion formation in mice. *Circ. Res.*, **2003**, *93*, 884-888.

[258] Kishi T.; Hirooka Y.; Masumoto A.; Ito K.; Kimura Y.; Inokuchi K.; Tagawa T.; Shimokawa H.; Takeshita A.; Sunagawa K. Rho-kinase inhibitor improves increased vascular resistance and impaired vasodilation of the forearm in patients with heart failure. *Circulation,* **2005**, *111*, 2741-2747.

[259] Fukumoto Y.; Matoba T.; Ito A.; Tanaka H.; Kishi T.; Hayashidani S.; Abe K.; Takeshita A.; Shimokawa H. Acute vasodilator effects of a Rho-kinase inhibitor, fasudil, in patients with severe pulmonary hypertension. *Heart,* **2005**, *91*, 391-392.

[260] Arita R.; Hata Y.; Nakao S.; Kita T.; Miura M.; Kawahara S.; Zandi S.; Almulki L.; Tayyari F.; Shimokawa H.; Hafezi-Moghadam A.; Ishibashi T. Rho kinase inhibition by fasudil ameliorates diabetes-induced microvascular damage. *Diabetes,* **2009**, *58*, 215-226.

[261] Pearson J. T.; Jenkins M. J.; Edgley A. J.; Sonobe T.; Joshi M.; Waddingham M. T.; Fujii Y.; Schwenke D. O.; Tsuchimochi H.; Yoshimoto M.; Umetani K.; Kelly D. J.; Shirai M. Acute Rho-kinase inhibition improves coronary dysfunction in vivo, in the early diabetic microcirculation. *Cardiovasc. Diabetol.*, **2013**, *12*, 111.

[262] Yin H.; Ru H.; Yu L.; Kang Y.; Lin G.; Liu C.; Sun L.; Shi L.; Sun Q. Targeting of Rho kinase ameliorates impairment of diabetic endothelial function in intrarenal artery. *Int. J. Mol. Sci.*, **2013**, *14*, 20282-20298.

[263] Arita R.; Nakao S.; Kita T.; Kawahara S.; Asato R.; Yoshida S.; Enaida H.; Hafezi-Moghadam A.; Ishibashi T. A key role for ROCK in TNF-alpha-mediated diabetic microvascular damage. *Invest. Ophthalmol. Vis. Sci.*, **2013**, *54*, 2373-2383.

[264] Garnock-Jones K. P. Ripasudil: first global approval. *Drugs,* **2014**, *74*, 2211-2215.

[265] Satoh K.; Fukumoto Y.; Shimokawa H. Rho-kinase: important new therapeutic target in cardiovascular diseases. *Am. J. Physiol. Heart Circ. Physiol.*, **2011**, *301*, H287-296.

[266] Sharma A.; Bernatchez P. N.; de Haan J. B. Targeting endothelial dysfunction in vascular complications associated with diabetes. *Int. J. Vasc. Med.*, **2012**, *2012*, 750126.

CHAPTER 5

Pyridinium Oximes as Cholinesterase Reactivators. An Update of the Structure-Activity Relationship and Efficacy in the Treatment of Poisoning with Organophosphorus Compounds

Milan Jokanović[1,*] and Radica Stepanović Petrović[2]

[1]*Faculty of Medicine, University of Nish, Nish, Serbia,* [2]*Faculty of Pharmacy, University of Belgrade, Belgrade, Serbia*

Abstract: During more than five decades, pyridinium oximes have been developed as therapeutic agents used in the medical treatment of poisoning with organophosphorus compounds. Their mechanism of action is reactivation of acetylcholinesterase (AChE) inhibited by organophosphorus agents. Organophosphorus compounds (OPC) are used as pesticides and developed as warfare nerve agents such as tabun, soman, sarin, VX and others. Exposure to even small amounts of an OPC can be fatal and death is usually caused by respiratory failure resulting from paralysis of the diaphragm and intercostal muscles, depression of the brain respiratory center, bronchospasm, and excessive bronchial secretions. The mechanism of OPC poisoning involves phosphorylation of the serine hydroxyl group at the active site of AChE leading to the inactivation of this essential enzyme, which has an important role in neurotransmission. AChE inhibition results in the accumulation of acetylcholine at cholinergic receptor sites, producing continuous stimulation of cholinergic fibers throughout the central and peripheral nervous systems. Presently, a combination of an antimuscarinic agent, *e.g.* atropine, AChE reactivator such as one of the standard pyridinium oximes (pralidoxime, trimedoxime, obidoxime, asoxime) and diazepam are used for the treatment of organophosphate poisoning in humans.

Despite enormous efforts devoted to synthesis and development of new pyridinium oximes as potential antidotes against poisoning with OPC, only four compounds are used in human medicine so far. However, they differ in their activity in poisoning with warfare nerve agents and pesticides and there is still no universal broad-spectrum oxime capable of protecting against all known OPC. This article reviews the latest data on structure-activity relationship of pyridinium oximes including their efficacy in treatment of poisoning with organophosphorus compounds.

Keywords: Acetylcholinesterase, asoxime, atropine, cyclosarin, diazepam, HI-6, HLö-7, LüH-6, methoxime, MMB-4, obidoxime, organophosphorus compounds, PAM-2, pralidoxime, pyridinium oximes, sarin, soman, tabun, TMB-4, trimedoxime, VX.

***Corresponding author Milan Jokanović:** Faculty of Medicine, University of Nish, Nish, Serbia; E-mail: milan.jokanovic@gmail.com

Atta-ur-Rahman, M. Iqbal Choudhary and Allen B. Reitz (Eds.)

INTRODUCTION

Organophosphorus compounds (OPC) have been used as pesticides and developed as warfare nerve agents such as tabun, soman, sarin, VX and others. Certain OP insecticides have found its application in human and veterinary medicine. Malathion is used topically in the treatment of pediculosis (lice) and scabies infestations, while dichlorvos and several other OP insecticides are used as topical ectoparasiticides in veterinary practice [1]. Exposure to even small amounts of an OPC could be fatal; death is usually caused by respiratory failure resulting from paralysis of the diaphragm and intercostal muscles, depression of the brain respiratory center, bronchospasm, and excessive bronchial secretions. The mechanism of OPC poisoning involves phosphorylation of the serine hydroxyl group at the active site of acetylcholinesterase (AChE), causing inactivation of this essential enzyme which has an important role in neurotransmission, leading to accumulation of acetylcholine at cholinergic receptor sites, producing continuous stimulation of cholinergic fibers throughout the central and peripheral nervous system. Presently, a combination of an antimuscarinic agent, such as atropine, AChE reactivator such as one of the standard pyridinium oximes (pralidoxime, trimedoxime, obidoxime, asoxime) and diazepam are used for the treatment of organophosphate poisoning in humans. This article reviews the mechanisms of OPCs action, structure-activity relationships for pyridinium oximes and their role as AChE reactivators in medical treatment of OPC poisonings.

INTERACTION OF CHOLINESTERASES WITH OPC INHIBITORS

There are different types of cholinesterases in the human body, which differ in respect to their tissue distribution, substrate affinity, and physiological functions. The principal ones are acetylcholinesterase (EC 3.1.1.7, AChE), found in the nervous system and also present in the outer membrane of red blood cells, and plasma cholinesterase (EC 3.1.1.8, ChE), which are a group of enzymes present in plasma, liver, cerebrospinal fluid and glial cells. Under normal physiological conditions, AChE performs the breakdown of acetylcholine (ACh), which is the chemical mediator responsible for conduction of nerve impulses at the sites of cholinergic transmission. However, its physiological role in blood is not completely understood. On the other hand, ChE is a circulating plasma glycoprotein synthetized in the liver and other tissues, which does not serve any known physiological function. This enzyme can be found in more than 40 genetic variants differing in their susceptibility to chemicals. It seems that ChE may have roles in cholinergic neurotransmission [2]. Also, it may be involved in other nervous system functions (cellular proliferation and neurite growth during the development of the nervous system) and in neurodegenerative disorders [3].

The function of AChE is the termination of action of ACh at the junctions of the various cholinergic nerve endings with their effector organs or post-synaptic sites. OPC are the most important AChE inhibitors and often called anticholinesterases. In the presence of inhibitors, AChE becomes progressively inhibited and is no longer capable of hydrolyzing ACh to choline and acetic acid [4]. Consequently, ACh accumulates at cholinergic receptor sites and thus produces effects equivalent to excessive stimulation of cholinergic receptors throughout the central and peripheral nervous system.

ChE is also an esterase that can react with most of the compounds that react with AChE in a similar way. ChE also reacts with nerve warfare agents, but while the effects of inhibition of AChE can be fatal (depending on the dose taken), inhibition of ChE does not have any known consequences on the normal body functions. ChE has an important role in detoxification of OPC since it can bind a significant amount of the compounds decreasing their concentration in blood and subsequently the toxic effects of OPC poisoning. For this reason, purified ChE was investigated as an exogenously administrated prophylactic agent and a bioscavenger. In further text, regarding the effects of pyridinium oximes, we will refer just to AChE.

Both substrate and inhibitors react covalently with the esterases in essentially the same manner, because acetylation of the serine residue at AChE catalytic site is analogous to phosphorylation. In contrast to the acetylated enzyme, which rapidly separates acetic acid and restores the catalytic site, the phosphorylated enzyme is stable. Inhibited enzyme can be spontaneously reactivated at different rates depending on chemical structure of the inhibitor, while for warfare nerve agents having branched alkyl groups, it may occur at a very slow rate or not at all. AChE inhibited with OP insecticides having dimethyl phosphate structure can spontaneously reactivate with half-times of 1.3 hours *in vitro* and 2.1 hours *in vivo* [5, 6]. As the result, dimethyl phosphates are recognized as reversible AChE inhibitors, while other OPC having branched alkyl groups are practically irreversible AChE inhibitors.

The variations in the acute toxicity of OPC are result of their different chemical structures and rates of spontaneous reactivation and aging (Fig. **1**, reaction 3). Aged form of phosphorylated AChE is resistant to both spontaneous and oxime-induced reactivation. For example, when complexed with VX, AChE in erythrocytes spontaneously reactivates at a rate of about 1% per hour for about the first 48 hours, while the VX-AChE complex ages very little during this period [7, 8]. The soman-AChE complex does not spontaneously reactivate since the half-time for aging is about 2-4 minutes [7-9]. The half-time for aging of the sarin-AChE complex is between 5 and 12 hours (depending on experimental conditions) and only about 5% of the enzyme undergo spontaneous reactivation [9]. The half-time for aging of the tabun-AChE complex is about 46 hours [7-9]. For OP pesticides containing dimethyl phosphate groups the half-time of spontaneous reactivation of phosphorylated AChE *in vitro* is about 1 hour and the half-time of aging is 3.7 hours, while these for diethyl phosphates are slower being about 31-57 hours and 31 hours, respectively [6, 10, 11].

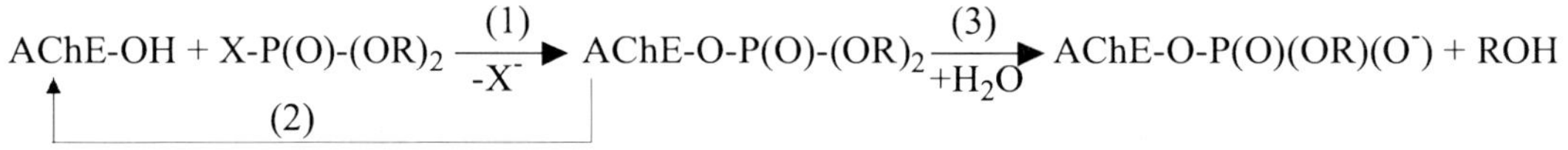

Figure 1: Interaction of acetylcholinesterase (AChE-OH) with organophosphorus compounds. Reaction 1 shows interaction of organophosphate molecule with the serine hydroxyl group at the active site of AChE. Inhibited AChE cannot further serve its physiological function, which causes the accumulation of acetylcholine at the nerve endings. Reaction 2 is spontaneous reactivation of inhibited AChE which occurs very slowly for most OP compounds. Reaction 3, called "aging", represents non-enzymatic time-dependent loss of one alkyl group (R) bound to the phosphorus. The aging reaction depends on the chemical structure of the inhibitor and leads to a stable non-reactivatable form of phosphorylated AChE. X stands for acyl radical (*i.e.* Cl^-, F^-, CN^-, *p*-nitrophenol *etc.*).

The rate of spontaneous reactivation (Fig. **1**, reaction 2) can be accelerated by pyridinium oximes that have a chemical structure which 'fits' the structure of the inhibited AChE. The oximes can only be of benefit as long as inhibited AChE is not completely converted to its aged form. The aging reaction, although appearing with many phosphorylated AChE complexes, has the only major clinical importance and is an imperative problem particularly in the treatment of soman poisoning. Aging with soman occurs so fast that no clinically relevant spontaneous reactivation of AChE may occur before aging has taken place. Hence, recovery of function depends on resynthesis of AChE which is a relatively slow process [9]. After soman exposure, it is important to immediately administer atropine and oximes so that some extent of AChE reactivation occurs before all AChE has aged. Even though aging occurs slowly and reactivation relatively rapidly in the case of other nerve agents and OPC, early oxime administration is clinically important particularly in patients poisoned with these agents.

CLINICAL ASPECTS OF ACUTE OPC POISONING

Signs and symptoms of acute poisoning with anticholinesterase agents are predictable from their biochemical mechanism of action and are directly related to the levels of AChE activity. In cases of human intoxication, general acute symptoms of peripheral nicotinic and muscarinic intoxication are clearly apparent [12]. These symptoms include miosis (unreactive to light); sweating, rhinorrhea, lacrimation, and salivation; abdominal cramps and other gastrointestinal symptoms; respiratory difficulties and cough; dyspnea, constriction sensation in the chest, wheezing; twitching of facial muscles and tongue, tremors, and fasciculations; bradycardia and ECG changes, pallor, and cyanosis; anorexia, nausea, vomiting, diarrhea, and involuntary urination and defecation. These signs and symptoms are accompanied by central effects such as dizziness, tremulousness, and confusion; ataxia; headache, fatigability, and paresthesia. Finally, seizures, convulsions, twitching, coma, and respiratory failure may occur. If the subject survives past the day of poisoning, there are personality changes, mood swings, aggressive events and psychotic episodes including schizoid reactions, paranoid delusions, and exacerbations of preexisting psychiatric problems. Sleep is poor from nightmares and hallucinations; disturbances or deficits in memory and attention, and additional delayed effects also occur [12-16].

The first four to six hours are the most critical in acute poisoning with OP pesticides. If there is improvement in symptoms after initial treatment then the patient is very likely to survive if adequate treatment is continued [14, 16]. Duration of effects is determined mainly by the properties of the compound: its liposolubility, the stability of the OP-AChE complex and whether it is reactivatable after the use of cholinesterase reactivators (such as oximes). It is important to note that only OP containing P=O bond (known as direct inhibitors) are potent AChE inhibitors; those having a P=S group (indirect inhibitors) must be metabolically activated to P=O group [17]. The signs and symptoms of poisoning with direct inhibitors appear quickly during or after exposure, while those with indirect inhibitors appear slowly and last longer, even up to several days after cessation of exposure.

Clinical diagnosis is relatively simple and is based on medical history, circumstances of exposure, clinical presentation, and laboratory tests. Confirmation of diagnosis can be made by measurement of erythrocyte AChE or plasma ChE. Activities of these enzymes are accepted as biomarkers of exposure and/or toxicity of OP. Erythrocyte AChE is identical to the enzyme present in the target synapses and its levels are assumed to reflect the effects in target organs. Thus, erythrocyte AChE is regarded as biomarker of toxicity of these compounds. AChE in brain is restored by *de novo* synthesis more rapidly than in erythrocytes where AChE activity is recovered *via* erythropoesis. The level of activity of ChE should be carefully interpreted since the normal range in healthy subjects is relatively wide, rendering interpretation in individual patients difficult unless the results of previous estimations in the patient are available. Inhibition of ChE does not provide accurate information related to clinical severity of the poisoning. Many OP insecticides, for example chlorpyrifos, demethon and malathion, are apparently more potent inhibitors of ChE than they are of erythrocyte AChE and, consequently, ChE inhibition might occur to a greater extent than AChE inhibition [16].

ANTIDOTES IN THE TREATMENT OF OPC POISONING

Atropine

Atropine acts by blocking the effects of excess concentrations of acetylcholine at muscarinic cholinergic synapses following OPC inhibition of AChE. Atropine is the initial drug of choice in acute OPC poisoning and can relieve the following symptoms of OPC poisoning: sweating, salivation, rhinorrhoea, lacrimation, nausea, vomiting and diarrhea. It can also help control of bradycardia and circulatory depressions, dilate the bronchi and abolish bronchorrhoea. It does not bind to nicotinic receptors and cannot relieve nicotinic effects of OPC. In addition, there is evidence regarding the anticonvulsant properties of atropine in poisoning with soman and VX [18, 19]. Atropine in combination with one of the standard pyridinium oximes has been used in the traditional therapy for OPC intoxications.

The standard dosing of atropine depends on the severity of OPC poisoning. The initial dose is usually 2 mg in an adult (0.02 mg/kg in a child) given every 5-10 min until hyperatropinization (flushing, dryness of the mouth, nose, lungs and the skin, heart rate 80100/min, normal blood pressure, mydriasis) is reached, which should be maintained during further treatment. The dose may be increased as required. Patients poisoned with OPC appear to be resistant to the toxic efects of atropine and may require relatively large doses of atropine administered during prolonged periods of time. According to IPCS [20], in severe OPC poisoning total dose of atropine given during five weeks of treatment can be as high as 30000 mg.

Abedin *et al.* [21] reported the results of an open-label randomized clinical trial undertaken to determine the optimal dosing regimen for atropine in the treatment of OPC poisoning. Out of 156 patients with OPC poisoning, 81 patients received conventional bolus dose atropine (group A) and 75 patients received rapidly incremental doses of atropine followed by infusion (group B). It was found that mortality in group A was 22.5% and in group B 8%. The mean duration of atropinization in group A was 151.74 min compared to 23.90 min for group B. More patients in group A experienced atropine toxicity than in group B (28.4% versus 12.0%). Respiratory support was required more often for patients in group A than in group B (24.7% versus 8%). It was concluded that rapid incremental dose atropinization followed by atropine infusion reduces mortality and morbidity from OPC poisoning and shortens the length of hospital stay and recovery suggesting that incremental atropine and infusion should become the treatment of choice for OPC poisoning.

Diazepam

Diazepam is a well-known benzodiazepine most frequently used for the treatment of convulsions that appear in OPC poisonings. The main site of action of diazepam is at the gamma-aminobutyric acid (GABA) receptor. The combination of atropine and diazepam is more effective in reducing mortality in OPC poisonings than atropine or oxime alone. Diazepam enhanced the clinical efficacy of low doses of atropine. In the cholinergic nervous system, diazepam appears to decrease the synaptic release of ACh. The main consequence of the action of benzodiazepines in CNS is hyperpolarization of neurons. This makes neurons significantly less susceptible to cholinergically-induced depolarization. The ultimate result is cessation of convulsions propagation [22, 23].

In patients poisoned with OPC, benzodiazepines may have a beneficial effect in reducing anxiety and restlessness, reducing muscle fasciculation, preventing seizures, convulsions, controlling apprehension and agitation and possibly reducing morbidity and mortality when used in combination with atropine and pyridinium oxime. Diazepam should be given to patients poisoned with OPC whenever convulsions or pronounced muscle fasciculation are present. In severe poisoning, diazepam administration should be considered even before these complications develop. The recommended dose of diazepam in cases of OPC poisoning is 5-10 mg i.v. in the absence of convulsions and 10-20 mg i.v. in cases with convulsions, which may be repeated as required [24].

It appears that other anticonvulsant benzodiazepines, such as midazolam, avizafone, lorazepam, are just as effective in stopping nerve agent-induced seizure as diazepam in hospital setting. In animal studies midazolam was even more effective than diazepam [25]. FDA and French army considered approval of midazolam as treatment of nerve agent-induced seizures [26, 27]. In addition to diazepam, Delacour and Dorandeu [28] recommended administration of clonazepam in the management of severe OP poisoning.

Detailed therapeutic protocols used in the treatment of poisoning with OPC are presented in several excellent reviews [12, 15, 16, 24, 29-33].

Oximes

Extensive studies over the past decades have investigated the mechanism of oxime action. There is convincing evidence that the antidotal potency of oximes is primarily attributed to their abilities to reactivate the phosphorylated AChE. Oximes reactivate phosphorylated AChE by displacing the phosphoryl moiety from the enzyme due to their high affinity for the enzyme and powerful nucleophilicity. The rate of reactivation depends on the structure of the phosphoryl moiety bound to the enzyme, the source of the enzyme, the structure and concentration of oxime present at the active site, the rate of post-inhibitory dealkylation known as aging, and the steric hindrance effects between oxime molecule and phosphoryl moiety attached to the active site of AChE [34, 35]. Phosphorylated oximes are formed during reactivation reaction and some of them appear to be potent inhibitors of AChE [36-38]. It was shown that phosphorylated oximes of 2-substituted pyridinium compounds (*e.g.* pralidoxime, asoxime) are rather unstable while those of 4-pyridinium aldoximes are markedly stable [37, 39, 40].

Oximes bind to AChE as reversible inhibitors and form complexes with AChE either at the acylation (catalytic) site, at the allosteric site, or at both sites of the enzyme and protect AChE from phosphorylation [41]. When the reversible inhibitor binds to the catalytic site, the protection is due to a direct competition between OPC and reversible inhibitor. Binding of a reversible inhibitor to the allosteric site induces indirect protection of the active site. Differences in the mechanisms of enzyme reactivation and protection, demonstrate how stereochemical arrangements of oximes can play a role in the potency of their therapeutic efficacy.

Mono- and bis-pyridinium oximes are effective against OP-inhibited AChE in the peripheral nervous system, but have a limited penetration potential across the blood-brain barrier. Limited penetration of asoxime was due to its pharmacokinetic profile and the two quaternary nitrogen atoms in its structure. However, pyridinium oxime penetration through blood-brain barrier was apparently underestimated since soman can cause seizure-related opening of the blood-brain barrier [42, 43] enabling passage of higher oxime concentrations into the brain. Shrot and coworkers [44] suggested that OP poisoning may have a favorable impact on brain penetration of oximes through the blood-brain barrier due to the induction of local inflammatory processes, the increase of brain blood flow and persistent seizures found at high doses of OPCs. Abdel-Rahman and colleagues [45] have shown that 0.5-1.0 LD_{50} of sarin caused a dose-dependent increase in permeability of blood-brain barrier in midbrain, brainstem, cerebrum and cerebellum in rats 24 hours after poisoning. Sakurada and coworkers [46] have determined the amount of PAM-2 passing accross the blood-brain barrier at approximately 10% of the given dose which may be effective in reactivation of OPC-inhibited AChE in brain. Obidoxime concentrations in brain were estimated to 3-5% of plasma levels in rodents [47] and those of HI-6 to 10% of the serum highest concentration [48]. Several studies [44, 49, 50] suggested that such oxime concentrations in brain are apparently sufficient to produce biochemical and physiological effects in soman poisoning, particularly in specific brain regions, such as the medulla oblongata. These effects can contribute to survival following OP intoxication, while others claim that it is uncertain whether this amount is sufficient to reactivate nerve agent-inhibited AChE [51]. When AChE reactivation was achieved in pontomedullar region good therapeutic effects of oximes were observed and survival of poisoned animals correlated with the AChE activity in this region [50]. In addition, it was suggested that in pralidoxime and obidoxime penetration through the blood-brain barrier a carrier-mediated transport mechanism might be involved [53].

STRUCTURE-ACTIVITY RELATIONSHIP FOR PYRIDINIUM OXIMES

Since the discovery that monopyridinium oximes were effective reactivators of OPC-inhibited AChE [54], a great number of monopyridinium and bispyridinium oximes had been synthesized and tested for their efficacy [35, 55-57]. However, no oxime has been found that is effective against all OP nerve agents [32, 35, 38, 49, 52, 58, 59], but methoxime appears to be a promising candidate [60, 61] (Fig. **2**). Several oximes, such as pralidoxime (PAM-2), obidoxime (LüH-6), trimedoxime (TMB-4) and asoxime (HI-6), have been adopted by different countries for treatment of OPC poisoning [34, 62]. Structure-activity relationships for the efficacy of pyridinium oximes are poorly understood, since oxime induced reactivation depends on many factors such as the structure of the phosphoryl moiety bound to the enzyme, the source of the enzyme, the structure and concentration of oxime which is present at the enzyme active site, the rate of aging and the steric hindrance effects between oxime molecule and phosphoryl moiety attached to the active site of AChE [32, 34, 36].

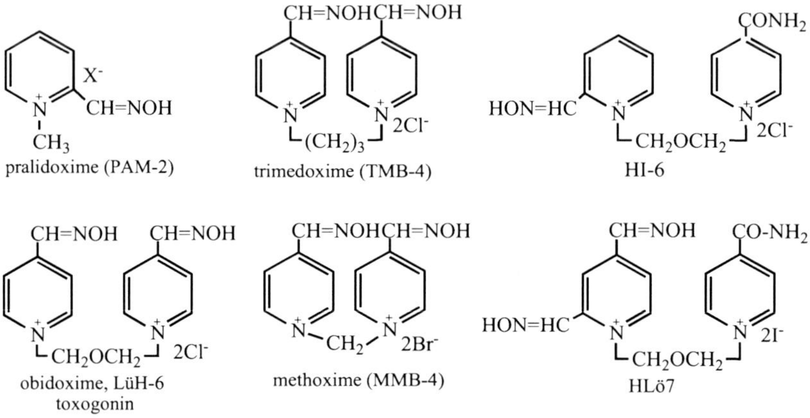

Figure 2: Chemical structure of pyridinium oximes used in treatment of OPC poisoning. In PAM-2 X stands for an anion, it can be Cl^-, I^- or methanesulfonate.

Structure-activity relationship related to pyridinium oximes depends on three most important factors that influence the affinity of the AChE reactivators toward inhibited AChE [32, 57, 63-65]:

a) The presence of quaternary nitrogen in the molecule of reactivator is the first well-known requirement. Due to the presence of the quaternary nitrogen in the oxime structure their affinity for AChE is increased allowing its approach to the AChE active site. It is also known that bispyridinium oximes which have two quaternary nitrogens achieved higher affinity and better reactivation of inhibited AChE than monopyridinium compounds.

b) It has been accepted that there should be one or two oxime group(s) placed at position 2 and/or 4 of the pyridinium ring(s). It is known that positions 2 and 4 are more suitable compared to the position 3 possibly due to the differences in pKa [32, 57, 63-69]. The great majority of the pyridinium oximes synthesized and studied during more than two decades fulfilled this condition.

c) The length of the connecting linker between bispyridinium nitrogens plays an important role in its potency. The linkers contained in currently used bispyridinium oximes are $-CH_2-CH_2-CH_2-$(in TMB-4), $-CH_2-O-CH_2-$(in HI-6, LüH-6 and HLö-7) or $-CH_2-$(in MMB-4).

Bisquaternary oximes have higher affinity towards both intact and inhibited AChE and, therefore, higher potency to reactivate OPC-inhibited AChE compared to monoquaternary oximes [70]. The above noted structural features, particularly the position of the oxime group in the pyridinium ring, depend on the chemical structure of AChE reactivators. While the potency of reactivators to reactivate tabun inhibited AChE with the oxime group in position 4 is higher compared to reactivators with the oxime group in other positions, the AChE reactivators with the oxime group in position 2 were the best reactivators of soman or cyclosarin-inhibited AChE [70]. The number of aldoxime groups does not seem to be so important since HI-6 has only one oxime group, but it is significantly more effective in reactivation of soman-and cyclosarin-inhibited AChE than bispyridinium oximes with two aldoxime groups such as obidoxime, methoxime or trimedoxime [70]. The chain linking the two quaternary nitrogens in bispyridinium oximes exerts a great effect on the reactivating efficacy, although this part of the oxime molecule does not play any role in the dephosphorylation process. It is a major factor in influencing oxime access to the active site of AChE and its reactivation rates. Tri- or tetra-carbon linker appears to be the most suitable for the sufficient potency of oximes to reactivate soman, tabun or cyclosarin-inhibited AChE [70], but one carbon linker (in methoxime) has also shown considerable efficacy against nerve agents.

In previous paper [32] the most interesting findings from different laboratories working on the synthesis and development of new oximes as AChE reactivators over the period of a few decades were reviewed. However, there appears to be a significant increase in the number of published scientific papers in this field in the period 2009-2015. In further text on new pyridinium oximes only the papers published after 2009 will be discussed.

Acharya and coworkers [71-74] have tested 47 new bispyridinium oximes for reactivation potential on electric eel or human AChE inhibited by sarin, VX or DFP and their efficacy was compared with PAM-2 and LüH-6. Among new oximes 12 symmetrical bispyridinium analogs of TMB-4 and LüH-6 having an aryl (*p*-C6H4 or *m*-C6H4) moiety or a carbon- carbon double or triple bond embedded in a diether linkage between two pyridinium rings with oxime groups at positions 2, 3 or 4. In 9 oximes the pyridinium rings were connected with methoxy alkane linkers, 8 were connected with xylene linkers while 18 were carbamoyl bis-pyridinium monooximes. None of the new oximes described in these studies proved to be better or promising AChE reactivators when compared with pralidoxime and obidoxime.

Bharate *et al.* [75] reported the reactivation potency of a series of 34 newly synthesized monoquaternary pyridinium oximes with heteroatomic side chains against paraoxon-inhibited electric eel AChE and recombinant human AChE of which thiophene-containing analogs provided some AChE reactivation. In a later paper [76] the same authors tested a series of 25 new bisquaternary pyridinium oximes bearing heterocyclic linkers in the same experimental model and compared their reactivation potential with pralidoxime, trimedoxime and obidoxime. The new oximes did not improve AChE reactivation when compared to the well-known oximes.

Eighteen monoquaternary reactivators of AChE with modified side chain (attached to quaternary nitrogen) were developed in an effort to extend the properties of pralidoxime. The known reactivators (pralidoxime, asoxime, obidoxime, trimedoxime, methoxime) and the new oximes were tested *in vitro* on a model of tabun and paraoxon-inhibited AChE. Monoquaternary reactivators were not able to exceed the reactivation potential of the standard oximes [77].

Musilek *et al.* [78] described 14 novel pyridinium oximes and presented the results of evaluation of their reactivation potential against human AChE inhibited by tabun, paraoxon, methylparaoxon and DFP in comparison with standard pyridinium oximes (pralidoxime, asoxime, obidoxime, trimedoxime, methoxime) and their previously tested oximes K027 and K203. The new bispyridinium oximes were structural analogs of K027 and K203. All new bispyridinium oximes contained oxime group in the position 4. Non-oxime moiety was attached to the position 4 of the second pyridinium ring. The linker between two nitrogen atoms was the prop-1,3-diyl as in K027, where linkers analogous to 3-4 C–C bonds were found to be the most promising for tabun or OP pesticides. It was found that three of the novel bispyridinium oximes provided certain degree of AChE reactivation.

Musilek *et al.* [57] reviewed the chemical structure and AChE reactivating potential of 330 oximes including those discussed in this study. There was no significant progress with the newly synthesized oximes in reactivation of AChE inhibited with different OPs when compared to the standards ones (pralidoxime, asoxime, obidoxime, trimedoxime, methoxime).

A series of nine bisquaternary pyridinium derivatives of 2-(hydroxyimino)-*N*-(pyridin-3-yl) acetamide having an amide group in conjugation to the oxime moiety have been synthesized [79]. These compounds were evaluated *in vitro* for their reactivation efficacy against sarin- and VX-inhibited human erythrocyte ghost acetylcholinesterase (*h*AChE) and compared with the reactivation efficacy of pralidoxime and obidoxime. Some of the compounds have shown better reactivation efficacy than pralidoxime and obidoxime against sarin- and VX-inhibited AChE.

Kliachyna *et al.* [80] synthesized a series of new uncharged functional AChE reactivators including heterodimers of tetrahydroacridine with 3-hydroxy-2-pyridine aldoximes and amidoximes. These novel molecules have shown some *in vitro* reactivation potential towards VX-, tabun- and paraoxon-inhibited human AChE that were equal or superior to those of the mono- and bis-pyridinium aldoximes currently used against nerve agent and pesticide poisoning, both in terms of reactivation efficiency and broadness of their activity spectrum.

Wei *et al.* [81] described a new class of uncharged non-pyridinium reactivators for nerve agent-inhibited acetylcholinesterase (AChE). Based on a dual site binding strategy, the authors conjugated the imidazolium aldoxime to different peripheral site ligands of AChE through alkyl chains. Compared with the known quaternary pyridinium reactivators, two of the resulting conjugates were effective non-pyridinium oxime conjugates exhibiting similar or superior ability to reactivate sarin-, VX- and tabun-inhibited AChE.

Among hundreds of oximes synthesized and studied during the past three decades, there were no major advances in designing compounds that would be effective against the nerve agents and other OPCs. The „universal" oxime that would be effective against all nerve agents and OPC has not yet been found [32, 59]. The monopyridinium oxime pralidoxime and bispyridinium oximes trimedoxime, obidoxime, asoxime and methoxime still remain the most effective oximes. Whether such an "universal" broad spectrum oxime, that would be equally effective against nerve agents and other OPCs, will be discovered is not presently known.

Until then, the following facts will remain. Asoxime is an effective antidote against toxic fluorophosphonates, especially soman, while trimedoxime and obidoxime are effective in the treatment of acute poisoning with OP insecticides. Trimedoxime and obidoxime are more effective oximes for the treatment of acute tabun poisonings than asoxime, but their potency to eliminate tabun-induced acute lethal effects is relatively low when administered at low human-relevant doses [11, 34, 64, 82-84]. More recent studies have indicated methoxime as a possible candidate to replace pralidoxime in the treatment of nerve agent poisoning [60, 61].

EFFICACY OF PYRIDINIUM OXIMES IN THE TREATMENT OF POISONING WITH WARFARE NERVE AGENTS

Among the many classes of oximes investigated so far, those with clinical application can be divided in two groups - the monopyridinium and bispyridinium oximes. Currently, the only used monopyridinium oxime is pralidoxime, while the most significant bispyridinium oximes comprise trimedoxime, obidoxime, asoxime and methoxime [32, 58].

It is important to say that none of the oximes discussed in this article were fully investigated in clinical trials possibly due to the problems of obtaining substances of GMP quality and ethical considerations. Despite of this, pralidoxime had been in used in clinical practice since 1950's. HI-6 was well tolerated in human volunteers [85-87], and in OP-poisoned patients [86] were it was shown to be effective against most OPCs except dimethoate and phosphamidon. As the result, asoxime was registered as a medicinal product for human use in former Yugoslavia in 1987.

Pralidoxime (PAM-2)

Pralidoxime was synthesized in the USA in 1955 [54]. Its four salts: chloride (PAM-2 Cl), methiodide, methylsulphate and mesylate (P2S) were investigated and introduced into the practice. Pralidoxime is effective in reactivating of AChE inhibited with sarin or VX [88-93], but not in the reactivation of the tabun- or soman-inhibited enzyme [94, 95].

Since pralidoxime, as quaternary pyridinium salt, penetrates the blood-brain barrier only to about 10% of its concentration in blood, its prodrug pro-2-PAM (*N*-methyl-1,6-dihydropyridine-2-carbaldoxime hydrochloride), apparently rapidly oxidized in brain to pralidoxime, was synthesized and tested in experimental animals. Pro-2-PAM provided some reactivation of sarin- and VX-inhibited AChE in the brain, blood and peripheral tissues of the guinea pigs, which was reflected by a limited ability to block or terminate seizures elicited by these agents. Pro-2-PAM provided marginal reactivation of blood, heart, and spinal cord AChE inhibited by cyclosarin, but was not effective against cyclosarin-induced seizures [51]. Pro-2-PAM was effective in suppression and elimination of seizures/status epilepticus in DFP-treated guinea pigs. In addition, pro-2-PAM provided a significant reduction of neurological damage in DFP-poisoned guinea pigs. In both cases pralidoxime was ineffective [96]. Clement [97] reported that pro-2-PAM was effective in protecting mice poisoned with DFP and sarin, but very slightly effective against soman in guinea pigs concluding that pro-2-PAM was not a significant improvement over pralidoxime with regard to prophylaxis against noted OPC. However, pro-2-PAM was less effective than pralidoxime against experimental poisoning with paraoxon in mice [98].

When given together atropine and pralidoxime act synergistically involving different mechanisms and the net effect is protection/treatment of animals poisoned with very high doses of certain OPs. Studies with nonhuman primates proved that the combination of atropine and pralidoxime protected against up to five times the LD_{50} of all known nerve agents except soman [9] and against 128 LD_{50} of paraoxon [99].

Pralidoxime may be given by slow intravenous injection over 5 to 10 minutes, by intravenous infusion over 15 to 30 minutes, or by subcutaneous or intramuscular injection; it has also been given orally [100]. The minimum therapeutic concentration of pralidoxime in plasma is 4 μg/mL and this level is reached in about 16 minutes after a single injection of 600 mg pralidoxime chloride [101].

In the treatment of OPC poisoning atropine and pralidoxime should be given as soon as possible and maintained as long as needed. When the effects of atropine become apparent, 1 to 2 g of pralidoxime chloride, iodide, or mesilate, should be given intramuscularly or intravenously and repeated after one hour and then every 8 to 12 hours if necessary. Alternatively, pralidoxime chloride may be given in an initial dose of 30 mg/kg given by intravenous infusion over 20 minutes, or by intravenous injection over at least 5 minutes if pulmonary oedema is present or infusion cannot be given; the initial dose is then followed by intravenous infusion at a rate of 8 mg/kg per hour [100, 102, 103]. Pralidoxime is relatively short acting (the apparent half-life of pralidoxime chloride is 74-77 minutes) and repeated doses may be needed, especially where there is any evidence of continuing absorption of the OPC [101]. The treatment should be continued until obviously no longer needed and the decision about this can be made on the basis of clinical status of the patient, relatively high AChE activity in erythrocytes when compared to control values, and the absence of OPCs and/or OP metabolites in urine [103].

In some countries, auto-injectors are available for emergency use containing pralidoxime, either alone or combined with atropine and/or avizafone, a water-soluble prodrug of diazepam [100]. French army considered inclusion of water soluble benzodiazepine midazolam in its autoinjector [27]. Typical doses are 600 mg of pralidoxime chloride or 500 mg of pralidoxime mesilate given intramuscularly up to 3 times, depending on symptoms. In severe poisoning pralidoxime may be given as a continuous infusion of 200 to 500 mg/hour, titrated against response. A maximum dose of 12 g in 24 hours has been suggested [100].

Pralidoxime has been very well tolerated in most cases [104]. Adverse effects of pralidoxime in volunteers include dizziness, drowsiness, blurred vision, occasional diplopia, impaired accomodation, nausea, headache, tachycardia, hyperventilation, increased systolic and diastolic blood pressure and muscular weakness. Tachycardia, laryngospasm, and muscle rigidity have been attributed to giving pralidoxime intravenously at too rapid a rate. Large doses of pralidoxime may cause transient neuromuscular blockade [6, 100, 101, 104, 105]. Forty to 60 minutes after intramuscular injection, mild to moderate pain may be experienced at the site of injection. Elevations in AST and/or ALT enzyme levels in blood were observed in 1 of 6 normal volunteers given 1200 mg of pralidoxime chloride intramuscularly, and in 4 of 6 volunteers given 1800 mg intramuscularly. Levels returned to normal in about two weeks [101]. Pralidoxime should be used cautiously in patients with renal impairment and a reduction in dosage may be necessary. Caution is also required in giving pralidoxime to patients with myasthenia gravis as it may precipitate a myasthenic crisis [100].

The clinical experience with the use of pralidoxime iodide, given with atropine and diazepam, in the treatment of the victims of Tokyo sarin attack victims in 1995 was very favourable [26, 106]. However, pralidoxime should not be recommended as the drug of choice for management of poisoning with all OPC due to its lack of efficacy against tabun and soman [50].

Tang *et al.* [107] conducted a retrospective cohort study of 152 cases aimed to clarify the efficacy of two treatments for patients with severe acute OP pesticide poisoning. Patients admitted to the hospital for OPC poisoning were divided into two groups with different therapeutic regimens: group A was administered a repeated pulse intramuscular injection of pralidoxime chloride, and group B received the same initial dosage of atropine and pralidoxime chloride, but pralidoxime chloride intravenous therapy was administered for only 3 days, regardless of the length of atropine therapy. The total dose of atropine was 57.40±15.14 mg in group A and 308.26±139.16 mg in group B; group A received less atropine than did group B. The length of intensive care unit stay in group A was reduced, and group A had fewer atropine adverse effects. It was concluded that in patients with severe poisoning, group A used less atropine, had fewer atropine adverse effects, and had a shorter stay in hospital. The authors suggested that therapy should be started as early as possible using a sufficient amount of pralidoxime chloride started intramuscularly in combination with atropine and that the drugs should not be prematurely discontinued.

Trimedoxime (TMB-4)

Trimedoxime was synthesized in the USA in 1957 [108]. It is the only one of the major bispyridinium oximes with a propylene linker between the two pyridinium rings. It was shown that trimedoxime is a more potent reactivator of the DFP-inhibited AChE than pralidoxime [109] and a better reactivator than obidoxime in the case of the tabun-inhibited enzyme [110].

Trimedoxime was the first oxime that was efficient in the treatment of animals intoxicated with tabun [67]. It could also protect the animals poisoned with sarin or VX, but not the ones intoxicated with soman [67, 94, 111]. However, trimedoxime was a considerbly toxic oxime since it was shown in mice that its LD_{50} is 3, 4 and 8 times less than that for pbidoxime, pralidoxime and HI-6, respectively [112].

Obidoxime (LüH-6, Toxogonin)

After its discovery in 1964, obidoxime showed a significant potential as an antidote in poisonings with OPCs [113]. Given with atropine, obidoxime efficiently protected experimental animals against poisoning with tabun [94, 114], sarin [94, 114] and VX [114]. Obidoxime was more effective than trimedoxime as an antidote against intoxication with tabun [115]. Similarly to pralidoxime and trimedoxime, obidoxime was also inefficient in soman poisoning in mice [67], guinea-pigs [94] and primates [116]. Obidoxime was more efficient than HI-6 against tabun intoxication. Similarly, obidoxime also was more effective as a medical counter-measure following intoxication with most cholinesterase inhibiting insecticides, while asoxime was considered to be a better antidote against soman inhibited AChE [62]. However, obidoxime was not a potent reactivator of tabun-inhibited human AChE [61, 117]. In contrast to trimedoxime, obidoxime, when administered with atropine in pyridostigmine-pretreated guinea-pigs, could provide some protection against soman as well [94].

When administered to human volunteers by intramuscular route obidoxime 5 mg/kg produced a plasma concentration >4 mg/L, from 5 min after injection to 3 hours [118]. Adverse effects of obidoxime in male volunteers were described as pallor, nausea, burning sensation, headache, generalized weakness, sore throat, and paresthesia of the face [6, 15, 119]. Transient hepatotoxic effects of obidoxime were reported in about 10% of patients [6, 120].

Obidoxime can be given with atropine in the treatment of OP poisoning in a usual initial dose of 250 mg (4 mg/kg) by slow intravenous injection. This may be followed by intravenous infusion of 750 mg over 24 hours, continued until the concentration of organophosphate is below critical levels; alternatively, repeated doses of 4 to 8 mg/kg may be given at intervals of 2 to 4 hours. It has also been given by intramuscular injection [121].

Asoxime (HI-6)

The first oxime that could reactivate soman-inhibited AChE and provide some protection of the animals experimentally poisoned with this nerve agent was synthesized in 1966. It has been shown that asoxime is more potent than obidoxime and HS-6 in the protection of various rodent species from intoxication with soman [61, 91, 94, 122], as well as sarin and particularly VX [67, 94]. Asoxime could not reactivate tabun-inhibited AChE [70, 123, 124] and it was inefficient, when used as the only oxime, against poisoning with tabun [59, 67, 91, 94]. However, other studies found that asoxime can provide a similar degree of protection against tabun and soman intoxication in primates [115]. It is also suggested that sufficiently high doses of asoxime could even protect rats from the multiple lethal doses of tabun [125]. Asoxime given with atropine protected guinea-pigs from poisoning against 5 LD_{50} of soman or cyclosarin [126]. It was suggested that asoxime was superior to the other oximes in treatment of cyclosarin-poisoned animals [50].

The acute toxicity of asoxime is the lowest among the six oximes (asoxime, pralidoxime, trimedoxime, obidoxime, HLö-7, methoxime) [67, 112, 127, 128], with LD_{50} value after i.m. administration to rats of 781 mg/kg.

Asoxime was introduced into clinical practice in Serbia in the early 1980s. It was administered to more than 200 patients poisoned with OP insecticides and volunteers. A clinical study with asoxime, administered intramuscularly at doses up to 500 mg to 22 healthy volunteers, revealed no adverse effects, whereas in patients poisoned by several OP insecticides, asoxime provided rapid reactivation of AChE in almost all cases except in dimethoate and phosphamidon poisoning [33, 83, 85, 129]. Similar results regarding asoxime tolerance were later found in a double-blind, placebo-controlled, single-dose study in 24 volunteers [87]. Safety of asoxime administration was in correlation with the recent report of Pohanka and his team [130] who found that therapeutic doses of asoxime are safe in dogs adding that tenfold therapeutic dose can cause hyperglycemia.

Clinical studies have shown that asoxime dosed at either 250 or 500 mg by intramuscular route reached plasma concentrations > 4 mg/L in 4-6 min. This concentration was maintained for 125 min following the lower dose (250 mg) and 200 min following the higher dose (500 mg) [83, 85].

Asoxime is considered to be a very promising bispyridinium oxime in medical treatment following exposure to most nerve agents. For these reasons asoxime is involved in the equipment of the Czech, Slovak, Swedish and Canadian Armies as antidote against nerve agent intoxication [27, 59] and under development in other countries [27, 61]. A disadvantage of asoxime compared to other available oximes is the lack of stability in aqueous solutions [62, 85].

In addition to the four pyridinium oximes that were introduced to clinical practice described above, the oximes HLö-7 and methoxime (MMB-4) require attention after proving to be efficient against warfare nerve agents in animal studies.

HLö-7

HLö-7 was synthesized in 1986 [131]. Only HLö-7 could reactivate AChE inhibited by any of the four major nerve agents *in vitro* and *in vivo* [131-133], as well as AChE inhibited by cyclosarin [134]. In addition, HLö-7 more efficiently restored the neuromuscular transmission impaired by *in vitro* superfusion of the neuromuscular preparation with tabun, sarin, soman or cyclosarin compared to pralidoxime, obidoxime and asoxime [135]. It was found that HLö-7 induced a significant reactivation of AChE in mice diaphragm inhibited with tabun, sarin, soman and cyclosarin [136]. Although both asoxime and HLö-7 can antagonize sarin-induced hypothermia (proving that, when given with atropine, they can pass the blood-brain barrier and gain access to the CNS) [136, 137], the toxicity of HLö-7 was 2.5 times higher than that of asoxime [137]. The cardiovascular tolerability of HLö-7 was similar, but still not as good as of asoxime, at least not in anaesthetised guinea-pigs [138]. HLö-7 appears to be more effective than asoxime against tabun and VX poisoning and less effective against sarin, soman and cyclosarin intoxication [134-138]. HLö-7-induced protection against tabun poisoning in guinea pigs was significantly better compared to asoxime, while asoxime was only slightly more efficient than HLö-7 in soman poisoning [140, 141]. However, Kuča *et al.* [142, 143] reported that HLö-7 was not a good reactivator of tabun-inhibited AChE *in vitro* possibly due to the presence of the lone electron pair on the tabun amide group.

The pharmacokinetic profile of HLö-7 was similar to that of asoxime. The mean absorption half-time of HLö-7 was about 14 min after intramuscular administration. Maximum HLö-7 concentration in plasma was reached after 30 min and the half-time of elimination was about 45 minutes [139].

Methoxime (MMB-4)

Methoxime is a structural analog of toxogonin showing potent antimuscarinic activity [144]. It is a bispyridinium 4,4'-bisaldoxime first described by Hobbiger and Sadler [145]. Methoxime is currently being proposed as a leading candidate for the replacement of pralidoxime as an oxime reactivator of nerve agent-inhibited AChE activity that can prevent lethality after warfare nerve agent exposure [27, 60, 61]. So far, there are no reports regarding clinical studies of methoxime [61].

Methoxime was less toxic in mice (LD_{50}=441 mg/kg, i.m.) than obidoxime (188 mg/kg, i.m.) but more toxic compared to asoxime (671 mg/kg i.m.) [128]. It was shown that methoxime was inferior to asoxime in reactivation of peripheral AChE from soman-poisoned animals, although both oximes were better than pralidoxime. In soman-poisoned rats methoxime was not successful in protection or reactivation of soman-inhibited AChE [146]. Luo and coworkers [147] found that methoxime was not a good reactivator of soman-inhibited AChE obtained from human and three monkey species being less effective compared to asoxime. This is consistent with the findings reported by other authors [61].

Methoxime was also less effective in reactivation of cyclosarin-inhibited rat brain AChE and sarin inhibited human or monkey AChE compared to asoxime [56, 61]. Methoxime was found to be a weak reactivator of tabun inhibited human AChE [109]. However, experimental data indicate that methoxime was superior to pralidoxime in reactivating OP-inhibited AChE and in preventing lethality in OP poisoned animals [60]. It was also reported that methoxime was the most effective oxime in reactivation of ChE in blood and peripheral tissues in guinea pigs poisoned by sarin, cyclosarin, VX and VR [148]. In addition, methoxime was reported to be better AChE reactivator than pralidoxime in paraoxon- and methylparaoxon-poisoned rats [149, 150].

COMBINATION OF OXIMES

The idea of using a combination of two already tested oximes in order to improve their antidotal efficacy in OP poisoning was initiated by Maksimović *et al.* [114] and Kovačević *et al.* [151], supported by Worek *et al.* [40] and further evaluated by Czech scientists. Kassa and his team have tested a few combinations of existing bispyridinium oximes against soman, sarin, cyclosarin and tabun poisoning *in vivo*. In soman poisoning, they have tested the ability of two combinations of asoxime+trimedoxime and asoxime+K203 to reduce neurotoxic signs and symptoms using a functional observational battery [152, 153]. It was found that both combinations with atropine improved the neuroprotective effecacy of antidotal treatment of acute soman poisonings when compared to the individual oximes. The same combinations of oximes were tested in sarin- or cyclosarin-poisoned rats and compared to the potency of individual oximes but no significant differences were found [153, 154].

Three combinations of bispyridinium oximes were tested in tabun-poisoned rats. The combinations of asoxime obidoxime and asoxime+K203 provided beneficial effects in reduction of acute toxicity of tabun and reactivation of tabun-inhibited AChE [155]. The combination of asoxime+trimedoxime showed improvement in reactivation of tabun-inhibited AChE in blood and diaphragm but also in pontomedullar area and frontal cortex of the brain [59].

EFFICACY OF PYRIDINIUM OXIMES IN POISONING WITH OP INSECTICIDES

The efficacy of pyridinium oximes has been demonstrated in experimental conditions and in many cases of patients poisoned with OP insecticides who have been treated in European clinics. In those cases the recommendations for pralidoxime dosing proposed by the WHO were followed [14].

Contrary to these findings, reports from Asia indicate that pralidoxime treatment was not sufficiently effective in their patients. However, in their studies pralidoxime was not used as recommended by the WHO [156-160]. In addition, these studies were apparently poorly designed due to suboptimal dose, short duration of treatment, long delay between patient exposure and pralidoxime administration, the chemical structure of OP pesticides not being taken in consideration, and so on [16, 33].

A particular problem in interpreting the beneficial role and efficacy of pyridinium oximes in clinical practice is a relative lack of published data, especially those evaluated in controlled clinical trials. Studies related to the efficacy of oximes in clinical setting showed the heterogeneity of therapeutic approaches (*i.e.*, dosing regimen, oxime choice and final outcome of the treatment) [16].

Jokanović and coauthors [33] reviewed the experience gained in clinical management of 296 patients poisoned with OP insecticides at Clinic of Toxicology of the National Poison Control Center in Military Medical Acedemy in Belgrade during the period of 1998-2007. Poisoning with dimethyl OP was confirmed in 246 patients (malathion 153, dimethoate 69, dichlorvos 4, fenitrothion 6, monocrotophos 8), with diethyl OP in 38 patients (diazinon 21, parathion 9, phorate 1, phoxim 1, phosalone 3, chlorpyriphos 2, quinalphos 1) and in 12 patients OP was not fully identified. The majority of poisonings (92%) was due to deliberate ingestion of OP insecticides. Accidental inhalational poisonings were registered in 3.8% patients, accidental ingestion in 4.1% patients and there were two cases of deliberate intramuscular and intravenous application of OPs. Suicidal poisonings were usually oral, and the severity of poisonings depended on the toxicity, the amount of OP insecticides taken, and the time from exposure to admission to the Clinic of Toxicology and initiation of treatment. In this study 15.9% of OP-poisoned patients died, which was consistent with the findings reported in other studies [161], where case fatality generally greater than 15%. About 42% of deceased patients died within seven days, and lethal outcome occurred within hours in 18% patients due to the severity of poisoning.

In this study 74 patients were treated with pralidoxime methylsulphate (Contrathion). The dosage regimen for severely poisoned patients included a loading dose of 400 mg (two bottles) by slow intravenous infusion, subsequently 200 mg/h continuous intravenous infusion administered in the first day, and continued as long as clinical signs of poisoning and OP insecticide and/or its metabolite were present in urine and blood. In moderate poisoning 400 mg of Contrathion was followed by 400 mg/3-4 h. However, due to the economic crisis in Serbia, the Clinic of Toxicology did not have any oxime available for several years and the patients were treated with atropine and diazepam only. In patients treated with atropine and oxime or atropine only there were no differences in mortality. Atropine consumption was lower in the group of patients treated with an oxime, but the difference was significant only in the group of severely poisoned patients where it was reduced by 30%.

Eddleston and colleagues [162] conducted a prospective study on 802 patients self-poisoned with chlorpyrifos, dimethoate, or fenthion. Compared with chlorpyrifos (8.0%), the proportion of fatalities was significantly higher with dimethoate (23.1%) or fenthion (16.2%) as was the proportion requiring endotracheal intubation (chlorpyrifos 15.0%, dimethoate 35.2%, fenthion 31.3%). Patients poisoned by the diethyl OP pesticide (chlorpyrifos) responded well to pralidoxime, whereas those to the two dimethyl OP pesticides (dimethoate, fenthion) responded poorly. The poor efficacy of pralidoxime in the treatment of human dimethoate and fenthion poisonings was in agreement with experimental studies reported by Jokanović and Maksimović [83] who found that the antidotal efficacy of obidoxime, trimedoxime, pralidoxime and asoxime (given with atropine and diazepam) in rats dosed with 2 LD_{50} of the dimethoate was low. However, there was a discrepancy between fenthion-poisoned patients and animals in that pralidoxime was ineffective as an antidote in patients, whereas the four oximes showed considerable efficacy in rats.

In a randomized controlled trial, Pawar and colleagues [163] studied the effects of very high doses of pralidoxime iodide (2 g loading dose, then 1 g either every hour or every 4 h for 48 h, then 1 g every 4h until recovery) in 200 patients with moderate OP poisoning (excluding severely ill patients). The OP pesticides involved were chlorpyrifos (diethyl OP) and dimethoate (dimethyl OP). The dosing regimen was associated with reduced case fatality, fewer cases of pneumonia, and reduced time on mechanical ventilation. This study suggests that large doses of pralidoxime could have a positive effect on patients if they are treated early and have good supportive care.

An interesting clinical trial where PAM-2 chloride was investigated in 235 patients self-poisoned with OP insecticides was reported by Eddleston and coworkers [164]. The patients were randomized to receive pralidoxime (n=121) or saline placebo (n=114). Pralidoxime produced substantial and moderate erythrocyte AChE reactivation in patients poisoned by diethyl and dimethyl compounds, respectively. Mortality was nonsignificantly higher in patients receiving pralidoxime: 30/121 (24.8%) receiving pralidoxime died, compared with 18/114 (15.8%) receiving placebo. The authors concluded that, despite clear reactivation of erythrocyte AChE in diethyl OP insecticide poisoned patients, they found no evidence that treatment with pralidoxime improves survival or reduces need for intubation in patients with OP insecticide poisoning. However, this study suffered from a major methodological problem of randomization as the authors wrote that "more severely poisoned patients were allocated to pralidoxime" than to the placebo treatment. Based on this statement it was expected that more patients would die in the pralidoxime group despite the oxime treatment. This was supported by another sentence present in this report that "patients died sooner in the pralidoxime group". In addition, there are other problems with this study such as the lack of atropine inclusion in both groups as World Health Organization [12] recommended treatment of OP poisoned patients with atropine, pralidoxime and diazepam, and not only with prlaidoxime. There were also relatively small number of patients included in the study, relatively long and variable period between self-poisoning and administration of pralidoxime (mean 4.3 hours, intervals 2.9-7.8 hours) probably having an impact on aging of phosphorylated AChE, ethical issues such as why the patients in placebo group were allowed to die without the best possible treatment according to the WHO recommendations. Because of these and other problems, the results of this study shall be considered with caution.

Kušić *et al.* [86] have tested asoxime in OP pesticide poisoning in 60 patients. Asoxime was administered four times a day as a single 500 mg i.m. injection with atropine and diazepam treatment. Oxime therapy was started on admission and continued for 2 to 7 days. Most patients were treated with asoxime and nine severely poisoned patients with quinalphos were treated pralidoxime chloride (1000 mg four times per day). Asoxime rapidly reactivated human erythrocyte AChE inhibited by diethoxy OPCs (phorate, pyridaphenthion, quinalphos) as well as that inhibited by dichlorvos, a dimethoxy OPC. AChE inhibited with other dimethoxy OPCs (dimethoate and phosphamidon) was reported to be resistant to asoxime treatment, whereas reactivation with malathion was slow (reactivation half-time 10 hours). Both asoxime and pralidoxime successfully reactivated AChE in quinalphos-poisoned patients, with asoxime acting as a faster AChE reactivator than pralidoxime.

Willems *et al.* [165] reported that ethyl parathion and methyl parathion poisoning could be effectively treated with pralidoxime methylsulfate (plasma concentrations 4 mg/L) and atropine when pesticide concentrations in plasma were relatively low. In severe poisoning with pesticide levels in plasma above 30 μg/L, high pralidoxime concentrations in plasma (14.6 mg/L) did not provide any improvement. In addition, pralidoxime at concentrations of 6.3 mg/L was not effective in AChE reactivation in dimethoate poisoning where AChE was inhibited with its active metabolite omethoate.

It was reported that in parathion poisoning, obidoxime dose of 250 mg i.v. administered as a bolus followed by infusion of 750 mg per day was effective, but AChE reactivation following severe poisoning did not occur until the concentration of paraoxon in plasma became low [166, 167]. Oxydemeton methyl poisoning responded to obidoxime therapy only when the oxime was instituted shortly after poisoning. In cases when obidoxime treatment was started too late, there was no reactivation of erythrocyte AChE and one out of six patients died.

Nine patients intoxicated with OP pesticides were treated with pralidoxime methylsulfate (Contrathion) using a dose of 4.42 mg/kg as a bolus injection followed by continuous infusion of 1.14 mg/kg/h. In patients poisoned with ethylparathion or methylparathion AChE reactivation could be obtained at low oxime concentrations (2.88 mg/L). In others, however, an oxime concentrationas high as 14.6 mg/L was ineffective. The therapeutic effect of the oxime apparently depended on the plasma concentrations of the OP pesticides. Due to AChE reinhibition, reactivation was absent as long as these concentrations remained above 30 μg/L [168].

In a clinical study involving 63 patients poisoned with organophosphorus insecticides, patients were divided into three groups: one was treated with atropine only, while the other two received atropine and either pralidoxime or obidoxime. Initial and maintenance intravenous doses for pralidoxime were 30 mg/kg and 8 mg/kg/h, respectively, and 8 mg/kg and 2 mg/kg/h, respectively, for obidoxime. The major clinical findings or AChE activities at the time of admission did not show statistically significant differences among the groups. Although the severity of intoxications, based on respiratory complications and duration of hospitalization, was higher in the atropine plus oxime groups, 12% and 50% of patients in the atropine and atropine plus obidoxime groups died, respectively. No mortality was found in the pralidoxime plus atropine group. Incidence of recurrent twitching and convulsions, repeated respiratory arrest, required mechanical respiration, required intensive care unit therapy and duration of hospitalization were smaller in the atropine plus obidoxime group than in the atropine plus pralidoxime group. Three of the patients who received the obidoxime combination therapy developed hepatitis and two of them died due to hepatic failure, which may indicate overdosage of obidoxime [169].

In the study reported by Tsai and coworkers [170] pralidoxime was administered to 56 patients within 48 hours of poisoning with OP pesticides. The average duration of pralidoxime treatment was 5.5±4.9 days and the mean total dose was 16.4±12.4 g. The mortality and length of stay did not improve with pralidoxime treatment. Days spent on ventilator support were decreased in patients treated with pralidoxime (6.7±1.9 days) compared to those who were not (23.0±4.8 days). However, there are many limitations in this study (relatively small number of treated patients, lack of data regarding the OP pesticides involved, dose taken, and AChE inhibition).

AChE inhibited by several OP pesticides, including dimethoate, demethon, triamiphos, ethoprophos, profenofos, fenamiphos, and pyridafenthion, was resistant to any attempt of reactivation with any oxime, probably due to the variations in phosphoryl moiety and distribution of electronic charge [16, 30, 33, 83].

CONCLUDING REMARKS

Despite of enormous efforts devoted to synthesis and development of new pyridinium oximes as potential antidotes against poisoning with OPCs in the past few decades, there were no major advances in designing compounds that would be effective against the nerve agents and other OPCs. Only four compounds so far are applicable in human medicine. However, they differ in their activity in poisoning with warfare nerve agents and OP insecticides. There is still no universal broad spectrum oxime capable of protecting against all known organophosphorus compounds. Asoxime seems to be a promising choice for nerve agent poisoning because it was able to protect experimental animals from toxic effects and improve survival of animals poisoned with sublethal doses especially in the cases of soman and cyclosarin. Asoxime was also effective in poisoning with OP insecticides. Methoxime is currently being proposed as a leading candidate for replacement of pralidoxime as an oxime reactivator of nerve agent-inhibited AChE activity that can prevent lethality after warfare nerve agent exposure. Available experimental evidence suggests that there are no clinically important differences between pralidoxime, obidoxime and asoxime in the treatment of nerve agent poisoning. On the other hand, obidoxime appears to be the best choice for the treatment of poisoning with OP insecticides. However, it is necessary to rigorously test these statements in randomized clinical trials.

CONFLICT OF INTEREST

The authors confirm that this article content has no conflict of interest.

ACKNOWLEDGEMENTS

This research was supported by a grant from the Ministry of Education, Science and Technological Development of the Republic Serbia (Project 175045).

DISCLOSURE

This is an updated Chapter of the earlier review article [by Jokanovic, Milan, and Milica Prostran. "Pyridinium oximes as cholinesterase reactivators. Structure-activity relationship and efficacy in the treatment of poisoning with organophosphorus compounds." *Current Medicinal Chemistry* 16.17 (2009): 2177-2188], with additional text and references.

REFERENCES

[1] Pesticides and Repelents. In *Martindale The Complete Drug Reference*, S.C. Sweetman, Ed.; Pharmaceutical Press, 36th edition, London, **2009**; pp. 2034-2051.

[2] Mesulam, M.M.; Guillozet, A.; Shaw, P.; Levey, A.; Duysen, E.G.; Lockridge, O. Acetylcholinesterase knockouts establish central cholinergic pathways and can use butyrylcholinesterase to hydrolyze acetylcholine. *Neuroscience*, **2002**, *110*(4), 627-639.

[3] Darvesh, S.; Hopkins, D.A.; Geula, C. Neurobiology of butyrylcholinesterase. *Nat. Rev. Neurosci.*, **2003**, *4*(2), 131-138.

[4] Jokanović, M.; Maksimović, M. Abnormal cholinesterase activity: understanding and interpretation. *Eur. J. Clin. Chem. Clin. Biochem.*, **1997**, *35*(1), 11-16.

[5] Reiner, E.; Pleština, R. Regeneration of cholinesterase activities in humans and rats after inhibition by O,O-dimethyl-O-dimethyl-2,2-dichlorvinyl phosphate. *Toxicol. Appl. Pharmacol.*, **1979**, 49(3), 451-454.

[6] Eyer, P. The role of oximes in the management of organophosphorus pesticide poisoning. *Toxicol. Rev.*, **2003**, *22*(3), 165-190.

[7] Karczmar, A.G. Acute and long lasting central actions of organophosphorus agents. *Fundam. Appl. Toxicol.*, **1984**, *4*(2Pt 2), S1-S17.

[8] Sidell, F.R. Nerve agents. In: *Textbook of Military Medicine. Medical Aspects of Chemical and Biological Warfare*; Zajtchuk, R. Ed.; Office of The Surgeon General Department of the Army, USA, **1997**, pp. 187.

[9] Stojiljković, M.P.; Pantelić, D.; Maksimović, M. *Tabun, sarin, soman and VX poisoning in rats: kinetics of inhibition of central and peripheral acetylcholinesterase, ageing, spontaneous and oxime-facilitated reactivation.* Proceedings of the 7th International Symposium on Protection against Chemical and Biological Agents. Stockholm, Sweden June 15-19, **2001**, pp. 1-12.

[10] Worek, F.; Diepold, C.; Eyer, P. Dimethylphosphoryl-inhibited human cholinesterases: Inhibition, reactivation, and aging kinetics. *Arch. Toxicol.*, **1999**, *73*(1), 7-14.

[11] Mason, H.J.; Sams, C.; Stevenson, A.J.; Rawborne, R. Rates of spontaneous reactivation and aging of acetylcholinesterase in human erythrocytes after inhibition by organophosphorus pesticides. *Hum. Exp. Toxicol.*, **2000**, *19*(9), 511-516.

[12] World Health Organization. *Organophosphorus insecticides: a general introduction.* Environmental Health Criteria, Geneva, **1986**; vol. 63.

[13] Karchmar, A.G. Anticholinesterases and war gases. In *Exploring the vertebrate central cholinergic nervous system*, Karchmar A.G. Ed.; Springer, **2007**; pp. 237-310.

[14] International Programme on Chemical Safety Poisons Information. *Organophosphorus pesticides.* Monograph G001. World Health Organization, Geneva, **1989**, Updated 1998. Available at: http:// www.inchem.org/documents/pims/chemical/ pimg001.htm

[15] Marrs, T.C.; Vale, J.A. Management of organophosphorus pesticide poisoning. In: *Toxicology of Organophosphorus and Carbamate Compounds*, Gupta R.C., Ed.; Elsevier Academic Press, Amsterdam, **2006**, pp.715-733.
[16] Jokanović, M. Medical treatment of acute poisoning with organophosphorus and carbamate pesticides. *Toxicol. Lett.*, **2009**, *190*(2), 107-115.
[17] Jokanović, M. Biotransformation of organophosphorus compounds. *Toxicol.*, **2001**, *166*(3), 139-160.
[18] McDonough, J.H.Jr.; McLeod, C.G.Jr.; Nipwoda, M.D. Direct micro-injection of soman or VX into the amygdala produces repetitive limbic convulsions and neuropathology. *Brain Res.*, **1987**, *435*(1-2), 123-137.
[19] Zilker, T. Medical management of incidents with chemical warfare agents. *Toxicol.*, **2005**, *214*(3), 221-231.
[20] International Programme on Chemical Safety Evaluation (IPCS). *Antidotes for poisoning by organophosphorus pesticides. Monograph on atropine.* World Health Organization, Geneva, **2002**.
[21] Abedin, M.J.; Sayeed, A.A.; Basher, A.; Maude, R.J.; Hoque, G.; Faiz, M.A. Open-label randomized clinical trial of atropine bolus injection versus incremental boluses plus infusion for organophosphate poisoning in Bangladesh. *J. Med. Toxicol.*, **2012**, *8*(2), 108-117.
[22] Sellström, A. Anticonvulsants in anticholinesterase poisoning. In: *Clinical and experimental toxicology of organophosphates and carbamates,* Ballantyne B, Marrs T.C., Eds.; Butterworth-Heinemann, Oxford-Baltimore, **1992**; pp. 578-586.
[23] Marrs, T.C. *Diazepam. Antidotes for Poisoning by Organophosphorus Pesticides.* International Programme on Chemical Safety, World Health Organization, **2004**.
[24] Johnson, M.K.; Vale, J.A. Clinical management of acute organophosphate poisoning: an overview. In: *Clinical and Experimental Toxicology of Organophosphates and Carbamates,* Ballantyne B, Marrs T.C., Eds.; Butterworth-Heinemann, Oxford-Baltimore, **1992**; pp. 528-535.
[25] Bokonjić, D.; Rosić N. Anticonvulsive and protective effects of diazepam and midazolam in rats poisoned by highly toxic organophosphorus compounds. *Arh. Hig. Rada Toksikol.*, **1991**, *42*(4), 359-365.
[26] Sidell, F.R.; Newmark, J.; Donough, J.H. Nerve agents. In *Medical aspects of chemical warfare*; Tuorinsky SD, Ed., Borden Institute (U.S.), Washington, **2009**, pp. 155-219.
[27] Masson, P. Evolution of and perspectives on therapeutic approaches to nerve agent poisoning. *Toxicol. Lett.*, **2011**, *206*(1), 5-13.
[28] Delacour, H.; Dorandeu, F. Nerve agent-induced acute poisoning: principles of medical management. *Réanimation*, **2014**, *23*(2), S458-S465.
[29] Lotti, M. Treatment of acute organophosphate poisoning. *Med. J. Aust.*, **1991**, *154*(1), 51-55.
[30] Bismuth, C.; Inns, R.H.; Marrs, T.C. Efficacy, toxicity and clinical use of oximes in anticholinesterase poisoning. In: *Clinical and Experimental Toxicology of Organophosphates and Carbamates*, Ballantyne B, Marrs T.C., Eds.; Butterworth-Heinemann, Oxford-Baltimore, **1992**; pp. 555-577.
[31] Johnson, M.K.; Jacobsen, D.; Meredith, T.J.; Eyer, P.; Heath, A.J.;Ligtenstein, D.A.; Marrs, T.C.; Szinicz, L.; Vale, A.J.; Haines, J.A. Evaluation of antidotes for poisoning by organophosphorus pesticides. *Emerg. Med.*, **2000**, *12*(1), 22-37.
[32] Jokanović, M.; Prostran, M. Pyridinium oximes as cholinesterase reactivators. Structure-activity relationship and efficacy in the treatment of poisoning with organophosphorus compounds. *Curr. Med. Chem.*, **2009**, *16*(17), 2177-2188.
[33] Jokanović, M.; Antonijević, B.; Vučinić, S. Epidemiological studies of anticholinesterase pesticide poisoning in Serbia. In: *Anticholinesterase Pesticides: Metabolism, Neurotoxicity and Epidemiology*, Satoh T, Gupta R.C., Eds.; John Wiley & Sons, **2010**; pp. 481-494.
[34] Jokanović, M.; Stojiljković, M.P. Current understanding of the application of pyridinium oximes as cholinesterase reactivators in treatment of organophosphate poisoning. *Eur. J. Pharmacol.*, **2006**, *553*(1), 10-17.
[35] Maxwell, D.M.; Koplowitz, I.; Worek, F.; Sweeney, R.E. A structure-activity analysis of the variation in oxime efficacy against nerve agents. *Toxicol. Appl. Pharmacol.*, **2008**, *231*(2), 157-164.
[36] Luo, C.; Saxena, A.; Smith, M.; Garcia, G.; Radić, Z.; Taylor, P.; Doctor, B.P. Phosphoryl oxime inhibition of acetylcholinesterase during oxime reactivation is prevented by edrophonium. *Biochem.*, **1999**, *38*(31), 9937-9947.

[37] Ashani, Y.; Bhattacharjee, A.K.; Leader, H.; Saxena, A.; Doctor, B.P. Inhibition of cholinesterases with cationic phosphonyl oximes highlights distinctive properties of the charged pyridine groups of quaternary oxime reactivators. *Biochem. Pharmacol.*, **2003**, *66*(2), 191-202.
[38] Worek, F.; Eyer, P.; Aurbek, N.; Szinicz, L.; Thiermann, H. Recent advances in evaluation of oxime efficacy in nerve agent poisoning by *in vitro* analysis. *Toxicol. Appl. Pharmacol.*, **2007**, *219*(2-3), 226-234.
[39] Kiderlen, D.; Eyer, P.; Worek, F. Formation and disposition of diethylphosphorylobidoxime, a potent anticholinesterase that is hydrolysed by human paraoxonase (PON1). *Biochem. Pharmacol.*, **2005**, *69*(12), 1853-1867.
[40] Worek, F.; Eyer, P.; Kiderlen, D.; Thiermann, H.; Szinicz, L. Effect of human plasma on the reactivation of sarin-inhibited human erythrocyte acetylcholinesterase. *Arch. Toxicol.*, **2000**, *74*(1), 21-26.
[41] Reiner, E. Inhibition of acetylcholinesterase by 4,4'-bipyridine and its effect upon phosphylation of the enzyme. *Croat. Chem. Acta*, **1986**, *59*(4), 925-931.
[42] Carpentier, P.; Delamanche, I.S.; Le Bert, M.; Blanchet, G.; Bouchard, C. Seizure-related opening of the blood-brain barrier induced by soman: possible correlation with the acute neuropathology observed in poisoned rats. *Neurotoxicol.*, **1990**, *11*(3), 493-508.
[43] Grange-Messent, V.; Bouchaud, C.; Jamme, M.; Lallement, G.; Foguin, A.; Carpentier, P. Seizure-related opening of the bloodbrain barrier produced by the anticholinesterase compound, soman: new ultrastructural observations. *Cell. Mol. Biol.*, **1999**, *45*(1), 1-14.
[44] Shrot, S.; Markel, G.; Dushnitsky, T.; Krivoy, A. The possible use of oximes as antidotal therapy in organophosphate-induced brain damage. *Neurotoxicol.*, **2009**, *30*(2), 167-173.
[45] Abdel-Rahman, A.; Shetty, A.K.; Abou-Donia, M.B. Acute exposure to sarin increases blood brain barrier permeability and induces neuropathological changes in the rat brain: dose-response relationships. *Neurosci.*, **2002**, *113*(3), 721-741.
[46] Sakurada, K.; Matsubara, K.; Shimizu, K.; Shiono, H.; Seto, Y.; Tsuge, K.; Yoshino, M.; Sakai, I.; Mukoyama H.; Takatori., T. Pralidoxime iodide (2-PAM) penetrates accross the blood-brain barrier. *Neurochem. Res.*, **2003**, *28*(9), 1401-1407.
[47] Falb, A.; Erdmann, W.D. Über die Permeation von ^{14}C-Obidoxim durch die sog. Blut-Hirnschranke von Mäusen und Ratten (Penetration of ^{14}C-Obidoxime through the so-called blood-brain barrier of mice and rats). *Arch. Toxicol.*, **1969**, *24*(2), 123-132.
[48] Cassel, G.; Karlsson, L.; Waara, L.; Ang, K.W.; Goransson-Nyberg, A. Pharmacokinetics and effects of HI 6 in blood and brain of soman-intoxicated rats: a microdialysis study. *Eur. J. Pharmacol.*, **1997**, *332*(1), 43-52.
[49] Kassa, J. Review of oximes in the antidotal treatment of poisoning by organophosphorus nerve agents. *J. Toxicol. Clin. Toxicol.*, **2002**, *40*(6), 803-816.
[50] Kassa, J. The role of oximes in the antidotal treatment of chemical casualties exposed to nerve agents. In: *Medical Aspects of Chemical and Biological Terrorism. Chemical Terrorism and Traumatism.* Monov A., Dishovsky C, Eds.; Publishing House of the Union of Scientists in Bulgaria, Sofia, **2005**, pp.193-208. Available at: http://www.jmedchemdef.org/archives.html.
[51] Shih, T.-M.; Guarisco, J.A.; Myers, T.M.; Kan, R.K.; McDonough, J.H. The oxime pro2-PAM provides minimal protection against the CNS effects of the nerve agents sarin, cyclosarin, and VX in guinea pigs. *Toxicol. Mech. Methods*, **2011**, *21*(1), 53-62.
[52] Bajgar, J.; Fusek, J.; Kuca, K.; Bartosova, L.; Jun, D. Treatment of organophosphate intoxication using cholinesterase reactivators: facts and fiction. *Mini Rev. Med. Chem.*, **2007**, *7*(5), 461-466.
[53] Lorke, D.E.; Kalasz, H.; Petroianu, G.A.; Tekes, K. Entry of oximes into the brain: a review. *Curr. Med. Chem.*, **2008**, *15*(8), 743-753.
[54] Wilson, I.B.; Ginsburg, S. A powerful reactivator of alkylphosphate-inhibited acetylcholinesterase. *Biochim. Biophys. Acta*, **1955**, *18*(1), 168-170.
[55] Schoene, K. Pyridinium salts as organophosphate antagonists. *Monogr. Neural Sci.*, **1980**, *7*(1), 85-98.
[56] Kuča, K.; Patočka, J. Reactivation of cyclosarin-inhibited rat brain acetylcholinesterase by pyridinium oximes. *J. Enzym. Inhib. Med. Chem.*, **2004**, *19*(1), 39-43.

[57] Musilek, K.; Dolezal, M.; Gunn-Moore, F.; Kuca, K. Design, evaluation and structure-activity relationship studies of the AChE reactivators against organophosphorus pesticides. *Med. Res. Rev.*, **2011**, *31*(4), 548-575.

[58] Dawson, R.M. Review of oximes available for treatment of nerve agent poisoning. *J. Appl. Toxicol.*, **1994**, *14*(5), 317-331.

[59] Bajgar, J. Optimal choice of acetylcholinesterase reactivators for antidotal treatment of nerve agent intoxication. *Acta Medica (Hradec Králové)*, **2010**, *53*(4), 207-211.

[60] Worek, F.; Wille, T.; Aurbek, N.; Eyer, P.; Thiermann, H. Reactivation of organophosphate-inhibited human, Cynomolgus monkey, swine and guinea pig acetylcholinesterase by MMB-4: A modified kinetic approach. *Toxicol. Appl. Pharmacol.*, **2010**, *249*(3), 231-237.

[61] Lundy, P.M.; Hamilton, M.G.; Sawyer, T.W.; Mikler, J. Comparative protective effects of HI-6 and MMB-4 against organophosphorous nerve agent poisoning. *Toxicol.*, **2011**, *285*(1), 90-96.

[62] Aas, P. Future considerations for the medical management of nerve agent intoxication. *Prehosp. Disaster Med.*, **2003**, *18*(3), 208-216.

[63] Kuča, K.; Musilek, K.; Jun, D.; Bajgar, J.; Kassa, J. Novel oximes. In: *Handbook of Toxicology of Chemical Warfare Agents*; R.C. Gupta Ed.; Elsevier, **2009**, pp. 997-1021.

[64] Jokanović, M. Structure-activity relationship and efficacy of pyridinium oximes in the treatment of poisoning with organophosphorus compounds: A review of recent data. *Curr. Top. Med. Chem.*, **2012**, *12*(16), 1775-1789.

[65] Worek, F.; Thiermann, H. The value of novel oximes for treatment of poisoning by organophosphorus compounds. *Pharmacol. Ther.*, **2013**, *139*(2), 249-259.

[66] Musilek, K.; Kuča, K.; Jun, D.; Doležal, M. Progress in synthesis of new acetylcholinesterase reactivators during the period 1990-2004. *Curr. Org. Chem.*, **2007**, *11*(2), 229-238.

[67] Maksimović, M.; Bošković, B.; Radović, Lj.; Tadić V.; Deljac, V.; Binenfeld, Z. Antidotal effects of bis-pyridinium-2-monoxime carbonyl derivatives in intoxication with highly toxic organophosphorus compounds. *Acta Pharm. Jugosl.*, **1980**, *30*(2), 151-160.

[68] Binenfeld, Z.; Deljac, V.; Knežević, M.; Pavlov, Lj.; Maksimović, M.; Markov, V.; Radović, Lj.; Rakin, D. Reactivating effects of pyridinium salts on acethylcholinesterase inhibited by organophosphorus compounds. *Acta Pharm. Jugosl.*, **1981**, *31*(1), 5-15.

[69] Kovacic, P. Mechanism of organophosphates (nerve gases and pesticides) and antidotes: electron transfer and oxidative stress. *Curr. Med. Chem.*, **2003**, *10*(24), 2705-2709.

[70] Kassa, J.; Jun, D.; Karasova, J.; Bajgar, J.; Kuča, K. A comparison of reactivating efficacy of newly developed oximes (K074, K075) and currently available oximes (obidoxime, HI-6) in soman, cyclosarin and tabun-poisoned rats. *Chem. Biol. Interact.*, **2008**, *175*(1-3), 425-427.

[71] Acharya, J.; Gupta, A.K.; Dubey, D.K.; Raza, S.K. Synthesis and evaluation of novel bis-pyridinium oximes as reactivators of DFP-inhibited acetylcholinesterase. *Eur. J. Med. Chem.*, **2009**, *44*(3), 1335-1340.

[72] Acharya, J.; Dubey, D.K.; Raza, S.K. *In vitro* evaluation of bispyridinium oximes bearing methoxy alkane linker as reactivators of sarin inhibited human acetylcholinesterase. *Toxicol. In Vitro*, **2010**, *24*(6), 1797-1802.

[73] Acharya, J.; Rana, H.; Kaushik, M.P. Synthesis and *in vitro* evaluation of xylene linked carbamoyl bis-pyridinium monooximes as reactivators of organophosphorus (OP) inhibited electric eel acetylcholinesterase (AChE). *Eur. J. Med. Chem.*, **2011**, *46*(9), 3926-3933.

[74] Acharya, J.; Dubey, D.K.; Srivastava, A.K.; Raza, S.K. *In vitro* reactivation of sarin-inhibited human acetylcholinesterase (AChE) by bis-pyridinium oximes connected by xylene linkers. *Toxicol. In Vitro*, **2011**, *25*(1), 251-256.

[75] Bharate, S.B.; Guo, L.; Reeves, T.E.; Cerasoli, D.M.; Thompson, C.M. New series of monoquaternary pyridinium oximes: Synthesis and reactivation potency for paraoxon-inhibited electric eel and recombinant human acetylcholinesterase. *Bioorg. Med. Chem. Lett.*, **2009**, *19*(17), 5101-5104.

[76] Bharate, S.B.; Guo, L.; Reeves, T.E.; Cerasoli, D.M.; Thompson, C.M. Bisquaternary pyridinium oximes: Comparison of *in vitro* reactivation potency of compounds bearing aliphatic linkers and heteroaromatic linkers for paraoxon-inhibited electric eel and recombinant human acetylcholinesterase. *Bioorg. Med. Chem.*, **2010**, *18*(2), 787-794.

[77] Musilek, K.; Kucera, J.; Jun, D.; Dohnal, V.; Opletalova, V.; Kuča, K. Monoquaternary pyridinium salts with modified side chainsynthesis and evaluation on model of tabun-and paraoxon-inhibited acetylcholinesterase. *Bioorg. Med. Chem.*, **2008**, *16*(17), 8218-8223.

[78] Musilek, K.; Komloova, M.; Holas, O.; Horova, A.; Pohanka, M.; Gunn-Moore, F.; Dohnal, V.; Dolezal, M.; Kuča, K. Mono-oxime bisquaternary acetylcholinesterase reactivators with prop-1,3-diyl linkage - Preparation, *in vitro* screening and molecular docking. *Bioorg. Med. Chem.*, **2011**, *19*(2), 754-762.

[79] Karade, H.N.; Valiveti, A.K.; Acharya, J.; Kaushik, M.P. Synthesis and in-vitro evaluation of bis-quaternary 2-(hydroxyimino)-*N*-(pyridin-3-yl) acetamide derivatives as reactivators against sarin and VX inhibited human acetylcholinesterase (*h*AChE). *Bioorg. Med. Chem.*, **2014**, *22*(9), 2684-2691.

[80] Kliachyna, M.; Santoni, G.; Nussbaum, V.; Renou, J.; Sanson, B.; Colletier, J.P.; Arboleas, M.; Loiosice, M.; Weik, M.; Jean, L.; Renard, P.Y.; Nachon, F.; Baati, R. Design, synthesis and biological evaluation of novel tetrahydroacridine pyridine- aldoxime and -amidoxime hybrids as efficient uncharged reactivators of nerve agent-inhibited human acetylcholinesterase. *Eur. J. Med. Chem.*, **2014**, *78*(2), 455-467.

[81] Wei, Z.; Liu, Y.Q.; Zhou, X.B.; Luo, Y.; Huang, C.Q.; Wang, Y.A.; Zheng, Z.B.; Li, S. New efficient imidazolium aldoxime reactivators for nerve agent-inhibited acetylcholinesterase, *Bioorg. Med. Chem. Lett.*, **2014**, *24*(24), 5743-5748.

[82] Kassa, J.; Kuča, K.; Bartosova, L.; Kunešova, G. The development of new structural analogues of oximes for the antidotal treatment of poisoning by nerve agents and the comparison of their reactivating and therapeutic efficacy with currently available oximes. *Curr. Org. Chem.*, **2007**, *11*(3), 267-283.

[83] Jokanović, M.; Maksimović, M. A comparison of trimedoxime, obidoxime, pralidoxime and HI-6 in treatment of oral organophosphorus insecticide poisoning. *Arch. Toxicol.*, **1995**, *70*(2), 119-123.

[84] Worek, F.; Kirchner, T.; Bäker, M.; Szinicz L. Reactivation by various oximes of human erythrocyte acetylcholinesterase inhibited by different organophosphorus compounds. *Arch. Toxicol.*, **1996**, *70*(8), 497-503.

[85] Kušić, R.; Bošković, B.; Vojvodić, V.; Jovanović, D. HI-6 in man: blood levels, urinary excretion, and tolerance after intramuscular administration of the oxime to healthy volunteers. *Fundam. Appl. Toxicol.*, **1985**, *5*(6 Pt2), S89-S97.

[86] Kušić, R.; Jovanović, D.; Ranđelović, S.; Joksović, D.; Todorović, V.; Bošković, B.; Jokanović, M.; Vojvodić, V. HI-6 in man: efficacy of the oxime in poisoning by organophosphorus insecticides. *Hum. Exp. Toxicol.*, **1991**, *10*(2), 113-118.

[87] Clement, J.G.; Bailey, D.G.; Madill, H.D.; Tran, L.T.; Spence, J.D. The acetylcholinesterase oxime reactivator HI-6 in man: Pharmacokinetics and tolerability in combination with atropine. *Biopharm. Drug Dispos.*, **1995**, *16*(5), 415-425.

[88] Johnson, D.D.; Stewart, W.C. The effects of atropine, pralidoxime and lidocaine on nerve-muscle and respiratory function in organophosphate-treated rabbits. Can. J. Physiol. Pharm., **1970**, *48*(9), 625-630.

[89] Sidell, F.R.; Groff, W.A. The reactivability of cholinesterase inhibited by VX and sarin in man. *Toxicol. Appl. Pharmacol.*, **1974**, *27*(2), 241-252.

[90] Harris, L.W.; Stitcher, D.L. Reactivation of VX-inhibited cholinesterase by 2-PAM and HS-6 in rats. *Drug Chem. Toxicol.*, **1983**, *6*(3), 235-240.

[91] Mesić, M.; Deljac, A.; Deljac, V.; Binenfeld, Z.; Kilibarda, V.; Maksimović, M.; Kovačević, V. Reactivations of acetylcholinesterase inhibited by organophosphorus compounds. Imidazole derivatives. II. *Acta Pharm. Jugosl.*, **1991**, *41*(2), 203-210.

[92] Masuda, N.; Takatsu, M.; Morinari, H.; Ozawa, T. Sarin poisoning on the Tokyo subway. *Lancet,* **1995**, *345*(8962), 1446.

[93] Nozaki, H.; Aikawa, N. Sarin poisoning on the Tokyo subway. *Lancet*, **1995**, *345*(8962), 1446.

[94] Inns, R.H.; Leadbeater, L. The efficacy of bispyridinium derivatives in the treatment of organophosphate poisoning in the guinea pig. *J. Pharm. Pharmacol.*, **1983**, *35*(7), 427-432.

[95] Koplovitz, I.; Stewart, J.R. A comparison of HI-6 and 2-PAM against soman, tabun, sarin and VX in the rabbit. *Toxicol. Lett.*, **1994**, *70*(3), 269-279.

[96] DeMar, J.C.; Clarkson, E.D.; Ratcliffe, R.H.; Campbell, A.J.; Thangavelu, S.G.; Herdman, C.A.; Leader, H.; Schulz, S.M.; Marek, E.; Medynets, M.A.; Ku, T.C.; Evans, S.A.; Khan, F.A.; Owens,

R.R.; Nambiar, M.P.; Gordon, R.K. Pro-2-PAM therapy for central and peripheral cholinesterases. *Chem. Biol. Interact.*, **2010**, *187*(1-3), 191-198.

[97] Clement, J.G. Efficacy of Pro-PAM (N-methyl-1,6-dihydropyridine-2-carbaldoxime hydrochloride) as a prophylaxis against organophosphate poisoning. *Toxicol. Appl. Pharmacol.*, **1979**, *47*(2), 305-311.

[98] Bošković, B.; Tadić, V.; Kušić, R. Reactivating and protective effects of pro-2-PAM in mice poisoned with paraoxon. *Toxicol. Appl. Pharmacol.*, **1980**, *55*(1), 32-36.

[99] Murphy, S.D. Toxic effects of pesticides. In: *Casarett and Doull's Toxicology - The Basic Science of Poisons*, Klaassen C.D., Amdur M.O., Doull J. Eds.; 3rd edition, Macmillan Publishing Company, New York, **1986**; pp. 519-581.

[100] Pralidoxime. In: *Martindale The Complete Drug Reference*, Sweetman, S.C. Ed.; Pharmaceutical Press, 36th edition, London; **2009**; pp.1460-1461.

[101] Pralidoxime. In: *AHFS Drug Information*, G.K. McEvoy, Ed.; American Society of Health-System Pharmacists, Bethesda, MD; **2008**, pp. 3665-3667.

[102] International Programme on Chemical Safety Poisons Information, *Organophosphorus pesticides.* Monograph G001; World Health Organization, Geneva; **1989**; updated 1998. Available at: http://www.inchem.org/documents/pims/chemical/ pimg001.htm

[103] Milatović, D.; Jokanović, M. Pyridinium Oximes as Cholinesterase Reactivators in the Treatment of Poisoning with Organophosphorus Compounds. In: *Handbook of the Toxicology of Chemical Warfare Agents*, Gupta R.C., Ed.; Academic Press/Elsevier, **2009**, pp. 985-996.

[104] Sidell, F.R.; Groff, W.A. Intramuscular and intravenous administration of small doses of 2-pyridinium aldoxime methochloride to man. *J. Pharm. Sci.*, **1971**, *60*(8), 1224-1228.

[105] Jager, B.V.; Stagg, G.N. Toxicity of diacetylmonoxime and of pyridine-2-aldoxime methiodide in man 15-30 mg/kg 2-PAM: no effects on EEG in a healthy man. *Bull. Johns Hopkins Hosp.*, **1958**, *102*(2), 203-211.

[106] Stojiljković, M.P.; Jokanović, M. AUM Shinrikyo and Terrorist use of Nerve Agents in Japan. In: *Medical Aspects of Chemical and Biological Terrorism: Chemical Terrorism and Traumatism*, Monov A., Dishovsky C, Eds.; Publishing House of the Union of Scientists in Bulgaria, Sofia, Bulgaria, **2005**, pp. 101-115. available at: http://www.jmedchemdef.org/archives.html

[107] Tang, X.; Wang, R.; Xie, H.; Hu, J.; Zhao, W. Repeated pulse intramuscular injection of pralidoxime chloride in severe acute organophosphorus pesticide poisoning. *Am. J. Emerg. Med.*, **2013**, *3*(6), 946-949.

[108] Poziomek, E.J.; Hackley, B.E.; Jr., Steinberg, G.M. Pyridinium aldoximes. *J. Org. Chem.*, **1958**, *23*(5), 714-717.

[109] Hobbiger, F.W.; Sadler, P.W. Protection by oximes of bispyridinium ions against lethal diisopropylphosphonofluoridate poisoning. *Nature*, **1958**, *182*, 1672.

[110] Hobbiger, F.W.; Vojvodić, V. The reactivating and antidotal actions of N, N'trimethylenbis (pyridinium-4-aldoxime) (TMB-4) and N,N'-oxydimethylenbis (pyridinium-4-aldoxime) (toxogonin), with particularly reference to their effect on phosphorylated acetylcholinesterase in the brain. Biochem. Pharmacol., 1966, 15(11), 1677-1690.

[111] Binenfeld, Z. Medical protection against nerve gas poisoning. Past, present and future trend. A critical appraisal. Proc. 2nd Int. Symp. Protect. Chem. Warfare Agents; Stockholm (Sweden), **1986**, pp. 315-321.

[112] Clement, J.G. Toxicology and pharmacology of bispyridinium oximes -insight into the mechanism of action *vs.* soman poisoning *in vivo. Fundam. Appl. Toxicol.*, **1981**, *1*(2), 193-202.

[113] Erdmann, W.D.; Engelhard, H. Pharmakologisch-toxikologische Untersuchungen mit dem Dichlorid des Bis-(4-hydroxyimminomethyl-pyridinium-(1)-methyl)-öthers, einem neuen Esterase-Reaktivator. *Arzneimittelforschung*, **1964**, *14*(1), 5-11.

[114] Maksimović, M.; Kovačević, V.; Binenfeld, Z. Protective and reactivating effects of HI-6-Toxogonin mixture in rats and guinea pigs poisoned by nerve agents. *Acta Pharm. Jugosl.*, **1989**, *39*(1), 27-33.

[115] Heilbronn, E.; Tolagen, B. Toxogonin in sarin, soman and tabun poisoning. *Biochem. Pharmacol.*, **1965**, *14*(1), 73-77.

[116] Hamilton, G.; Lundy, P.M. HI-6 therapy of soman and tabun poisoning in primates and rodents. *Arch. Toxicol.*, **1989**, *63*(2), 144-149.

[117] Worek, F.; Thiermann, H.; Szinicz, L.; Eyer, F. Kinetic analysis of interactions between human acetylcholinesterase, structurally different organophosphorous compounds and oximes. *Biochem. Pharmacol.*, **2004**, *68*(11), 2237-2248.

[118] Sidell, F.R.; Groff, W.A. Toxogonin: blood levels and side effects after intramuscular administration in man. *J. Pharm. Sci.*, **1970**, *59*(6), 793-797.

[119] Simon, F.A.; Pickering, L.K. Acute yellow phosphorus poisoning: "Smoking stool syndrome". *J. Am. Med. Assoc.*, **1976**, *235*(6), 1343-1344.

[120] Marrs, T.C. Toxicology of oximes used in treatment of organophosphate poisoning. *Adverse Drug React. Toxicol. Rev.*, **1991**, *10*(1), 61-72.

[121] Obidoxime chloride. In: *Martindale The Complete Drug Reference*, Sweetman, S.C. Ed.; Pharmaceutical Press, 36th edition, London, **2009**; p. 1456.

[122] Oldiges, H.; Schoene, K. Pyridinium und Imidazolinium-Salze als Antidote gegenü ber Soman-und Paraoxonvergiftungen bei Möuse. *Arch. Toxicol.*, **1970**, *26*(4), 293-305.

[123] Clement, J.G. HI-6: Reactivation of central and peripheral acetylcholinesterase following inhibition by soman, sarin and tabun *in vivo* in the rat. *Biochem. Pharmacol.*, **1982**, *31*(7), 1283-1287.

[124] Ćetković, S. Effect of PAM-2Cl, HI-6 and HGG-12 in poisoning by tabun and its thiocholine analog in the rat. *Fundam. Appl. Toxicol.*, **1984**, *4*(2 Pt2), S116-S122.

[125] Lundy, P.M.; Goulet, J.C.; Hand, B.T. Hormone-and dose schedule-dependent protection by HI-6 against soman and tabun poisoning. *Fundam. Appl. Toxicol.*, **1989**, *12*(3), 595-603.

[126] Lundy, P.M.; Hill, I.; Lecavalier, P.; Hamilton, M.G.; Vair, C.; Davidson, C.; Weatherby, K.L.; Berger, B.J. The pharmacokinetics and pharmacodynamics of two HI-6 salts in swine and efficacy in the treatment of GF and soman poisoning. *Toxicol.*, **2005**, *208*(3), 399-409.

[127] Rousseaux,C.G.; Dua, A.K. Pharmacology of HI-6, an H-series oxime. *Can. J. Physiol. Pharmacol.*, **1989**, *67*(10), 1183-1189.

[128] Bartosova, L.; Kuča, K.; Kunesova, G.; Jun, D. The acute toxicity of acetylcholinesterase reactivators in mice in relation to their structure. *Neurotox. Res.*, **2006**, *9*(4), 291-296.

[129] Jovanović, D.; Maksimović, M.; Joksović, D.; Kovačević, V. Oral forms of the oxime HI-6: a study of pharmacokinetics and tolerance after administration to healthy volunteers. *Vet. Hum. Toxicol.*, **1990**, *32*(5), 419-421.

[130] Pohanka, M.; Novotny, L.; Zdarova-Karasova, J.; Bandouchova, H.; Zemek, F.; Hrabinova, M.; Misik, J.; Kuca, K.; Bajgar, J.; Zitka, O.; Cernei, N.; Kizek, R.; Pikula, J. Asoxime (HI-6) impact on dogs after one and tenfold therapeutic doses: Assessment of adverse effects, distribution, and oxidative stress. *Environ. Toxicol. Pharmacol.*, **2011**, *32*(1), 75-81.

[131] Löffler, M. Quartäre Salze von Pyridin-2,4-dialdoxim als Gegenmittel für Organophosphat-Vergiftungen (dissertation). University of Freiburg, Germany, 1986.

[132] DeJong, L.P.A.; Verhagen, M.; Langenberg, J.; Hagedorn, I.; Löffler, M. The bispyridinium-dioxime HLo-7. A potent reactivator for acetylcholinesterase inhibited by the stereoisomers of tabun and soman. *Biochem. Pharmacol.*, **1989**, *38*(4), 633-640.

[133] Worek, F.; Kirchner, T.; Szinicz, L. Effect of atropine and HI-6 on respiratory and circulatory function in guinea-pigs poisoned by O-ethyl S-[2-(diisopropylamino) ethyl]methyl-phosphonothioate (VX). *Pharmacol. Toxicol.*, **1994**, *75*(5), 302-309.

[134] Lundy, P.M.; Hansen, A.S.; Hand, B.T.; Boulet, C.A. Comparison of several oximes against poisoning by soman, tabun and GF. *Toxicol.*, **1992**, *72*(1), 99-105.

[135] Alberts, P. A new H-oxime restores rat diaphragm contractility after esterase inhibition *in vitro*. *Eur. J. Pharmacol.*, **1990**, *184*(1), 191-194.

[136] Clement, J.G.; Hansen, A.S.; Boulet, C.A. Efficacy of HLö-7 and pyrimidoxime as antidotes of nerve agent poisoning in mice. *Arch. Toxicol.*, **1992**, *66*(3), 216-219.

[137] Clement, J.G. Central actions of acetylcholinesterase oxime reactivators. *Toxicol. Appl. Pharmacol.*, **1992**, *112*(1), 104-109.

[138] Worek, F.; Szinicz, L. Investigation of acute cardiovascular and respiratory toxicity of HLö-7 dimethanesulphonate and HI-6 dichloride in anaesthetized guinea-pigs. *Pharmacol. Toxicol.*, **1993**, *73*(2), 91-95.

[139] Eyer, P.; Hagedorn, I.; Klimmek, R.; Lippstreu, P.; Löffler, M.; Oldiges, H.; Spöhrer, U.; Steidl, I.; Szinicz, L.; Worek, F. HLö-7 dimethanesulphonate, a potent bispyridinium-dioxime against anticholinesterases. *Arch. Toxicol.*, **1992**, *66*(9), 603-621.

[140] Melchers, B.P.C.; van der Laaken, A.L.; Busker, R.W.; Bruijnzeel, P.L.B.; van Helden, H.P.M. Non-reactivating effects of HI-6 on hippocampal neurotransmission. *Arch. Toxicol.*, **1994**, *69*(2), 118-126.
[141] Melchers, B.C.P.; Philippens, I.H.C.H.M.; Wolthuis, O.L. Efficacy of HI-6 and HLö-7 in preventing incapacitation following nerve agent poisoning. *Pharmacol. Biochem. Behav.*, *1994*, *49*(4), 781-788.
[142] Kuča, K.; Cabal, J.; Kassa, J.; Jun, D.; Hrabinova, M. A comparison of the potency of the oxime HLö-7 and currently used oximes (HI-6, pralidoxime, obidoxime) to reactivate nerve agent-inhibited rat brain acetylcholinesterase by *in vitro* methods. *Acta Medica (Hradec Kralove)*, **2005**, *48*(2), 81-86.
[143] Kuča, K.; Musilek, K.; Karasova, J.; Jun, D.; Soukup, O.; Pohanka, M.; Ghosh, K.K.; Hrabinova, M. On the universality of oxime HLö-7 -Antidote for case of the nerve agent poisoning. *Mil. Med. Sci. Lett.*, **2011**, *80*(1), 80-84.
[144] Amitai, G.; Kloog, Y.; Balderman, D.; Sokolovsky, M. The interaction of bispyridinium oximes with mouse brain muscarinic receptor. *Biochem. Pharmacol.*, **1980**, *29*(4), 483-488.
[145] Hobbiger, F.; Sadler, P.W. Protection against lethal organophosphate poisoning by quaternary pyridine aldoximes. *Brit. J. Pharmacol.*, **1959**, *14*(2), 192-201.
[146] Shih, T-M. Comparison of several oximes on reactivation of soman-inhibited blood, brain and tissue cholinesterase activity in rats. *Arch. Toxicol.*, **1993**, *67*(9), 637-646.
[147] Luo, C.; Tong, M.; Maxwell, D.M.; Saxena, A. Comparison of oxime reactivation and aging of nerve agent-inhibited monkey and human acetylcholinesterases. *Chem. Biol. Interact.*, **2008**, *175*, 261-266.
[148] Shih, T-M.; Skovira, J.W.; O'Donnell, J.C.; McDonough, J.H. *In vivo* reactivation by oximes of inhibited blood, brain and peripheral tissue cholinesterase activity following exposure to nerve agents in guinea pigs. *Chem. Biol. Interact.*, **2010**, *187*(1-3), 207-214.
[149] Petroianu, G.A.; Nurulain, S.M.; Nagelkerke, N.; Al-Sultan, M.A.; Kuca, K.; Kassa, J. Five oximes (K-27, K-33, K-48, BI-6 and methoxime) in comparison with pralidoxime: survival in rats exposed to the organophosphate paraoxon. *J. Appl. Toxicol.*, **2006**, *26*(3), 262-268.
[150] Petroianu, G.A.; Nurulain, S.M.; Nagelkerke, N.; Shafiullah, M.; Kassa, J.; Kuca, K. Five oximes (K-27, K-48, obidoxime, HI-6 and trimedoxime) in comparison with pralidoxime: survival in rats exposed to methyl-paraoxon. *J. Appl. Toxicol.*, **2007**, *27*(5), 453-457.
[151] Kovačević, V.; Maksimović, M.; Pantelić, D.; Vojvodić, V.; Binenfeld, Z. Protective and reactivating effects of HI-6 PAM-2 mixture in rats with nerve chemical warfare agents (nerve CWA). *Acta Pharm. Jugosl.*, **1989**, *39*(2), 161-165.
[152] Kassa, J.; Zdarova Karasova, J.; Tesarova, S. A comparison of the neuroprotective efficacy of individual oxime (HI-6) and combinations of oximes (HI-6+trimedoxime, HI-6+K203) in soman-poisoned rats. *Drug Chem. Toxicol.*, **2011**, *34*(3), 233-239.
[153] Kassa, J.; Zdarova Karasova, J.; Sepsova, V.; Caisberger, F. The benefit of combinations of oximes for the reactivating and therapeutic efficacy of antidotal treatment of sarin poisoning in rats and mice. *Basic Clin. Pharmacol. Toxicol.*, **2011**, *109*(1), 30-34.
[154] Kassa, J.; Zdarova Karasova, J. A comparison of the neuroprotective efficaicy of individual oximes (HI-6, trimedoxime, K203) and their mixtures (HI-6 + trimedoxime, HI-6 + K203) in cyclosarin-poisoned rats. *Mil. Med. Sci. Lett.*, **2011**, *80*(1), 12-20.
[155] Kassa, J.; Zdarova Karasova, J.; Pavlikova, R.; Caisberger, F.; Bajgar, J.; Misik, J. The influence of combinations of oximes on the reactivating and therapeutic efficacy of antidotal treatment of tabun poisoning in rats and mice. *J. Appl. Toxicol.*, **2010**, *30*(2), 120-124.
[156] Cherian, A.M.; Peter, J.V.; Samuel, J.; Jaydevan, R.; Peter, S.; Joel, S.; Jeyasselan, L.; Thomas, K. Effectiveness of P2AM (P2AM - pralidoxime) in the treatment of organophosphorus poisoning (OPP) a randomized, double blind placebo-controlled clinical trial. *J. Assoc. Phys. India*, **1997**, *45*(1), 22-24.
[157] Cherian, M.A.; Roshini, C.; Peter, J.V.; Cherian, A.M. Oximes in organophosphorus poisoning. *Indian J. Crit. Care Med.*, **2005**, *9*(3), 155-163.
[158] De Silva, H.J.; Wijewickrema, R.; Senanayake, N. Does pralidoxime affect outcome of management in acute organophosphate poisoning? *Lancet*, **1992**, *339*(8802), 1136-1138.
[159] Johnson, S.; Peter, J.V.; Thomas, K.; Jeyaseelan, L.; Cherian, A.M. Evaluation of two treatment regimens of pralidoxime (1 gm single bolus dose vs 12gm infusion) in the management of organophosphorus poisoning. *J. Assoc. Phys. India*, **1996**, *44*(8), 529-531.

[160] Singh, S.; Batra, Y.K.; Singh, S.M.; Wig, N.; Sharma, B.K. Is atropine alone sufficient in acute severe organophosphate poisoning? Experience of a North West Indian hospital. *Int. J. Clin. Pharmacol. Ther.*, **1995**, *33*(11), 628-630.

[161] Eddleston, M.; Juszczak, E.; Buckley, N.A.; Senarathna, L.; Mohamed, F.; Dissanayake, W.; Hittarage, A.; Azher, S.; Jeganathan, K.; Jayamanne, S.; Sheriff, M.H.R.; Warrell, D.A. Multiple-dose activated charcoal in acute self-poisoning: a randomised controlled trial. *Lancet*, **2008**, *371*(9612), 579-587.

[162] Eddleston, M.; Eyer, P.; Worek, F.; Mohamed, F.; Senarathna, L.; von Meyer, L.; Juszczak, E.; Hittarage, A.; Azhar, S.; Dissanayake, W.; Sheriff, M.H.R.; Szinicz, L.; Dawson, A.H.; Buckley, N.A. Differences between organophosphorus insecticides in human selfpoisoning: a prospective cohort study. *Lancet*, **2005**, *3668*(9495), 1452-1459.

[163] Pawar, K.S.; Bhoite, R.R.; Pillay, C.P.; Chavan, S.C.; Malshikare, D.S.; Garad, S.G. Continuous pralidoxime infusion versus repeated bolus injection to treat organophosphorus pesticide poisoning: a randomised controlled trial. *Lancet*, **2006**, *368*(9553), 2136-2141.

[164] Eddleston, M.; Eyer, P.; Worek, F.; Juszczak, E.; Alder, N.; Mohamed, F.; Senarathna, L.; Hittarage, A.; Azher, S.; Jeganathan, K.; Jayamanne, S.; von Meyer, L.; Dawson, A.H.; Sheriff, M.H.R.; Buckley, N.A. Pralidoxime in Acute Organophosphorus Insecticide Poisoning - A Randomised Controlled Trial. *PLoS Med.*, **2009**, *6*(6): e1000104.

[165] Willems, J.L.; DeBisschop, H.C.; Verstraete, A.G.; Declerck, C.; Christiaens, Y.; Vanscheeuwyck, P.; Buylaert, W.A., Vogelaers, D.; Colardyn, F. Cholinesterase reactivation in organophosphorus poisoned patients depends on the plasma concentrations of the oxime pralidoxime methylsulphate and of the organophosphate. *Arch. Toxicol.*, **1993**, *67*(2), 79-84.

[166] Thiermann, H.; Mast, U.; Klimmek, R.; Eyer, P.; Hibler, A.; Pfab, R.; Felgenhauer, N.; Zilker, T. Cholinesterase status, pharmacokinetics, and laboratory findings during obidoxime therapy in organophosphate poisoned patients. *Hum. Exp. Toxicol.*, **1997**, *16*(8), 473-480.

[167] Thiermann, H.; Szinicz, L.; Eyer, F.; Worek, F.; Eyer, P.; Felgenhauer, N.; Zilker, T. Modern strategies in therapy of organophosphorus poisoning. *Toxicol. Lett.*, **1999**, *107*(1-3), 133-139.

[168] Aragao, I.; Soares, M.E.; Bastos, M.L., Lopes, M. Continuous infusion of obidoxime and the importance of plasma levels monitoring in organophosphate poisoning. *Intens. Care Med.*, **1996**, *22*(Suppl 3), S388.

[169] Balali-Mood, M.; Shariat, M. Treatment of organophosphate poisoning. Experience of nerve agents and acute pesticide poisoning on the effects of oximes. *J. Physiol. Paris*, **1998**, *92*(5-6), 375-378.

[170] Tsai, J-R.; Sheu, C-C.; Chong, I-W.; Cheng, M-H.; Huang, M-S.; Hung, J-Y.; Hwang, J-J. Organophosphate poisoning: 10 years of experience in southern Taiwan. *Kaohsiung J. Med. Sci.*, **2007**, *23*(3), 112-119.

CHAPTER 6

Calixarenes in Biotechnology and Bio-Medical Research

Roman V. Rodik, Vyacheslav I. Boyko and Vitaly I. Kalchenko*

Institute of Organic Chemistry, National Academy of Sciences of Ukraine, Ukraine

Abstract: Application of calixarene derivatives in biotechnology and bio-medical researches is reviewed in this article. Results on antiviral, bactericidal, antithrombothic, antituberculosis, anticancer activities of modified calixarenes are presented. Transfection ability, specific protein complexation, enzyme mimic, membranotropic properties and toxicity of modified calixarenes are described.

Keywords: Alkaline phosphatase, anticancer, anticoagulant and antithrombotic, antituberculosis activity, antiviral, bactericidal, calixarenes, chloride channel, enzyme mimic, fungicidal, magnetic resonance imaging, protein complexation, protein tyrosine phosphatase, sodium-potassium pump, toxicity, transfection.

INTRODUCTION

Modern trend in medicinal chemistry is design of new bio-active compounds on the base of structural information of the bio-targets. This way raises the opportunity to obtain drugs with great efficiency and selectivity due to high specificity of the receptor-substrate interaction. Wide classes of compounds are used as molecular platform for this purpose, namely: peptides [1, 2], carbohydrates [3, 4], cyclodextrines [5, 6], as well as calixarenes [7, 8].

Calixarenes – synthetically available macrocyclic compounds obtained by precise cyclo-condensation of *p*-substituted phenols and formaldehyde - possess intramolecular lipophillic cavities formed by aromatic rings of the macrocyclic skeleton. Calixarenes are easily amenable to the upper or lower rim functionalization of the macrocyclic skeleton.

Due to their ability to establish a variety of interactions, multi-site hydrogen bonding, specific stacking, and generalized electrostatic interactions, calixarene

***Corresponding author Vitaly I. Kalchenko:** Institute of Organic Chemistry, National Academy of Sciences of Ukraine, Murmanska Str., 5 02660, Kyiv-94, Ukraine; Tel: 380(44) 559 06 67;
E-mail: vik@ioch.kiev.ua

Atta-ur-Rahman, M. Iqbal Choudhary and Allen B. Reitz (Eds.)

derivatives can be exploited as valid recognition motifs in the development of synthetic supramolecular systems. They show unique ability to recognize and bind in the Host-Guest supramolecular complexes cations [9-11], anions [12, 13] and neutral molecules [14] of appropriate size and architecture. These properties open wide perspectives of practical application of the calixarenes in different branches of chemistry, physics and biology [7, 15, 16].

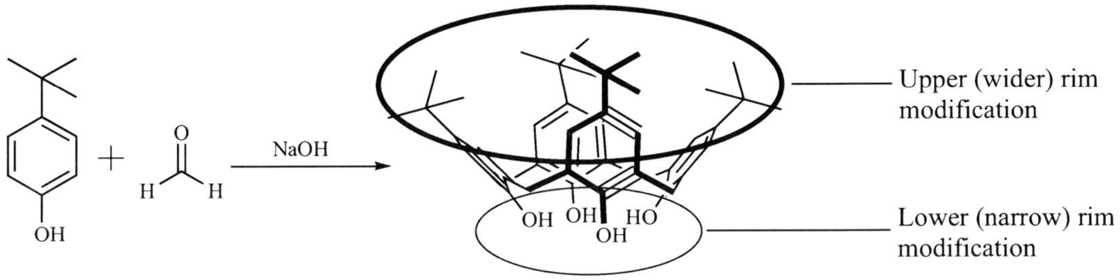

During the last 10 years, calixarene derivatives, especially water soluble and amphiphilic ones have been the subject of growing interest in the biological domain [17, 18]. Several review are devoted to sub-classes of calixarenes such as amino [19], or sulfonated ones [20, 21]. Some surveys dedicated potential application of calixarenes in different bio-medical fields [22, 23]. The highly diverse biomedical applications of these molecules now include antibacterial, anticancer, antiviral, antithrombotic, membranotropic activities, selective enzyme blocking and mimicking, as well as protein complexation [24-26]. Calixarenes have various properties which may be of importance for the remediation of many diseases. This review provides an overview of the studies that led to the elucidation of biological activity and biopharmaceutical application of the calixarenes.

1. CALIXARENE-BASED ENZYME MIMICS

1.1. Artificial Esterase of Phosphates and Carboxylates

In Nature, chemical transformations of phosphate esters by metalloenzymes are generally facilitated by the cooperative action of two metal ions, *e.g.*, in P1 nuclease, DNA polymerase I, phospholipase C, and alkaline phosphatase. These metal ions are usually Zn(II), Mg(II), Mn(II), or Fe(II) and are separated by 3-5 Å in the active site of these enzymes. Besides, *cis*-diaqua Cu(II) complexes, either mononuclear or dinuclear, have also been reported as active artificial catalysts for the cleavage of phosphate diesters [27].

An imidazolocalix[4]arene metallocomplex **1** with two *cis*-diaqua Cu(II) centers at the upper rim was investigated [28] as a model for dinuclear metalloenzymes that catalyze chemical transformation of phosphate esters. The calixarene **1** efficiently catalyzes the transesterefication of RNA model 2-nydroxypropyl-*p*-nitrophenyl phosphate (HPNP) and hydrolysis of the DNA model ethyl-*p*-nitrophenyl phosphate (EPNP) with turnover conversion, thereby exhibiting rate enhancement factors of $1\cdot10^4$ and $2.7\cdot10^4$, respectively [28].

Saturation kinetics and pH variation studies point to the formation of a Michaelis-Menten complex in which the phosphate group is doubly Lewis acid activated by coordination to the two Cu(II) centers. Moreover, the calixarene dinuclear complex **1** binds and cleaves the phosphate diesters HPNP and EPNP far more efficiently than mononuclear reference **2** [28].

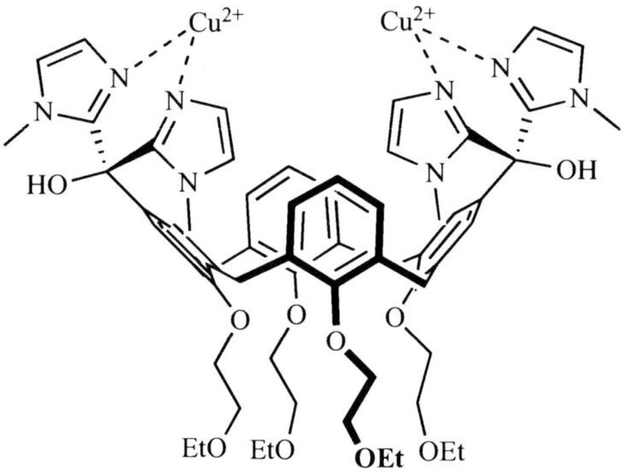

1

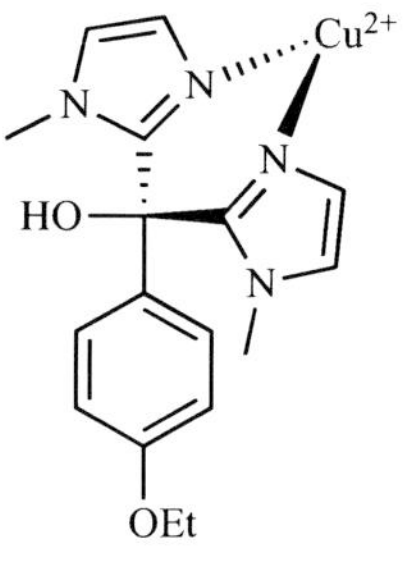

2

An important role of the calix[4]arene moiety in catalysis might be that rapid low energy conformational changes of the macrocyclic backbone which compensate for otherwise energetically unfavorable geometric transitions at the catalytic reaction center, mimicking in this way one of the functions of the enzyme peptide backbone [28].

The dinuclear Zn(II) calixarene complex **3** shows both binding to a phosphate diester substrate and high catalytic activity. The presence of 0.48 mM of **3** induces a 23000-fold rate enhancement in the catalytic cyclization of the RNA model substrate HPNP (pH 7.5). Comparison of **3** with monofunctionalized calix[4]arene complex **4** and reference pyridine complex **5** shows that the high catalytic activity of **3** can be attributed to a favorable contribution of the calix[4]arene moiety in substrate binding and catalytic synergic action of the two Zn(II) centers [29].

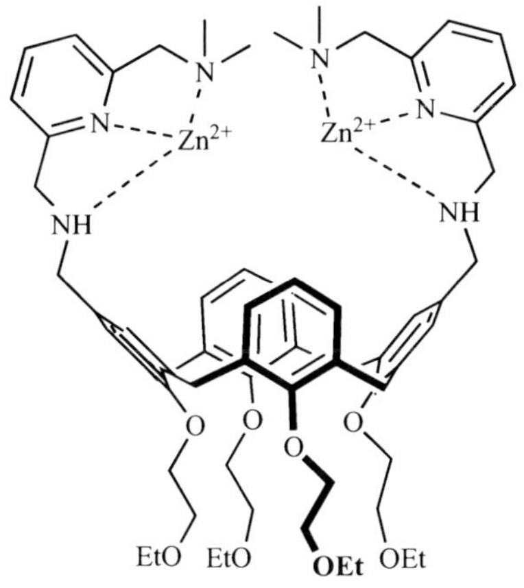

3

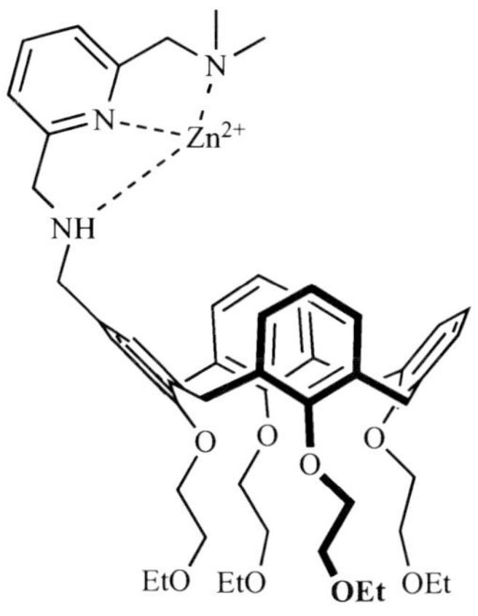

4

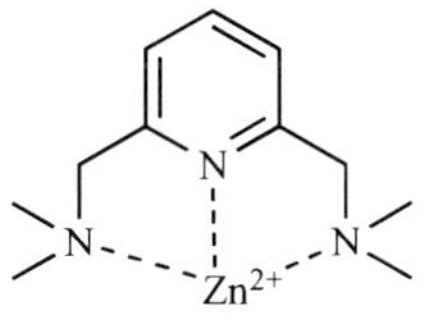

5

Calix[4]arene metallocomplex **3** is not catalytically active in the hydrolisys of diethyl *p*-nitrophenyl phosphate, ethyl *p*-nitrophenyl phosphate or *p*-nitrophenyl phosphate, what shows that the *β*-hydroxyl group HPNP is essential in the catalytic process. Therefore, the most likely mode of catalysis is a bifunctional mechanism in which one of the Zn(II) centers serves as Lewis acid activator of the phosphate group and the other is involved in activation of the nucleophilic *β*-hydroxyl group giving rise to relatively fast intramolecular reaction (Fig. **1**) [29].

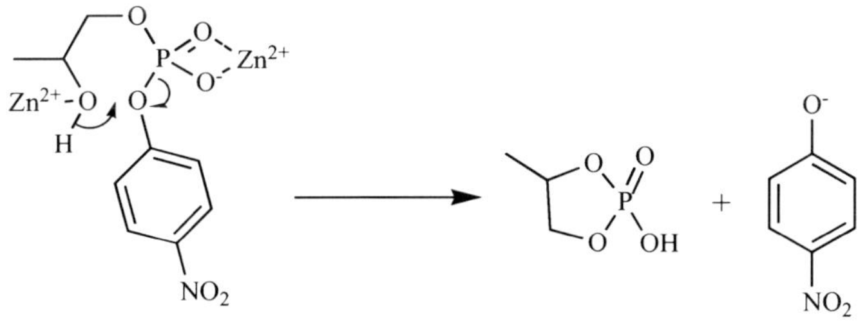

Figure 1: Scheme of catalytic cleavage of HPNP.

Metallocomplex of diaminopyridylcalix[4]arene **6** has the same catalytic activity as **3**, but pH-optimum was shifted from 7.5 to 6.8 [29]. It is assumed that one of aminogroups play a role of intramolecular catalyst in the deprotonation of hydroxyl group of substrate HPNP.

High catalytic activity of the flexible complex **3** originates from a very high substrate binding affinity, affording complex in which substrate is converted with a relatively moderate rate. A rigid analogue **7** exhibits both a lower substrate binding strength and lower catalytic rate. This demonstrates the importance of certain flexibility between cooperating catalytic centers [30].

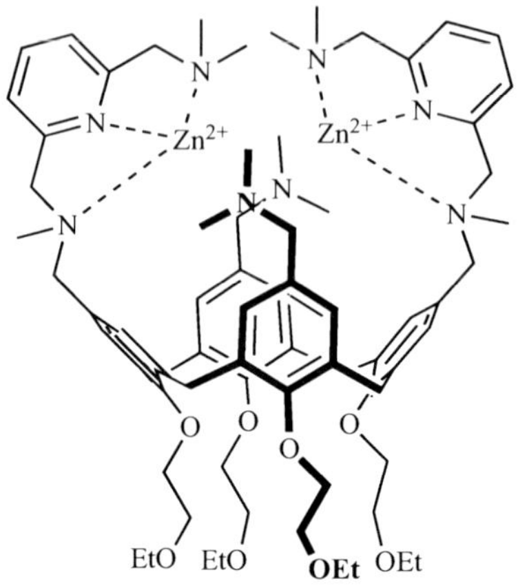

6

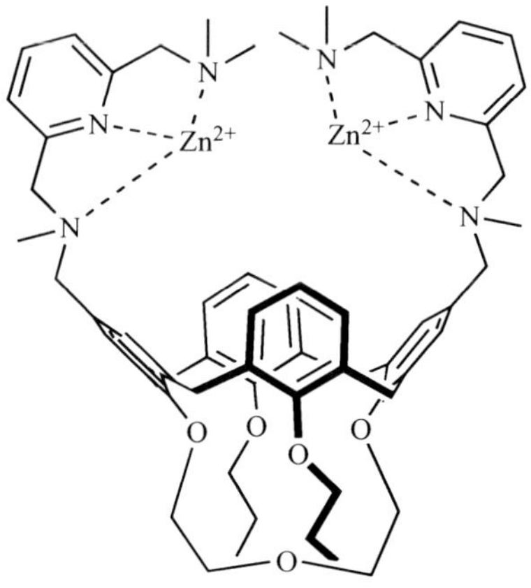

7

The presence of the third Zn(II) center in close proximity of the dinuclear Zn(II) cluster, *e.g.* complex **8**, results in a decreased substrate affinity and an increased catalytic rate to a factor 32000 over of the uncatalyzed reaction. This implies that **8** binds the pentacoordinated phosphorus transition state better than **3**. Participation of three Zn(II) ions in the catalytic cleavage of phosphate esters was proposed for the trinuclear enzymes phospholipase C and P1 nuclease, and efficiency is confirmed by using artificial trinuclear Zn(II) complex [30].

The next study of the same team's investigation [29] is concerned hydrolysis of RNA dinucleotides (3',5'-NpN) by the trinuclear Zn(II) Cu(II) and mixed 2Zn(II)+Cu(II) calixarene complexes **8-10** [31].

Catalytic cleavage of the RNA dinucleotides (0.09mM) by the complexes **8-10** (0.9mM) was carried out in the 35% ethanol with HEPES buffer at 50 °C. The process was identified as intramolecular transesterefication, affording cyclic ribonucleosides. The catalytic activity of Me(II) calixarene complexes measured for a series of GpG, UpU, CpC, GpA, ApG and ApA dinucleotides is shown in Table **1** [29].

The trinuclear calixarene metallocomplexes exhibit a very high catalytic activity; rate accelerations over uncatalyzed reactions are on the order 10^4–10^5. Surprisingly, for different nucleobases in the nucleotides large differences in rate were observed: GpG » UpU » ApA. The trinuclear complex **8** is a factor of 10, 19 and 160 times more active than dinuclear complex **7** for cleavage of CpC, UpU, and GpG respectively. These differences in activity are much higher than that observed in the transesterefication of HPNP.

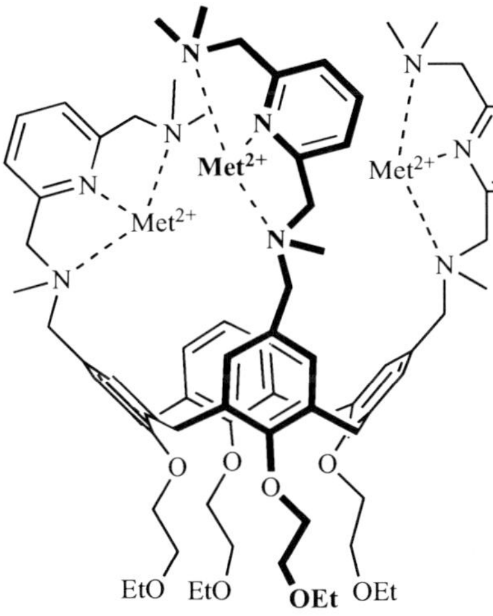

Met = Zn (**8**), Cu (**9**), 2Zn+Cu (**10**)

Table 1. Pseudo first order speed rate constants (k_{obs} 10^5 c^{-1}) for cleavage of dinucleotides RNA.

Substrate	3	4	8	9	10
GpG	0.45	-	72	28	88
UpU	0.45	0.56	8.5	1.2	13
CpC	0.58	-	6.1	1.9	7.1
ApA	0.28	0.31	0.44	0.47	0.46

In 35% ethanol, pH = 8 (20 mM buffer HEPES)

Whereas zinc complex **8** is superior to its analogue copper complex **9** (except the case of ApA), a statistical mixture calixarene, two equivalents of Zn(II) and one Cu(II) (the heterotrinuclear analogue **10** as the main species) shows a higher activity. This may be due to synergy of the favorable properties of Zn(II) in the binding (K_{ass}) and Cu(II) in the conversion (k_{cat}) of the phosphate diester substrate.

Preliminary experiments with a 24-mer RNA show that **8** and **10** exhibit catalytic activity in the cleavage of RNA oligonucleotides [31].

At the final stage a series of calix[4]arenes functionalized at the 1, 1,2-, 1,3-, and 1,2,3-positions of the upper rim with azacrown ligating units **11-14** were synthesized, and their bi- and trimetallic zinc(II) and copper(II) complexes were investigated as catalysts in the cleavage of phosphodiester as RNA models and as cleaving agents of 6-, 7-, and 17-meric oligoribonucleotides [32, 33].

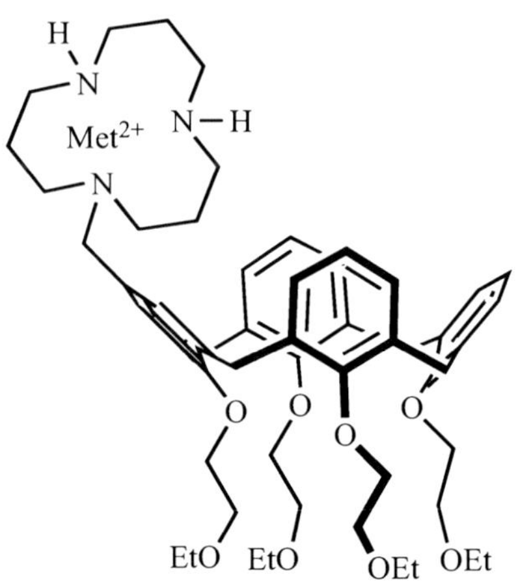

11 a, Met = Zn
b, Met = Cu

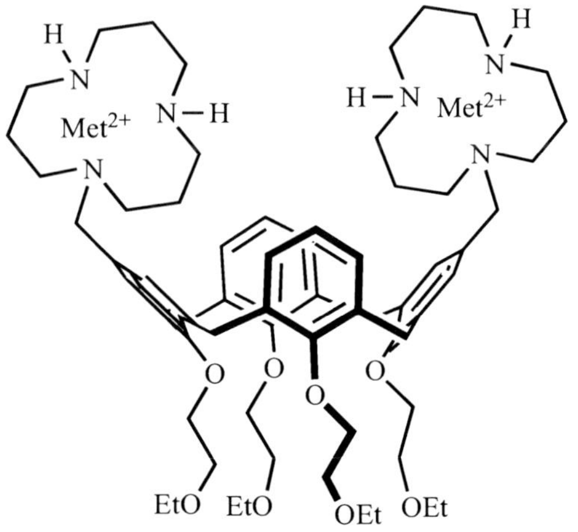

12 a, Met = Zn
b, Met = Cu

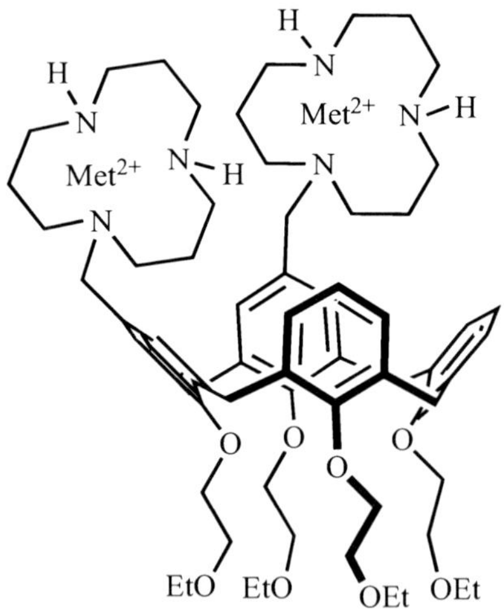

13 a, Met = Zn
b, Met = Cu

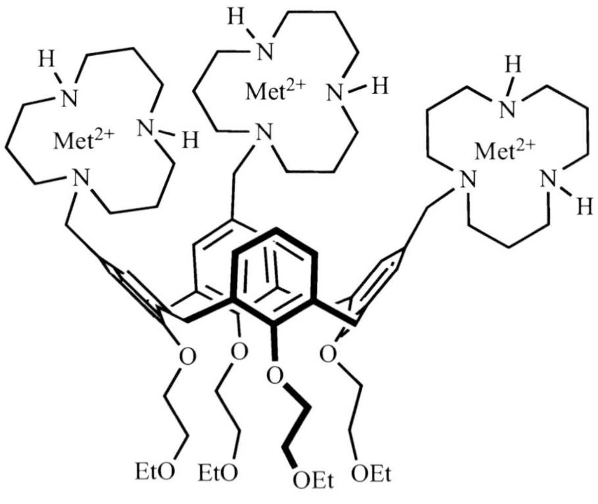

14 a, Met = Zn
b, Met = Cu

The results of comparative kinetic studies using monometallic controls indicate that the subunits of all of the zinc(II) complexes and of the 1,3-distal bimetallic copper(II) complex **12b** act as essentially independent monometallic catalysts. The lack of cooperation between metal ions in the above complexes is in marked contrast with the behavior of the 1,2-vicinal bimetallic copper(II) complex **13b**, which exhibits high catalytic efficiency and high levels of cooperation between metal ions in the cleavage of HPNP and of diribonucleoside monophosphates. A third ligated metal ion at the upper rim does not enhance the catalytic efficiency.

A kinetic investigation of the RNA-oligomer cleavage reactions by **12b, 13b, 14b** was carried out using gel electrophoresis to separate and analyze reactants and products having a radioactive phosphate label in the terminal 5′-position. The degree of cooperation was assessed on the basis of a comparison with rates of cleavage by mononuclear controls. A remarkable selectivity of cleavage of the CpA phosphodiester bond was observed for all metal complexes, in sharp contrast with the UpU and UpG selectivity previously observed in the cleavage of diribonucleoside monophosphates by the same metal complexes. The highest rate acceleration, brought about in the cleavage of the 5′-pCpA bond in hexanucleotide by 50 μM trinuclear complex **14b** (water solution, pH 7.4, 50 °C), amounts to 5×10^5-fold, as based on the estimated background reactivity of the CpA dimer. Selectivity in the cleavage of oligoribonucleotides by copper(II) complexes closely resembles that experienced by ribonuclease A and by a number of metal-independent RNase A mimicks.

Another example of mimicking of the natural enzymes action concerns carboxylate esterase activity. An imidazole substituted calix[4]arenes **15-18** are easily accessible acid-base catalysts with acyltransferase activity which show rate enhancements up to a factor of 10, as tested for release of *p*-nitrophenol from *p*-nitrophenolates [34].

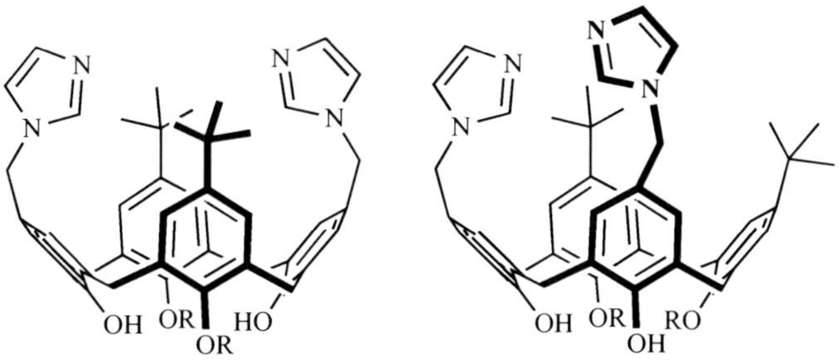

15 R = H
16 R = Bz

17 R = H
18 R = Bz

Proximal and distal substituted calix[4]arenes **15-18** improved the hydrolysis in similar way. Benzoylated species **16, 18** exhibited higher catalytic ability than tetrahydroxycalix[4]arenes **15** and **17**. This result is in favour of a mechanism for the catalysis involving the imidazole groups of the calix[4]arenes.

1.2. Carboxylcholine Esterase Mimic

Synthetic non-peptidic receptor displaying the binding properties of a whole antibody was designed by mimicking the key interaction found in a hapten-antibody complex [35]. Incorporation of suitable reactive functional groups into the binding site should open a new way for creation of enzyme mimics. Calix[6]arenes **19a,b** were synthesized accordingly to this idea [36].

Both **19a,b** catalyze methanolysis of choline *p*-nitrophenylcarbonate (PNPCC) in chloroform containing 1% methanol. The calixarene fragment strongly binds to trimethylammonium head group through cation-π interaction. The tetrahedral intermediate resulting from methoxide-ion addition to the ester carbonyl is stabilized through hydrogen bonding and electrostatic attraction by the guanidinium moiety.

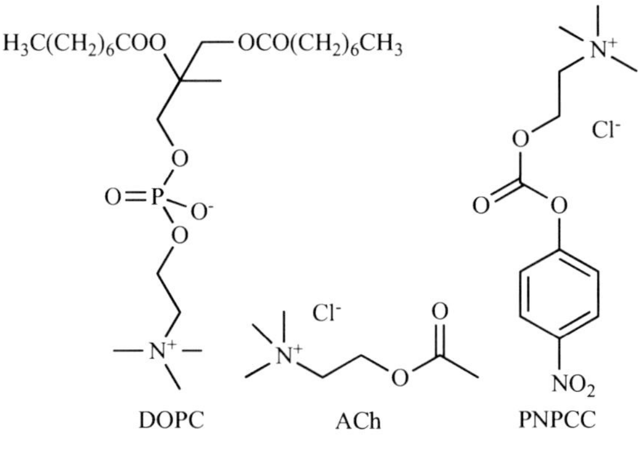

The observed cholinesterase activity of **19b** had been anticipated from ability **19a,b** bind dioctanoyl-L-α-phosphatidylcholine (DOPC) [34], which is a transition state analogue for the $B_{Ac}2$-type cleavage of choline esters. In neat chloroform calixarene **19b** forms complexes with DOPC and acetylcholine (ACh) with association constants $9.5 \cdot 10^4$ M^{-1} and $8.6 \cdot 10^2$ M^{-1} respectively. Complexation is weaker, but still significant in the mixture of solvents employed for catalysis studies: $CHCl_3/CH_3OH$, 99:1, v/v.

The overall mechanism of catalysis probably involves reversible formation of a substrate-catalyst complex (S·cat), followed by rate-limiting reaction with a solvent component (k_{cat}), as shown in the scheme below (Fig. **2**) [36].

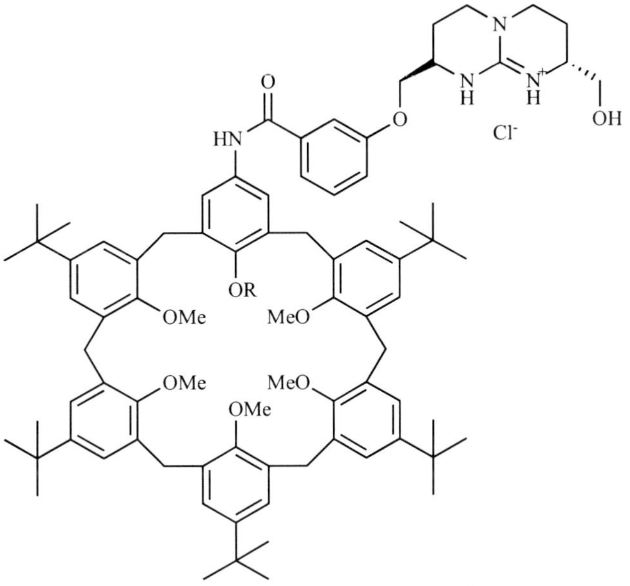

19 a R = H, b R = cyclohexyl

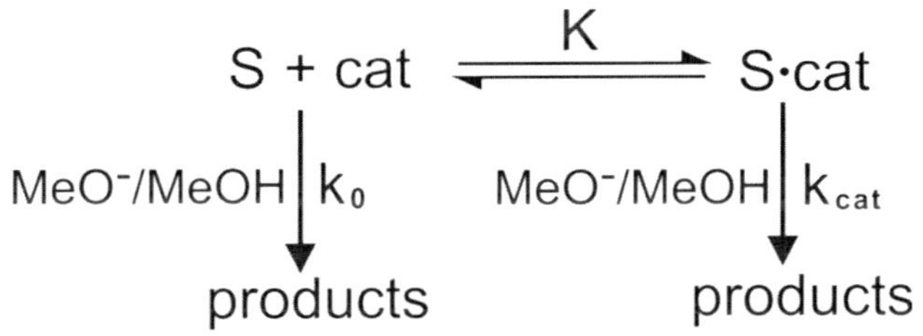

Figure 2: Proposed overall mechanism of DOPC metanolysis.

Assuming that PNPCC is bound by **19a,b** with equal affinities, the process of methanolysis was investigated on this substrate. The value of k_{cat}/k_0 is equal 600 for **19a**, 1000 for **19b** and surprisingly, 770 for MOPC167 antibody which is a well-known natural object for study of choline esters hydrolysis [37, 38].

1.3. Glyceraldehyde-3-Phopsphate Dehydrogenase Mimetic Compound

The object of enzyme mimicry is to design artificial systems which have efficiency and selectivity comparable to the natural ones and their catalytic transformations occur following Michaelis-Menten kinetics. Many macrocyclic compounds, especially water-soluble cyclodextrines and cyclophanes as well as calixarenes have been used as binding units for artificial enzymes [39, 40].

The hexasulfonated calix[6]arene derivatives **20, 21** are simple calixarene examples of enzyme mimics.

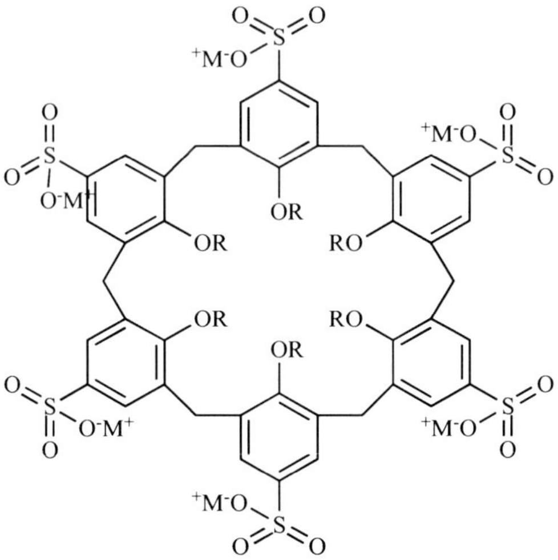

20 R = H, $M^+ = Na^+$
21 R = CH_2COOH, $M^+ = NH_4^+$

It was found that these calix[6]arenes efficiently accelerate acid-catalyzed hydration of 1-benzyl-1,4-dihydronicotinamide (synthetic model of NADH, Fig. **3**) and reaction proceeds accordingly to the Michaelis-Menten kinetics [41].

Compounds **20, 21** have acidic protons to catalyze the reaction and anionic sulfonates to stabilize the cationic intermediate at the two edges of the cavity. The rate constants were greater by 426-1220-fold than those for noncyclic analogues. The catalytic activity of **21** is further enhanced at the low-pH region (k equal $2.66 \cdot 10^5$ s^{-1} at pH = 6.3 and $3.28 \cdot 10^6$ s^{-1} at pH = 4) because of the increase in the number of the undissociated carboxylic groups. Also, **21** has a greater association constant (1340 M^{-1}) relatively to **20** (564 M^{-1}) that may be rationalized in terms of the elongated cavity size which would be favorable to interact with NADH-model molecule [41].

Comparison of the series of carboxyethylcalix[n]arenes **22-25** and model compound **26** in acceleration acid-catalyzed hydration of 1-benzyl-1,4-

dihydronicotinamide was carried out by other group [42]. The obtained results are shown in the Table **2**.

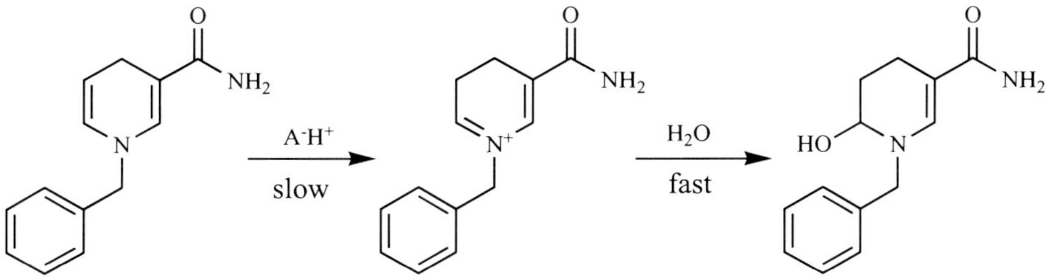

Figure 3: Scheme of catalytic hydration of NADH-model compound.

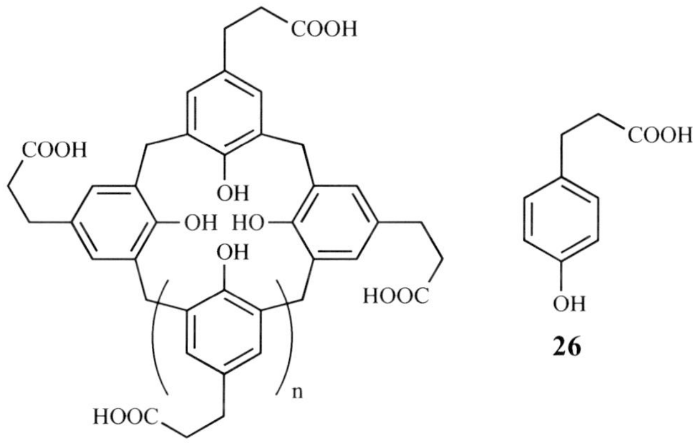

22-25

n = 2-5

26

Table 2. Kinetic parameters for the catalytic hydration of 1-benzyl-1,4-dihydro-nicotinamide (Fig. 3).

Catalyst	k_1 10^{-4}, s^{-1}	K_a M^{-1}
22 n = 2	11	3800
23 n = 3	17	880
24 n = 4	4.4	5000
25 n = 5	4.6	3700
26	0.2	
20	46.2	564

Data from the Table **2** show that *p*-(carboxyethyl)calix[6]arene **23** is a better catalyst than either the smaller or larger members of the series but that is considerably inferior to its sulfonic acid analog **20**. The calixarene effect is even greater if monomeric phenol was taken as the reference point. The ratio for the calixarene *vs.* monomer is 85 for **23** and 1215 for **20** [40].

1.4. ATP-Ase Mimetics

It has been mentioned that supramolecular catalysis involves two steps, the initial substrate bound by receptor molecule, and followed by the chemical transformation of the bound species within the supramolecular complex and finally release of the products with generation of the catalytic unit.

The catalytic effect of water-soluble calix[4]arene-tetrasulfonate **27** on hydrolysis of ATP in aqueous solution was investigated [43].

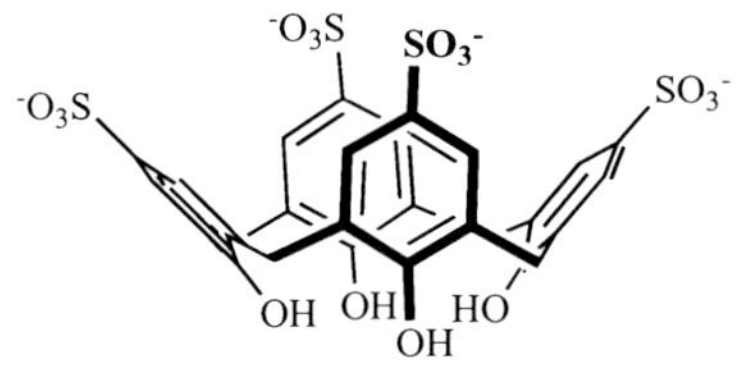

27

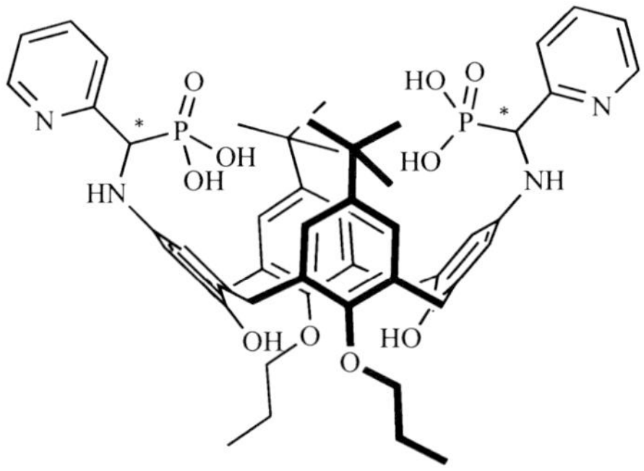

28

Both laser photolysis and pulse radiolysis give evidence of a supramolecular interaction between ATP and **27** species. The results of laser photolysis and pulse radiolysis are consistent with the supramolecular catalysis speculated from the HPLC experiment. The proposed mechanism is unclear and involves (as assumed) a supramolecular complex formation with intermolecular H-bonds between triphosphate group of ATP and phenolic hydroxyls of calixarene [41].

Calix[4]arene bis(aminomethyl)phosphonic acid **28** induces the ATP hydrolysis so can mimic ATPase action [44].

The velocity of calixarene-dependent hydrolysis of ATP exceeds the velocity of spontaneous hydrolysis of ATP at least 14–15 times. The kinetic curve of the ATP hydrolysis induced by calixarene **28** had a tendency to plateau (in the course of time) observing from 45-60 minutes of the incubation period when the reaction practically came to the end. The dissociation constants of the 1 : 1 Host–Guest complexes of ATP-Guest with the calixarene-Host within 197–231 μM were determined by HPLC from relationship of the capacity factor of the Guest and concentration of the calixarene Host in the mobile phase [44].

It was identified that univalent cations Na^+, K^+, Li^+, $choline^+$ and bivalent cations Ca^{2+} or Mg^{2+} did not influence the reaction of ATP hydrolysis. In the presence of the other bivalent cation the inhibition of the reaction occurred in line with the sequence: $Cu^{2+} > Ba^{2+} > Pb^{2+} > Sr^{2+} > Ni^{2+} = Zn^{2+} > Mn^{2+} > Co^{2+}$. The alkalization in the range of pH 6.0– 8.0 stimulated the ATP hydrolysis. The magnitude of activation energy of the reaction was 50.7 kilojoules per mole [45].

1.5. Candida Rugosa Lipase Improvements

Candida rugosa lipase was encapsulated within a chemically inert sol–gel support prepared by polycondensation with tetraethoxysilane and octyltriethoxysilane in the presence of N-methylglucamine based calix[4]arene **29** magnetic nanoparticles (Fig. **4**) [46]. The results indicate that the magnetic calix[4]arene based encapsulated lipase particularly has shown high conversion and enantioselectivity. The enantiomeric ratio equal 460 for encapsulated lipase and 166 for the free enzyme with an *ee* value of >98% for S-Naproxen hydrolysis.

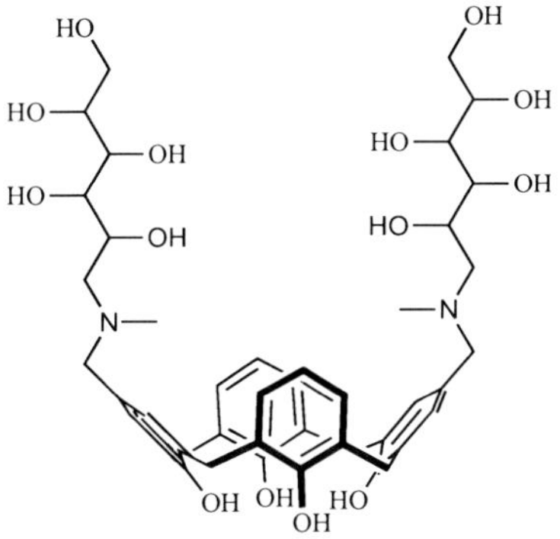

29

Fig. (4). contd....

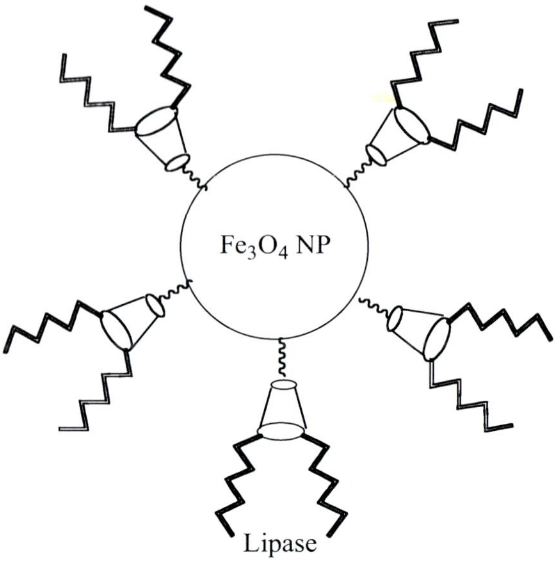

Figure 4: Structure of calixarene **29** and calixarene nanoparticle lipase associate.

2. CALIXARENE-BASED MIMETICS OF RECEPTORS, ANTIBODIES AND ENZYME INHIBITORS

2.1. Antibody McPC603 Mimetic

Biotechnological applications of catalytic antibodies meet with considerable problems such as high cost and lability at high temperatures and in nonaqueous solvents. Potentially, most of these problems could be obviated by using synthetic molecules with similar binding properties and predicted biological activities. Complexation of non-peptidic abiotic receptor **19a**, with DOPC was described [33]. The structure of DOPC·**19a** complex has a close resemblance to the antigen binding site in the McPC603·phosphoryl choline (PC) complex.

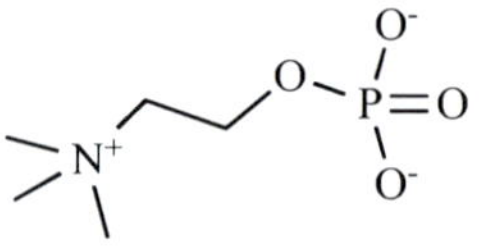

PC

A high binding constant, $K_{ass} = 7.3 \cdot 10^4$ M^{-1} ($\Delta G° = -6.7$ kcal/mol) obtained for the DOPC·**19a** complex in chloroform also is very similar to the complex

McPC603 with PC in water ($\Delta G°$ = - 6.0 kcal/mol). In both cases the phosphocholine head adopts a linear conformation, and orients in a similar way within the binding site, and gives rise to similar interactions:

1) the phosphate group interacts strongly with one of the nitrogens of a cationic guanidinium group (Arg 52H in McPC603, the bycyclic guanidinium fragment in **19a**);

2) a similarly oriented hydrogen bond is established between one of the phosphate oxygens and either the hydroxyl group of Tyr 33H or amide nitrogen of the receptor's side chain;

3) the aromatic rings are oriented toward the methyl groups on the choline head in such a way that each of the N–C vectors is perpendicular to the plane of the opposite ring.

So, synthetic non-peptidic receptor displays the binding properties of a whole antibody and incorporation of suitable reactive functional groups into the binding site should open a new way for creation enzyme mimics.

2.2. Cytochrome C Antibodies

The design of synthetic Hosts that can recognize protein surfaces and disrupt biologically important protein-protein interactions remains a major unsolved problem in bioorganic chemistry. In contrast, the immune system offers numerous antibodies that show high sequence and structural selectivity in binding to a wide range of protein surfaces. The diversity of recognition is even more remarkable, because all antibody (fragment antigen binding) FAB regions have a common structural motif of four-six hypervariable peptide loops.

The new type of synthetic Host **30** that mimics antibody combining sites in possession of several peptide loops arrayed around central binding region was synthesized and investigated [47].

The large surface area (near 400 Å^2) in the molecule allows strong binding to a complementary surface on cytochrome c (cyt c) and disrupts, in similar way to cytochrome c peroxidase (cyt c peroxidase), the approach of reducing agents to the active site of the cyt c. Thus, in phosphate buffer Fe^{III}·cyt c complex ($1.57 \cdot 10^{-5}$ M) is rapidly reduced by excess of ascorbate ($2.00 \cdot 10^{-3}$ M) with pseudo-first-order rate constant 0.109. In the presence **30** ($1.91 \cdot 10^{-5}$ M) the rate of cyt c reduction is diminished tenfold (k_{obs} = 0.01 s^{-1}), consistent with the calixarene-cyclopeptide binding to the protein surface and inhibiting approach of ascorbate to the heme edge. The importance of the four loops in **30** is seen by the minimal effect that even large concentrations of single loop **31** ($5.36 \cdot 10^{-4}$ M) had on the ascorbate reduction. A very similar inhibition of ascorbate reduction is seen with cyt c·cyt c peroxidase complex, further suggesting that **30** binds to a similar region as its natural partner. It has been found *via* fluorescence titration that **30** effectively compete with cyt c peroxidase for binding for cyt c, forming 1:1 cyt c·**30** complex with binding constant $3 \cdot 10^{8}$ M^{-1} [48] and can disrupt cyt c·Apaf-1 complex, key product in apoptosis.

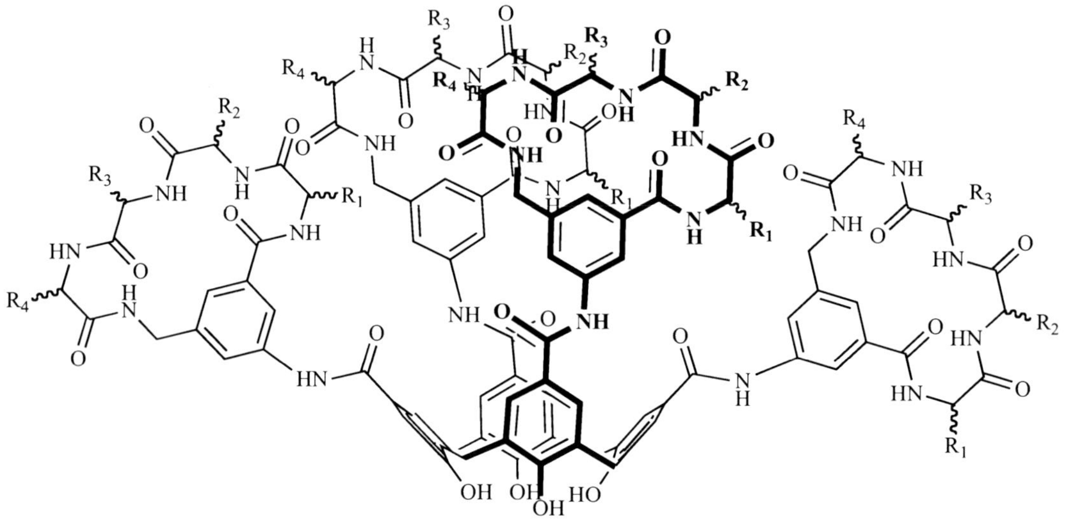

$R_1 = R_3 = H$

$R_2 = R_4 = CH_2COOH$

30

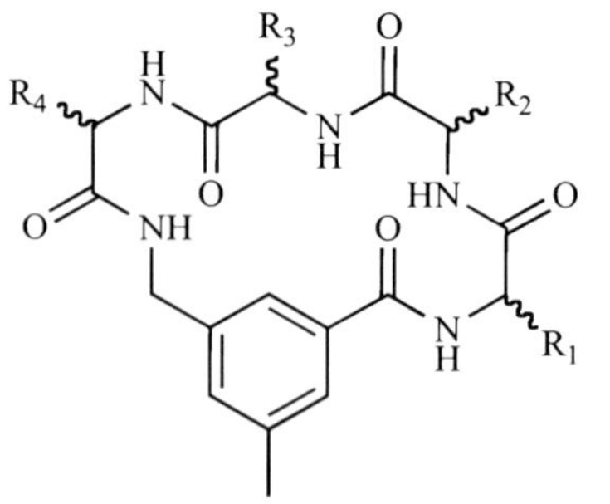

31

2.3. Heparin Neutralization

Polycationic calix[8]arenes **32a,b** derivatives able to neutralize UF and LMW heparin faster and more efficiently than protamine and polylysine, in aqueous solution and in blood were described [49].

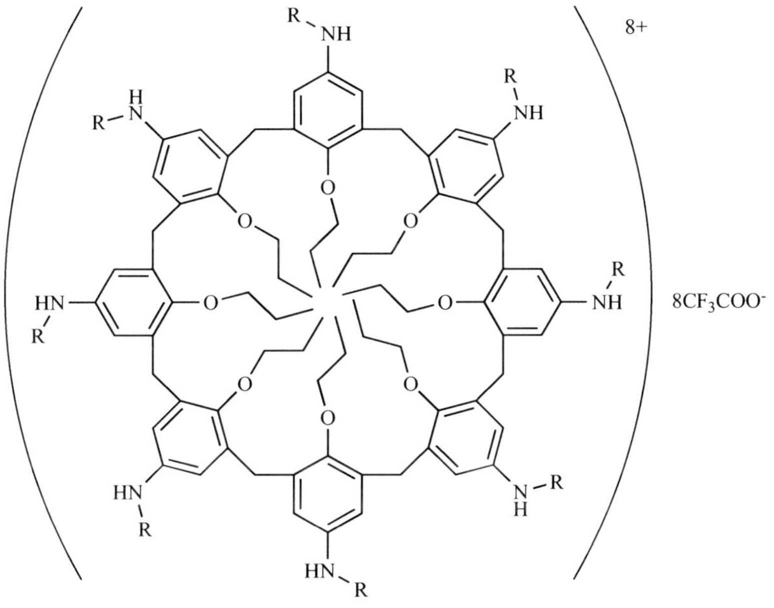

32 **a**: R = L-Lys
b: R = $CO(CH_2)_6NH_2$

The arrangement of multiple functional groups on the flexible macrocyclic scaffold of calix[8]arene, with respect to the conformationally rigid protamine and low ordered polylysine, allowed a mutual adaptability between calixarene polycations **32** and heparin, significantly enhancing the recognition performances. Fluorescence, NMR titration, and activated partial thromboplastin time (aPTT) experiments confirmed that **32a,b** have a very high specificity and affinity towards heparin neutralization as in aqueous solution so in blood. Analogous results were obtained with low molecular weight heparin (LMWH) whose effect on protamine is unable to completely reverse. The results highlight that even for this kind of recognition where the electrostatic interactions seem to be prevalent, other factors like mutual induced fit and conformational adaptability of the receptor could influence significantly the affinity and specificity of the Host–Guest association.

2.4. Inhibition of Acetylcholine Hydrolysis

The design of new selective inhibitors of enzymatic reactions is also of great interest. So, negatively charged calix[4]resorcinarene **33** inhibits a hydrolysis of neurotransmitter ACh.

It was shown [48] that the retardation of the hydrolysis rate amounts almost to a factor ten ($k_2 = 4.8 \cdot 10^{-4}$ s^{-1}; $k_1 = 4.1 \cdot 10^{-3}$ s^{-1}), it could be attributed to a general destabilization of the negatively charged $B_{Ac}2$-type transition state by the **33** charged cavity. An explanation of the speed retardation is based on electrostatic effect of the charged ammonium substituent. It is important that inhibition is occurred without the specific enzyme (acetylcholine esterase) blockade [50]. After the two Zn(II) cations are introduced in the tetraanion of the similar calix[4]resorcinarene **34** the complex becomes electro-neutral and inhibition of ACh is not observed [51].

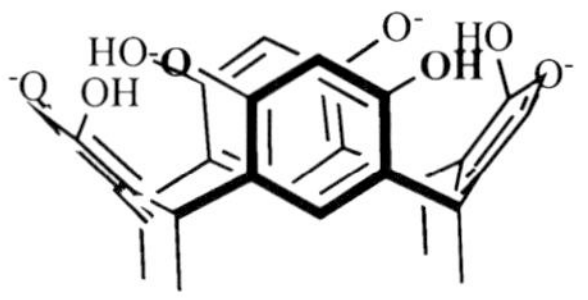

33

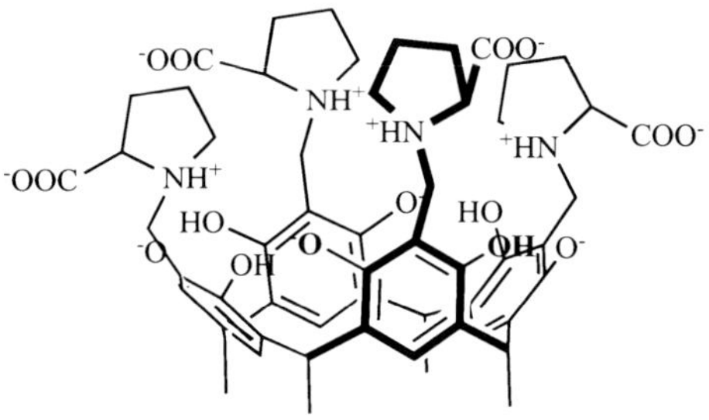

34

2.5. Myometrium Actomyosin and Myosin Inhibition

It was showed that the calix[4]arene **35** bearing one methylenebisphosphonic acid fragment inhibits the ATPase activity of myometrium actomyosin (inhibition coefficient $I_{0.5} = 84 \pm 2$ μM) and myometrium myosin subfragment-1 ($I_{0.5} = 83 \pm 7$ μM) [52]. The study of calixarene **35** effect on myometrium myosin subfragment-1

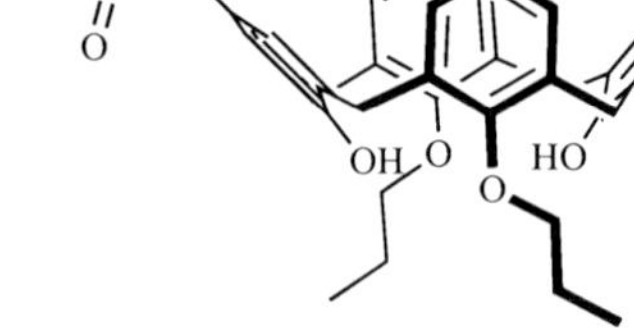

35

ATPase showed that in a concentration of 10 μM it almost did not influence ATPase activity. At a further increase of this calixarene concentration, it inhibited myosin head ATPase in dose-dependent mode, and at 100 μM concentration the inhibiting effect vs. control was 60 ± 5%.

According to the linearized graph in Hill coordinates (Fig. **5**), inhibition coefficient $I_{0.5}$=83 ± 7 μM.

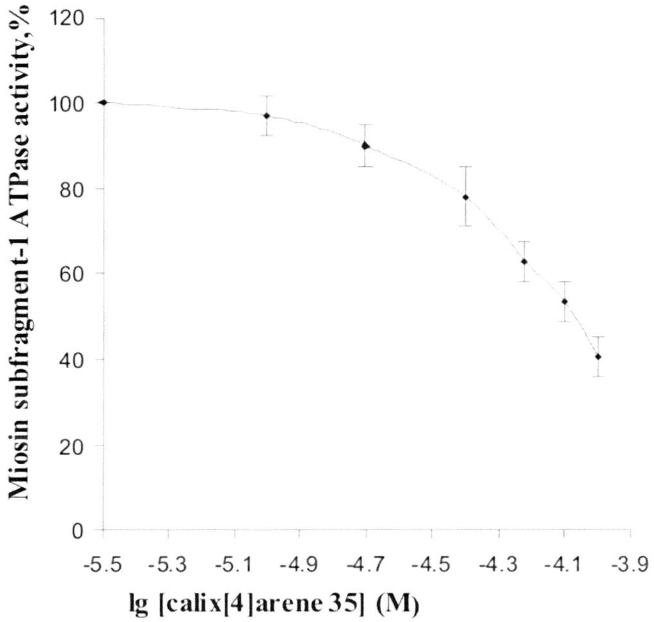

Figure 5: Catalytic titration of myosin subfragment-1 ATPase activity with calixarene **35**.

To understand the molecular mechanism of calixarene **35** inhibiting effect on myosin subfragment-1 ATPase the computer simulation for the interaction of calixarene **35** (as a separate ligand or coupled with ATP) with myometrium myosin subfragment-1 in the presence of Mg^{2+} as a cofactor (Fig. **6**). Mg cation takes part in the binding of ATP to the active site, and in the process of its hydrolysis.

Analysis of the docking of calixarene **35** as a separate ligand into the ligand binding site of myosin subfragment-1 suggests that residues Asp_{320}, Asn_{321}, Leu_{676} and Gln_{678} may be involved in the fixation of the calixarene phosphonate groups.

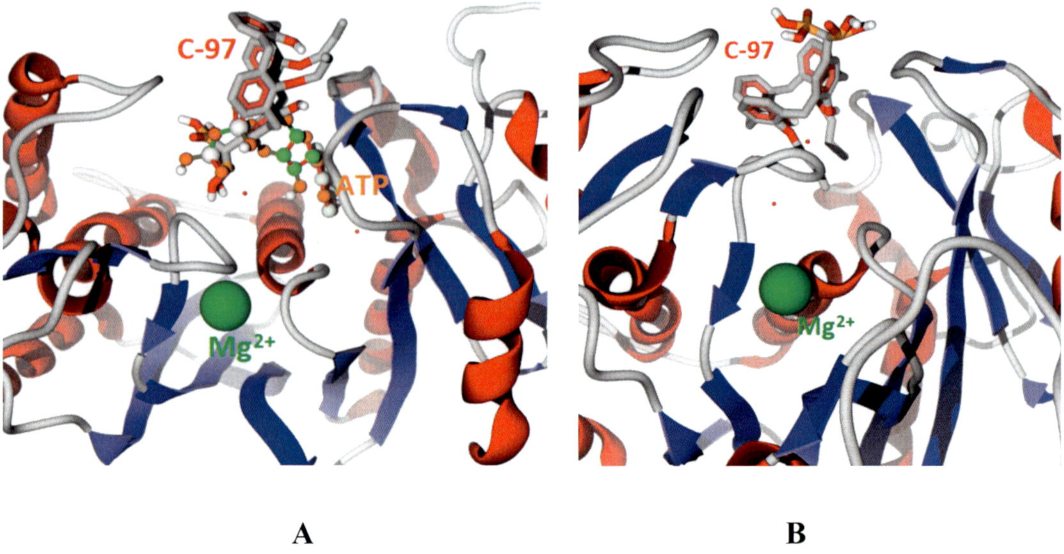

Figure 6: Docking of calixarene **35** as a separate ligand into the active site of myosin subfragment-1 (**A**) and docking of the paired ligand "**35** + ATP" (**B**) in the presence of Mg^{2+} as a cofactor.

2.6. Inhibitors of Alkaline Phosphatases

Inhibition of nonspecific alkaline phosphatases is of significant current medical interest, since these enzymes catalyze the hydrolysis of phosphate monoesters, which are involved in a number of important biochemical pathways.

Calix[4]arenes **35** and **36** bearing one or two methylenebisphosphonic acid fragments displayed stronger inhibition of calf intestine alkaline phosphatase than simple methylenebisphosphonic acid **37** or 4-hydroxyphenyl methylenebisphosphonic acids **38** [53].

The inhibitory activities of the compounds **35-38** were investigated in the calf intestine alkaline phosphatase catalyzed hydrolysis of *p*-nitrophenylphosphate. The inhibition constants K_i and K_i' for the calix[4]arene bis(methylenebisphosphonic) acid **36** in Tris- HCl buffer at pH 9 are 0.38 μM and 2.8 μM respectively. The influence of compounds **35-38** on the enzyme activity is in agreement with mixed-type or partial mixed-type inhibition. The mechanism of catalysis by alkaline phosphatase includes formation of the enzyme–substrate complex, phosphoryl transfer from this complex to a nucleophilic serine group, the hydrolysis of the covalent phosphorylated enzyme intermediate, the formation of a non-covalent enzyme–phosphate complex and the release of inorganic phosphate (Fig. **7**).

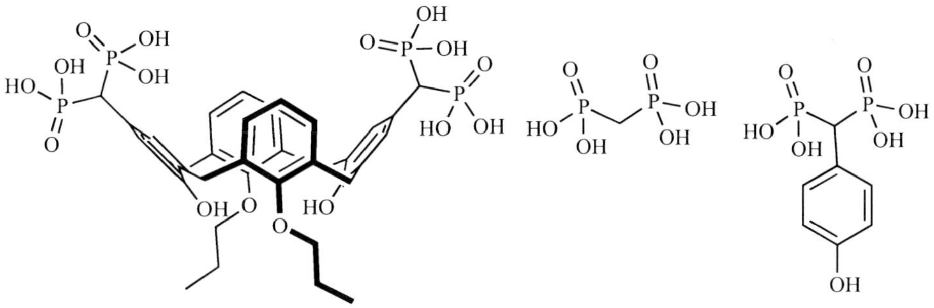

36 **37** **38**

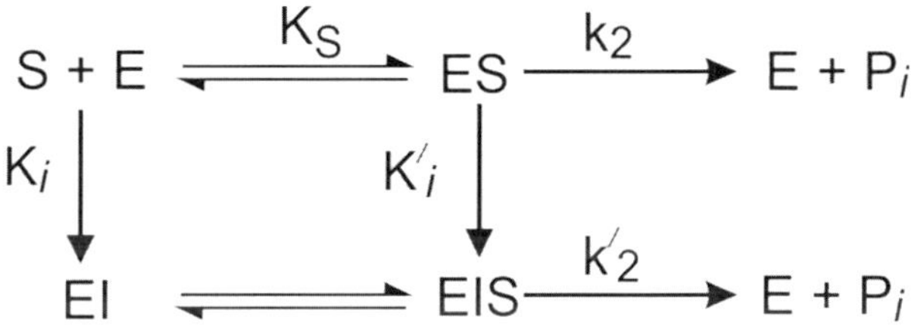

Figure 7: Schematic mechanism of alkaline phosphatase action and inhibition.

The dissociation constants of the enzyme inhibitor K_i and the enzyme–substrate inhibitor K_i' complexes with **35-38** are given in the Table **3**.

Table 3. Inhibition constants for the inhibitors 35-38 of calf intestine alkaline phosphatase.

Inhibitor	Ki, μM	Ki′, μM
35	2.5	46
36	0.38	2.8
37	67±5	750±100
38	22±4	290±110

The inhibitory activity of calixarene **36** is the consequence of the preorganization of two methylenebisphosphonic acid substructures. The efficient coordination of

36 with the metal cations in the enzyme's active site and the additional van der Waals interactions brought about by the scaffold itself are probably responsible for its high activity.

Further investigations of inhibitory effect of chiral calixarene-aminophosphonic acids **39, 40** to alkaline phosphatase of different domestic animals were carried out by the same team [54].

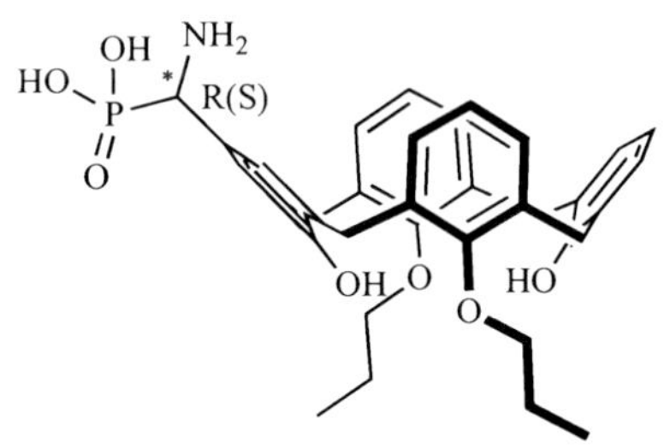

39 a R-isomer, b S-isomer

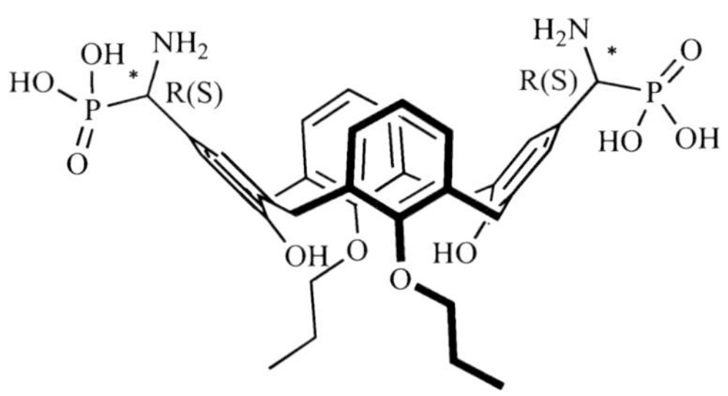

40 a RR- diastereomer
b SS - diastereomer

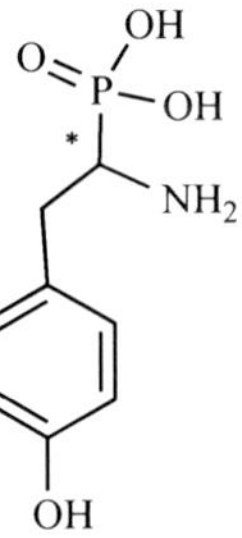

41 S-isomer

Compounds **39, 40** display almost identical affinities to bovine intestinal alkaline phosphatase (BIAP). But the inhibition of porcine kidney alkaline phosphatase (PKAP) is sensitive to the absolute configuration of the calixarene α-aminophosphonic acids. The K_i value for *S*-isomer **39b** is about two times smaller than for its *R*-counterpart **39a**. The enantioselectivity of the inhibition is drastically increased in bis-aminophosphonic acids **40a,b**. Namely, the *R,R* isomer **40a** binds to PKAP about 50 times stronger than the *S,S* isomer **40b**, which corresponds to a difference in free energy of 2.3 kcal/mol (Table **4**).

Table 4. Inhibition constants (μM) for the inhibition of PKAP by compounds 39, 40 and (S)-4-hydroxybenzyl-2-aminomethylphosphonic acid 41 Free energies (kcal/mol) for the formation of enzyme-inhibitor complexes of PKAP.

Inhibitor	K_I	K_I'	ΔG_I
39a (R)	73± 13	540± 90	-5.6
39b (S)	32± 5	780± 100	-6.1
40a (RR)	1.7± 3	130± 30	-7.8
40b (SS)	86± 11	610± 210	-5.5
41 (S)	580± 110	6800± 840	-4.4

0.1 M Tris-HCl buffer (pH 9), 23°C; $\Delta G = -RT \ln (K_i^{-1})$.

So, the strength of the enzyme-inhibitor interactions is determined both by cooperative effects and the goodness of the fit of the chiral inhibitor to the chiral space around the binding site. To the best of our knowledge compound **40a** is the first alkaline phosphatase inhibitor of R-aminophosphonic acid type that has an affinity for the enzyme in the micromolar range.

2.7. Inhibitors of Protein Tyrosine Phosphatases (PTPs)

Next stage of phosphatase survey was study of series (thia)calix[4]arene phosphonous acids in inhibition of protein tyrosine phosphatases (PTP). Named enzymes regulate a number of biochemical processes which proceed through dephosphorylation of phosphotyrosine residues in proteins, including cell growth, differentiation, metabolism, and movement. Because of importance of protein phosphorylation–dephosphorylation cycle in living cells, the activities of these enzymes are associated with diabetes, obesity, cancer and other human diseases.

Preliminary, an inhibition of *Yersinia* protein tyrosine phosphatase by calix[4] arene and thiacalix[4]arene tetrakis(methylphosphonic) acids have been investigated [55]. The kinetic studies revealed that **42** and **43** are potent competitive inhibitors of *Yersinia* PTP with inhibition constants 0.92 and 0.22 μM. It was found the phosphonate groups as well as macrocyclic structure are essential for the activities of these compounds. Model docking studies revealed similar binding mode for calixarene inhibitors covering the active site, with location of the phosphonate residues around the entry of the phosphotyrosine binding cavity.

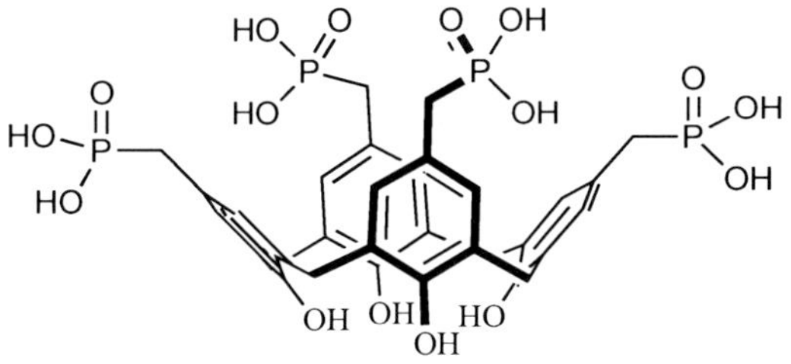

42

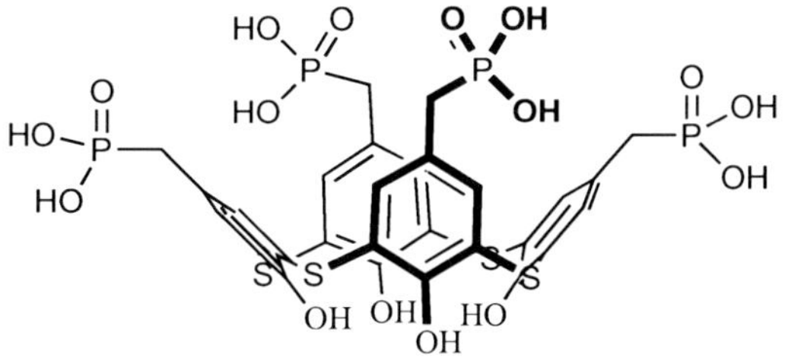

43

Protein tyrosine phosphatase 1B (PTP1B) belongs to the superfamily of PTPs and involves in insulin and leptin signal transduction by dephosphorylation of activated insulin receptor or leptin receptor-associated kinase, JAK2. As a negative regulator of insulin and leptin signalling, PTP1B is one of the most promising therapeutic targets for treatment of type 2 diabetes [56]. Calix[4]arenes bearing methylenebisphosphonic **35** and **36** or hydroxymethylenebisphosphonic **44** and **45** acid fragments at the wide rim of the macrocycle were studied as inhibitors of PTP1B [57]. Next part of the study operated with calix[4]arene mono- and bis-α-hydroxymethylphosphonic acids **46** and **47** [58]. The calixarene α-ketophosphonous acids **48** and **49** were also investigated and it was found that these macrocycles are able to inhibit PTP1B with IC_{50} values in the micromolar range displaying lesser efficiency but selectivity over TC-PTP, MEG1, MEG2, SHP2, LAR and PTP-β [59]. Some of the inhibitors showed IC_{50} values in the micromolar range and good selectivity in comparison with other protein tyrosine phosphatases such as TC-PTP, PTPb, LAR, and CD45 (Table **5**). Kinetic studies indicated that the calix[4]arene derivatives influence PTP1B activity as slow-binding inhibitors. The α -ketophosphonate moiety together with macrocyclic platform was supposed to bind to the active site of PTP1B by reversible slow-binding inhibition mechanism. Molecular docking performed on example of PTP1B indicated that hydroxymethylphosphonate fragment of the inhibitor **46** is able to mimic the binding of the substrate in the enzyme active site region. To clarify the mechanism of inhibition, the possible enzyme-inhibitor complexes have been considered using several crystal structures of PTP1B and all stereoisomeric forms of the inhibitors.

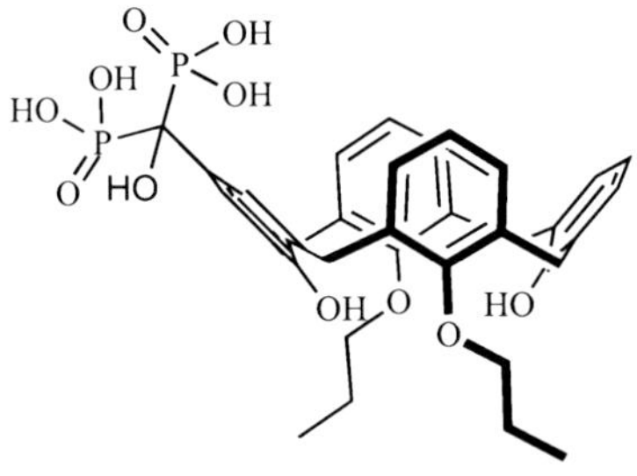

44

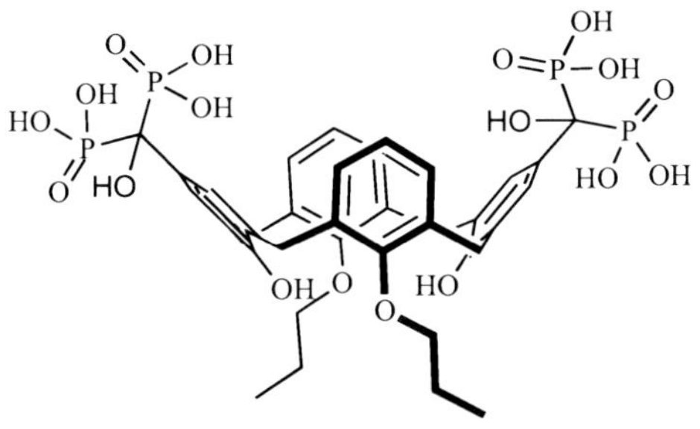

45

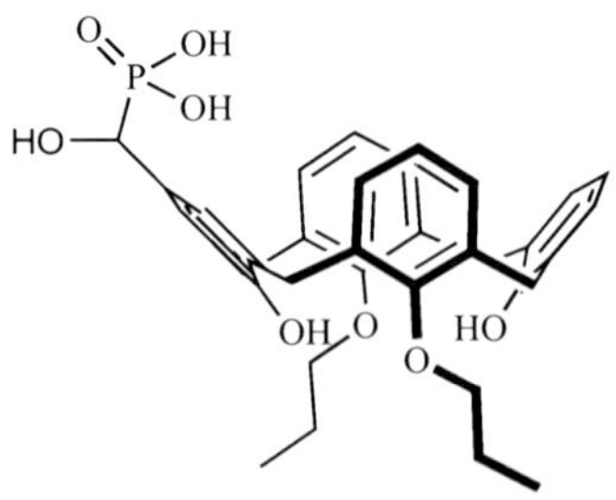

46

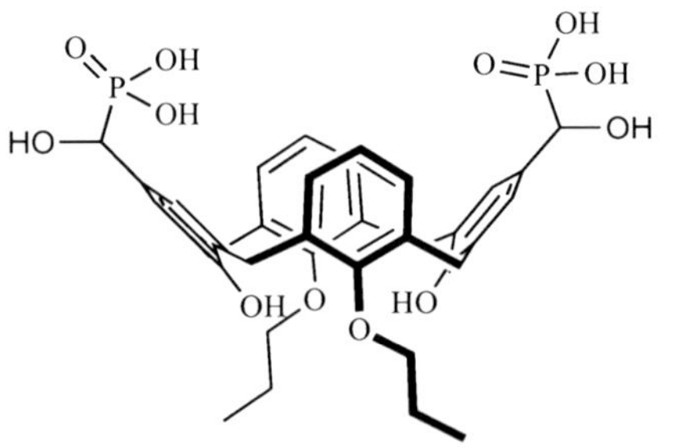

47

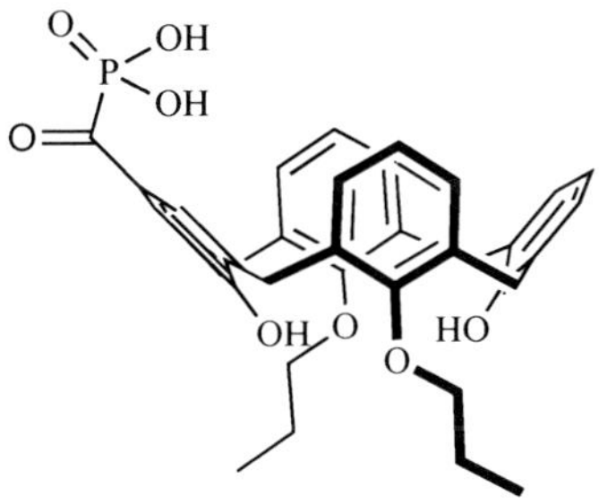

48

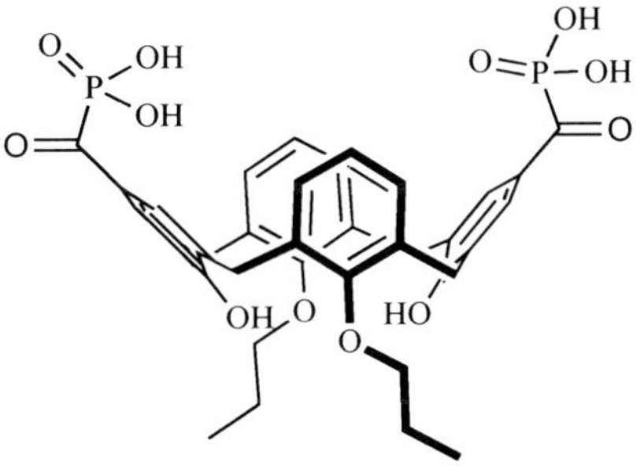

49

Table 5. Experimental inhibitory activity (IC_{50}, μM) of calix[4]arene phosphonous acids 35, 36, 44-49 as inhibitors of PTP1B and other protein tyrosine phosphatases.

Inhibitor	PTP1B	TC-PTP	PTPβ	CD45	SHP2	LAR
35	1.2 ± 0.2	9.3 ± 1.9	13 ± 1	5.7 ± 0.9	1.8 ± 0.5	>100
36	2.5 ± 0.2	4.9 ± 1.5	32 ± 7	22 ± 6	2.6 ± 0.6	>100
44	1.9 ± 0.2	27 ± 6	18 ± 1	8.4 ± 2.1	3.4 ± 0.3	≥100
45	8.3± 1.1	22 ± 2	12 ± 2	59 ± 10	9.2 ± 0.4	>100
46	1.6 ± 0.4	11±3	3.5±0.7	4.5±0.9	0.64±0.15	>100
47	8.4± 2.1	59±14	17±4	45±11	7.6±1.9	>100
48	3.2 ± 0.4	27 ± 3	32 ± 4	11 ± 3	40 ± 6	72 ± 5
49	6.7 ± 0.2	51 ± 10	30 ± 10	16 ± 3	53 ± 5	56 ± 3

In contrast to inhibition of alkaline phosphatases, better activity showed mono-substituted calixarene derivatives compared todi-substituted.

Last part of this systematic investigation involved tetrasubstituted thiacalix[4] arenes. It was demonstrated that the thiacalix[4]arene derivatives with covalently attached phosphonic acid monoester groups **50a,b** exhibit strong submicromolar inhibition of PTP1B and selectivity over some other PTPs [60]. Compound **50a** is an approximately 11-fold more potent inhibitor of PTP1B as compared to TC-PTP, Megl, SHP2, and 8-fold more active towards Meg2 (Table **6**). Molecular interactions between the phosphonate monoesters and HSA were detected to demonstrate a possible mechanisms of their bioavailability. The presented data provide new insight into properties of the phosphonate monoesters creating the basis for further development of potent macrocyclic inhibitors of PTP1B.

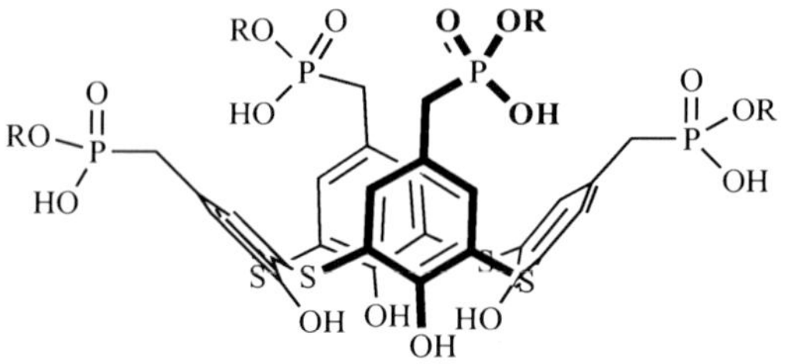

50 a R = n-C_2H_5
b R = n-C_4H_9

Table 6. Experimental inhibitory activity (IC_{50}, μM) of thicalix[4]arene phosphonous acids monoesters 50a,b as inhibitors of PTP1B and other protein tyrosine phosphatases.

Inhibitor	PTP1B	TC-PTP	MEG1	MEG2	SHP2
50a	0.34 ± 0.07	5.2 ± 1.4	2.8 ± 0.2	3.9 ± 1.5	3.6 ± 0.85
50b	0.32 ± 0.03	2.8 ± 0.5	1.9 ± 0.3	3.8 ± 0.9	3.4 ± 0.84

2.8. Inhibitors of Glutathione S Transferase

Overexpression of glutathione S-transferase (GSTase) isoenzymes is known to lead to formation of glutathione conjugates with chemotherapeutic agents, resulting in drug resistance. Dipropoxy- and tetrapropoxy-derivatives of calix[4]aren

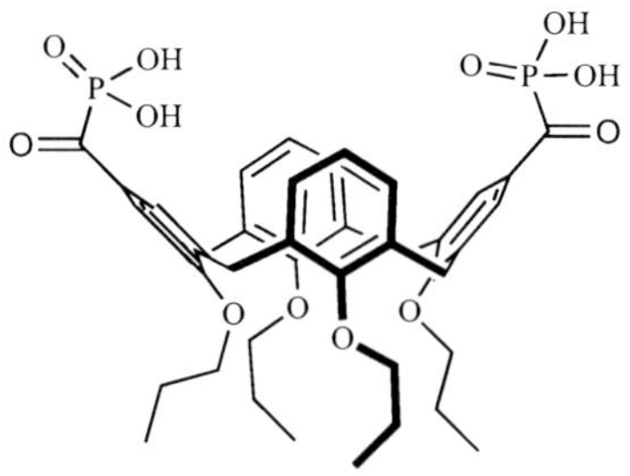

51

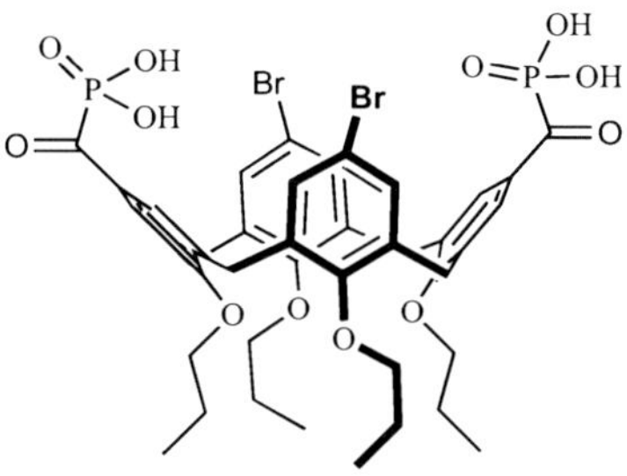

52

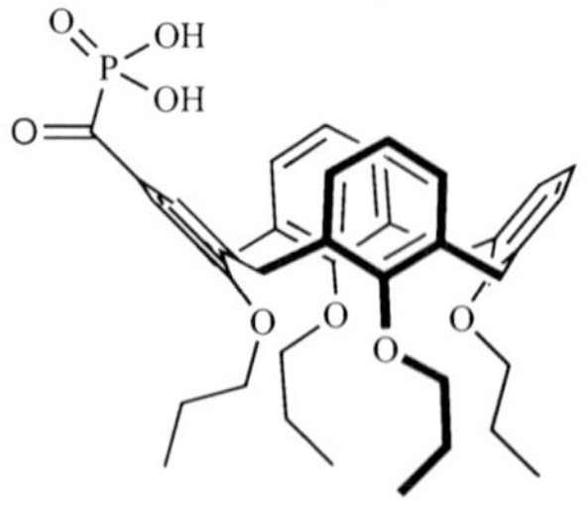

53

α-hydroxymethylphosphonic acids **46, 51-52** were designed to explore their inhibiting effects on activity of GSTase from equine liver and human placenta *in vitro* [61, 62]. These compounds were found to have affinity for the enzyme from equine liver in the micromolar range (Table **7**).

Table 7. Inhibition of glutathione S-transferases from equine liver (EL) and human placenta (HP) by calix[4]arene-α-hydroxymethylphosphonates.

Inhibitor	IC_{50}(EL), μM	IC_{50}(HP), μM
46	7.5 ± 2.3	No effect
51	6.1 ± 1.9	29 ± 13
52	4.5 ± 1.1	34 ± 6

The calix[4]arene **46** bearing one substituent at the upper rim was more effective in comparison with bis-substituted macrocycle **47**. Introduction of four propyl groups to the lower rim leads to the analogue **53** being less effective in comparison with **46**. At the same time, among tetrapropoxycalix[4]arenes, inhibiting properties of bisphosphonate derivatives **51** and **52** are higher than of monophosphonate **53**. Docking study proposed that calixarene inhibitors are bind to G-site of enzyme.

2.9. Inhibitors of Proteases

2.9.1. Chymotrypsin Inhibitor

Every protein contains a unique distribution of functional groups (charging, hydrogen bonding and hydrophobic) on its surface. In many cases, these regions are involved in important interactions to other proteins as part of multisubunit, reactivity modulating or signaling complexes. A calixarene platform is unique base for design of receptors by functionalization with high affinity groups. The calixarene **30** with four peptide loops which are arrayed around a central calix[4]arene core was identified [63] as a potent inhibitor of chymotrypsin that functions by binding to the surface of protein. Detailed kinetic analysis showed **30** to be a competitive inhibitor of α-chymotrypsin with K_i and K_i' values 0.81 and 0.11 μM, respectively and shows slow binding kinetics in an analogous manner to the several natural protein-proteinaceous inhibitors [64].

The observation of competitive inhibition of α-chymotrypsin by **30** lends support to the proposal that the receptor is binding close to the active site and blocking approach of the substrate. Surprisingly, **30** is able to displace soybean trypsin inhibitor from its 1:1 complex with α-chymotrypsin. This suggests that **30** is bind to a comparable region of α-chymotrypsin near the active site cleft. Incubation of α-chymotrypsin and **30** for 5 h followed by denaturation of the complex and reverse phase HPLC analysis showed that unreacted receptor was recovered, it is suggesting that no direct contact with the active site of enzyme occurred.

2.9.2. Tryptase Inhibitors

The appropriate functionalization of the calix[8]arene scaffold with basic amino acid fragments, furnished a new type of tryptase inhibitors **54a-d** [65]. Human tryptase was indirectly inhibited due to the antagonist effect of **54a-d** on the proteoglycan heparin.

Kinetic inhibition analysis on recombinant lung tryptase has shown a time-dependent competitive inhibition with both initial and steady-state rate constants in the nanomolar range. Trypsin, another serine protease, showed no inhibition by calixarene **54b**, underlining its selectivity for tryptase.

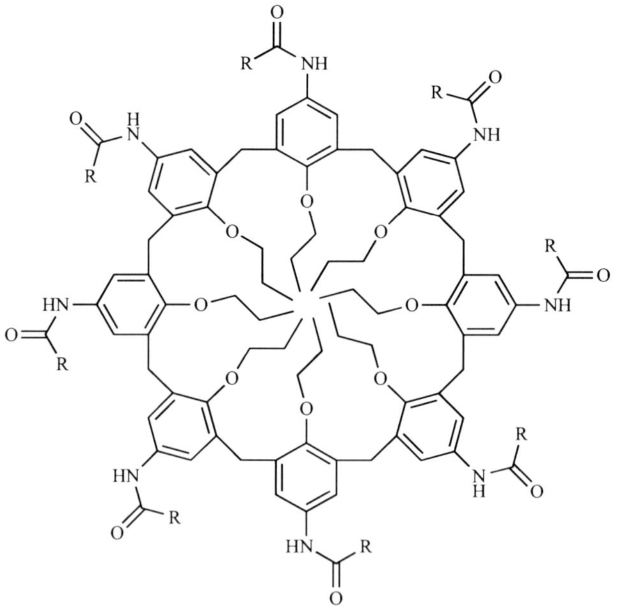

54a-d

a: R = $(CH_2)_2NH_2$
b: R = $CH(NH_2)(CH_2)_4NH_2$
c: R = $(CH_2)_5NH_2$
d: R = $(CH_2)_7NH_2$

3. MEMBRANOACTIVE CALIXARENES

3.1. Transport of Alkali Metal Cations (Carrier Mechanism)

Selective ion transport across phospholipid bilayer membranes is one of the crucial functions in a biological system. There are many studies related to model systems of selective ion transport across phospholipid bilayer membranes. Most have been performed using naturally-occurring antibiotic ionophores such as monensin, salinomycin and valinomycin. In contrast, the number of studies of synthetic ionophore-mediated ion transport is limited [65, 66]. The mechanism for ion transport through bilayer membranes can be divided into two categories: the channel and the carrier mechanism. For example, in the case of gramicidin, its monomer joins head-to-head by six intermolecular hydrogen bonds to form a channel sufficient to span a lipid bilayer. On the other hand, monensin and valinomycin act as mobile carriers in lipid bilayer membranes.

It was shown that tetrakis-ethoxycarbonylmethyl *p*-tertbutylcalix[4]arene **55** has capacity for selective Na^+ transport through phospholipid bilayer membranes [67].

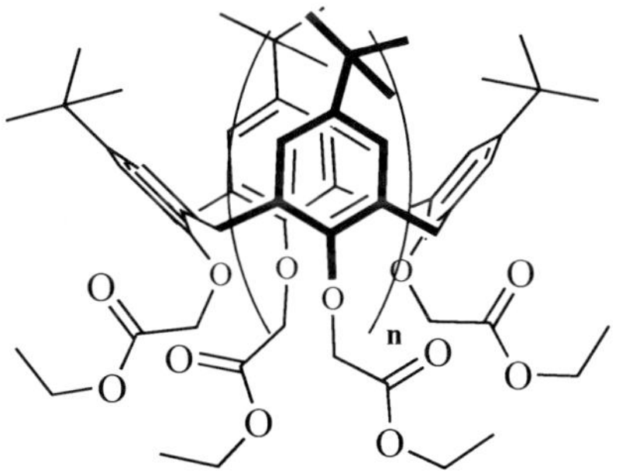

n = 1 (**55**), n = 2 (**56**), n = 3 (**57**), n = 4 (**58**), n = 5 (**59**)

When calixarene **55** was incorporated into a planar phospholipid bilayer membrane, electric currents, resulting from sodium ion fluxes across membrane were generated upon applying external voltages between two chambers. Also, incorporation of **55** into large unilamellar vesicles (LUVs) in NaCl aqueous solution gave rise to dynamic ^{23}Na NMR spectra arising from **55**-mediated Na^+ exchange across LUV bilayer membrane. The rates of Na^+ transport by **55** were 4-fold smaller than monensin and 16-fold higher than known sodium ionophore N,N-didodecyl-4,13-diaza-18-crown-6. The kinetics for **55**-mediated Na^+ transport follows a facilitated diffusion model in which one calix[4]arene molecule complexes with one sodium ion at the water-membrane interface and sodium ions are transported to the opposite interface by the diffusion of cationic Na^+·**55** complexes [65].

Investigation of alkali metal cation selectivity shows that permeability of sodium ions is larger by a factor 17 for K^+ or much larger for Li^+, Rb^+, Cs^+.

The further investigations were concerned with relationships between size of calixarene macrocycle and ionophoric activities in a phospholipid bilayer membrane system [68]. Calixarene esters **55-59** show capacity as ion carriers.

Ion selectivity for calix[n]arene-mediated ion transport in a phospholipid bilayer system are summarized as follows: for tetrameric ester **55,** $Na^+ \gg K^+ > Rb^+ > Cs^+ > Li^+$; for pentamer ester **56,** $K^+ > Rb^+ > Cs^+ \sim Na^+ > Li^+$; for hexamer ester **57**, $Cs^+ > Rb^+ > K^+ > Na^+ > Li^+$; for heptamer ester **58**, $Cs^+ > Rb^+ \sim K^+ > Na^+ > Li^+$. Both calix[n]arenes **57** and **58** display Cs^+ transport selectivity, while the ion transport ability of **57** is four times better than that of **58**. For octamer ester **59** no ion transport activity in phospholipid membrane was detected.

3.2. Formation of Membrane Channels for Alkali Metal Cations

Ion channels are naturally occurring molecular devices in the cells of all tissues of all animals that generate large ionic fluxes with on-off gate control. An ion channel with a typical conductance of 10 ps allows the passage of $5 \cdot 10^6$ ions per second through single molecular pore. These ion channels generally discriminate different ionic species, for example cation/anion and Na^+/K^+.

Two channel compounds **60a,b** having a calix[4]arene central unit in a 1,3-alternate conformation were described [69]. Passage of cations through the center of such a rigid framework has been demonstrated.

The overall length of the channel lies between 34 and 60 Å and is enough to span the lipid portion of a typical bilayer membrane (normally about 30 Å thick). Addition of **60a** or **60b** (1 μL of 1 μM solution) in the bilayer dramatically increased either the Na^+ or K^+ permeability of the previously nonpermeable membrane. Both compounds exhibited "burst" events of nearly 300 ps. These ion fluxes normally persisted less than a second, but the presumed overlap of several conductance events resulted in occasional episodes of longer than 3 s.

Compound **61**, the "control" structure having cone conformation, did not conduct cations under identical conditions.

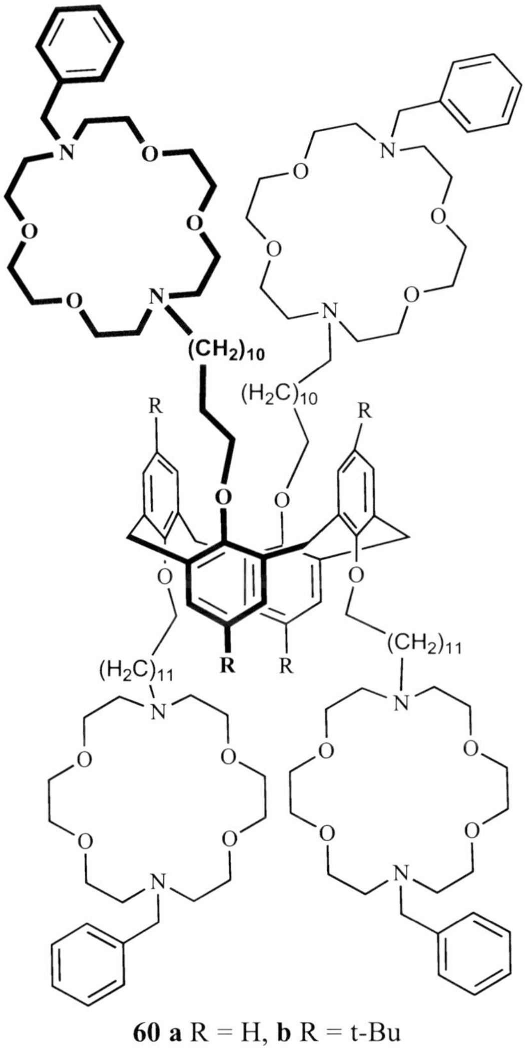

60 a R = H, **b** R = t-Bu

Authors propose such mechanism, in **60a,b** two crown ether groups on each side may rapidly exchange positions; this leads, on average, to a much larger opening in the membrane. If the sidearms sweep out a conical surface in each direction from the bilayer midplane, conductance should be quite high. But, further motion of cations is unclear and authors suggest that it is highly unlikely that either Na^+ or K^+ cations pass through the calix hole.

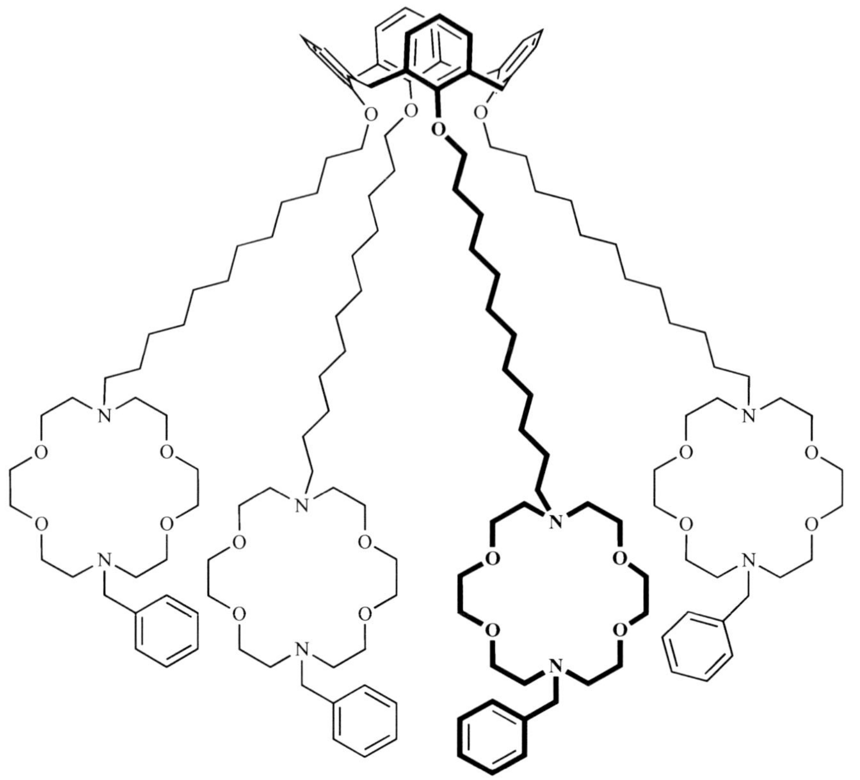

61

The first example of selective ion transport through a membrane bilayer by means of a channel mechanism with a simple and stable synthetic molecule **62a** was described [70]. In a membrane bilayer calixarene **62a** reproduces the essential biological properties of a single ion channel (Fig. **8**). This simplest model of ion channel may offer insight into origins of ion selectivity of ion channels. Under the conditions of single-salt gradient (0.5 M KCl/0.1 M KCl) the channel formed by **62a** showed a reversal potential of 35.9 mV, indicating that it has a distinct selectivity for cations over anions (ratio P_K/P_{Cl} of 20). Similar reversal potential measurements under gradients of K^+ and Na^+ revealed, that the channel discriminates K^+ from Na^+ ions with permeability ratio P_K/P_{Na} of roughly 3.

The next step was in combining of function of this non-peptidic channel **62a** with rigid methyl cholate **63** which has amphiphilic nature and shows a long-lasting open state [71]. The resulted macrocyclic calix[4]resorcinarene **62b** have four amphiphilic methyl cholate groups and as suggested two molecules are assembled in tail-to-tail coupled pair in bilayer membrane to afford a transmembrane channel [72].

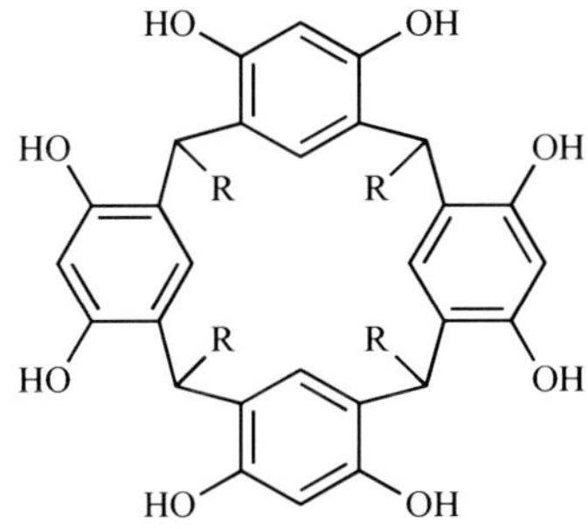

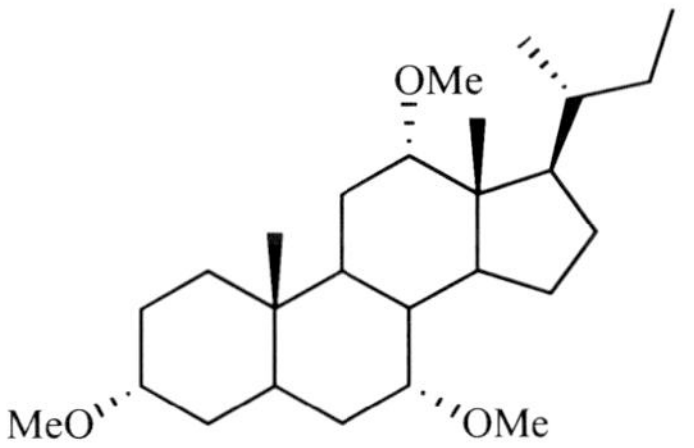

63

62a R = n-$C_{11}H_{23}$,

b R =

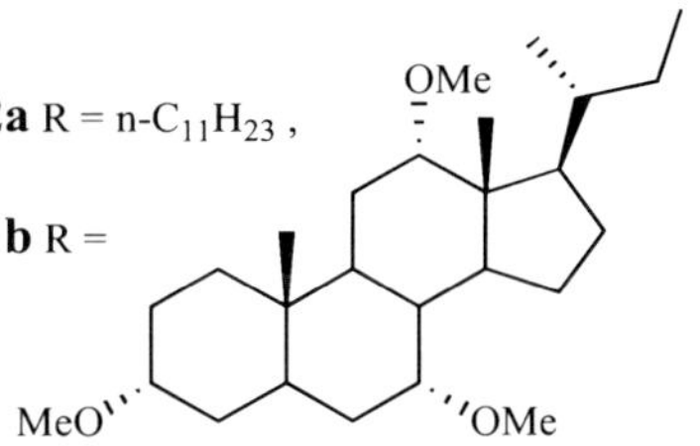

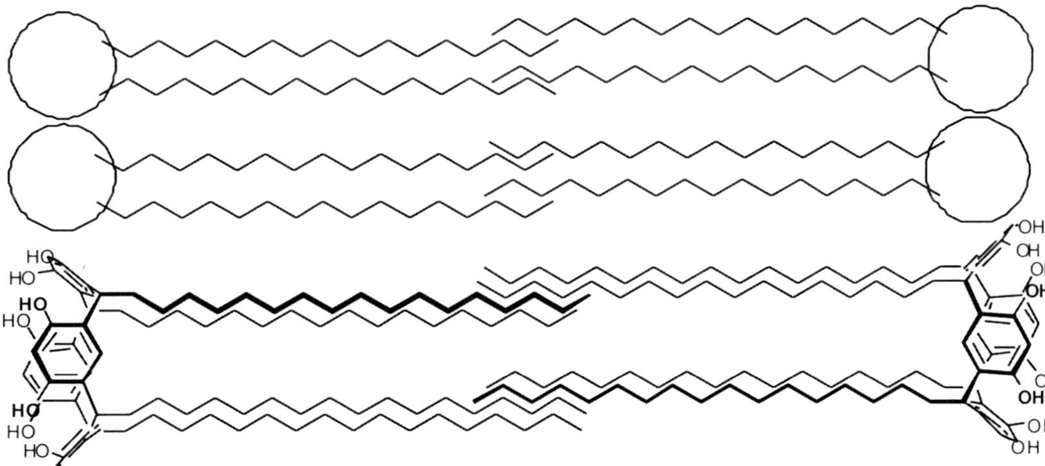

Figure 8: Formation of single ion channel by **62a** in soybean membrane bilayer.

In the symmetrical two chambers system with 0.5 M KCl (pH 7.2, HEPES buffer) at 70 mV, ion channels formed from **62b** showed long lasting states with single-channel currents. The ratio of open to closed states was about 95 to 5, and the average open times were in range 2.5-4.5 s. For channels formed by **62a** this index is 25-45 ms. Comparison of two systems suggests that the rapid transitions in **62a** are based on molecular motion of flexible alkyl side chains. In **62b** with rigid steroidal anchors in the membrane, these rapid open-closed transitions are almost quenched. The K^+/Na^+ selectivity was examined under asymmetric salt concentration conditions and the permeability ratio P_K/P_{Na} was determined as 2.8, which was similar to that of **62a**.

3.3. Calcium Pumps Modulators

The effect of the series of calixarenes on the activity of calcium pumps of smooth muscle cells was investigated [72, 73]. It was shown that calixarene-tetrasulfonylamidine **64** decreases the transport activity of the ruthenium red insensitive, oxalate-stimulated and Mg^{2+}, ATP-dependent calcium pump of sarcoplasmic reticulum by 75%. A similar result was obtained with the Mg^{2+}, ATP-dependent calcium pump purified from the plasma membrane ($I_{0,5}$ = 34.6 μM). On the other hand, calixarene **64** did not affect: the ruthenium red sensitive and thapsigargin-insensitive Ca^{2+}-transport in mitochondria; the catalytic activity of the plasma membrane Ca^{2+}-independent Mg^{2+}-ATPase; ouabaine-suppressed Na^+, K^+-ATPase. The results are represented in the Fig. **9** [74, 75].

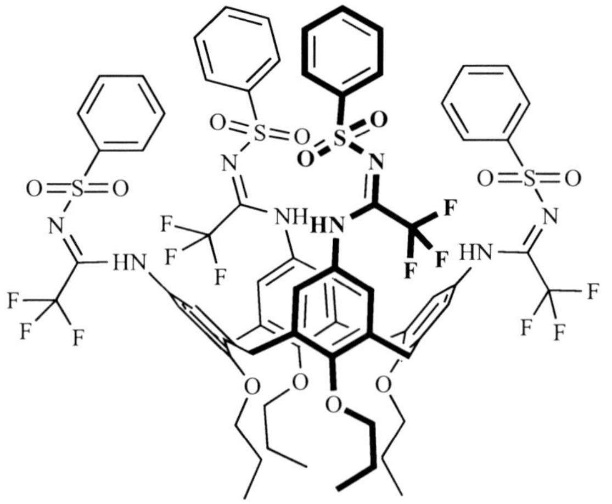

64

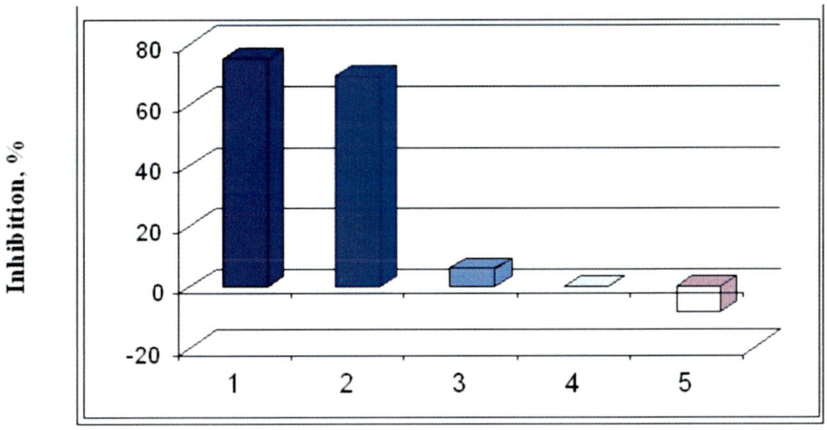

Figure 9: Inhibition of myometrium smooth muscle cells transporting ATP-ases by calixarene **64** (0.1 mM) (1) Mg^{2+}, ATP-dependent Ca^{2+} accumulation in the miometrial sarcoplasmic reticulum; (2) Mg^{2+}, ATP-dependent calcium pump of the plasma membrane; (3) ouabaine-suppressed Na^{+},K^{+}-ATPase; (4) Mg^{2+}, ATP-dependent Ca^{2+} accumulation in the mitochondria; (5) Ca^{2+}-independent Mg^{2+}-ATPase ("basal" Mg^{2+}-ATPase).

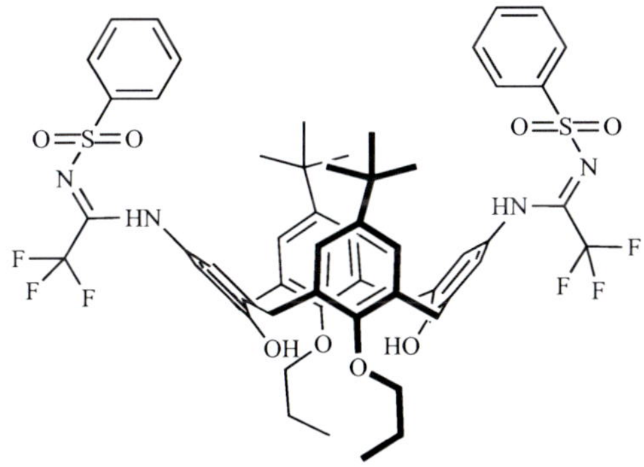

65

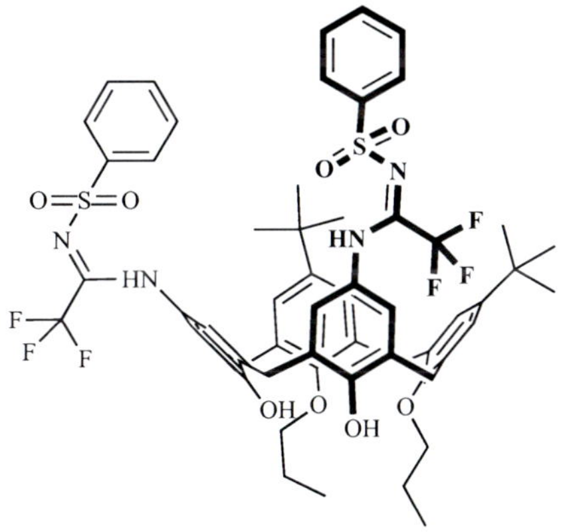

66

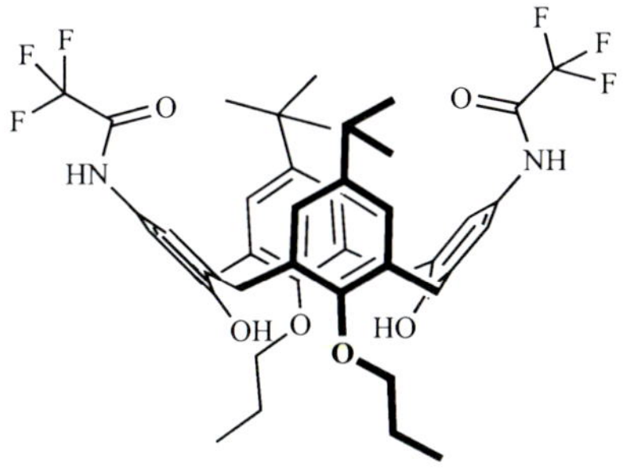

67

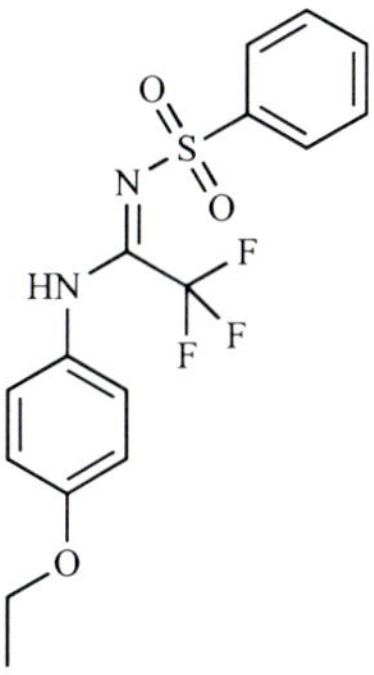

68

Further investigation of sulfonylamidines **65-66** revealed that di-substituted calixarenes as well as model compounds possess lesser activity (Table **8**). So, to efficiently inhibit the Ca^{2+},Mg^{2+}-ATPase activity of PM the calix[4]arene molecule has to have four phenylsulfonylamidine groups on the upper rim of the macrocycle. It should be noted that the inhibitory effects of the studied calix[4]arenes also depended on the mutual location of phenylsulfonylamidine groups on the upper rim of the calix[4]arene "bowl" [76].

Table 8. Inhibitory effect of the sulfonylamidine calix[4]arenes 64-66 as well as bis-amide calix[4]arene 67 and of the model compound 68 on specific enzymatic activity of Ca^{2+}, Mg^{2+}-ATPase of plasma membrane of myometrium cells.

Calix[4]arene	Inhibition coefficient, $I_{0,5}$, μM	The remaining enzymatic activity at concentration 100 μM, %
64	20.2 ± 0.5	25.0 ± 0.3
65	53.4 ± 3.6	38.5 ± 1.9
66	165 ± 20	54.1 ± 0.6
67	–	79.3 ± 0.9
68	–	90.7 ± 2.4

Calixarene amide **69** in concentration 100 μM acts multidirectionally on the functionally different Mg^{2+}-dependent ATP-hydrolase enzymatic systems. Calixarene **69** activates effectively the actomyosin ATPase (K_a = 52 ± 11 μM). Also, **69** reduces effectively basal Mg^{2+}- ATPase activity, insignificantly activating Na^{+},K^{+}- ATPase, but has no influence on transporting Ca^{2+},Mg^{2+}-ATPase activity of plasmatic membranes [73]. The results of the comparative investigation of **69** are shown in the Fig. (**10**).

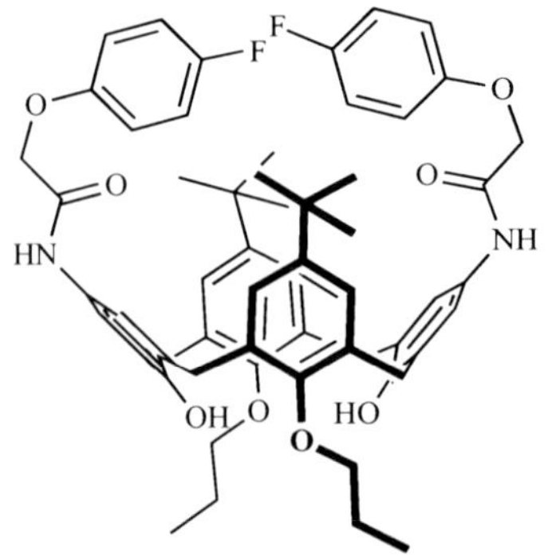

69

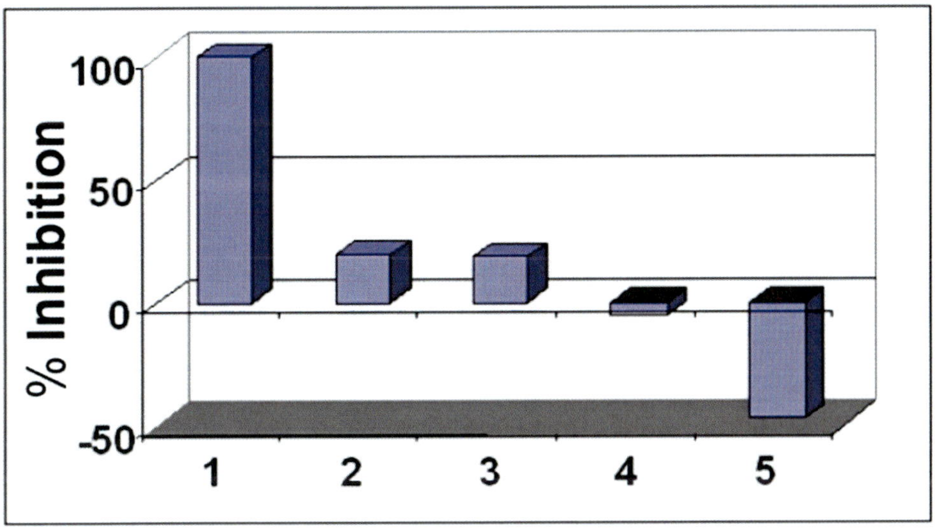

Figure 10: Influence of transporting ATP-ases of smooth muscle cells of miometrium by calixarene **69** (1) Mg^{2+}, ATP-dependent Ca^{2+} accumulation in the mitochondria; (2) Mg^{2+}, ATP-dependent Ca^{2+} accumulation in the miometrial sarcoplasmic reticulum; (3) Ouabaine-suppressed Na^{+},K^{+}-ATPase; (4) Mg^{2+}, ATP-dependent calcium pump of the plasma membrane; (5) Ca^{2+}-independent Mg^{2+}-ATPase ("basal" Mg^{2+}-ATPase).

3.4. Modulators of Sodium-Potassium Pump

Effect of calix[4]arenes **28, 35**, **47**, functionalized by fragments of hydroxyphosphonic, aminophosphonic and methylene-*bis*-phosphonic acid on enzymatic activity of oubaine-sensitive Na^{+}, K^{+}-ATPase and oubaine resistant basal Mg^{2+}ATPase was studied [75, 76]. The experiments were made on the suspension of myometrium cell plasma membranes. It was found that calixarene phosphonic acids **28, 35**, **47** in concentration of 100 μM inhibited enzymatic activity of Na^{+}, K^{+}-ATPase by 86–98% and did not practically affect activity of Mg^{2+}-ATPase. These calixarenes were more efficient than ouabaine in suppressing enzymatic activity of the sodium pump (Table **9**). In the presence of calixerenes **28, 35**, **47** the value of the appearance constant of inhibition $I_{0.5}$ was < 0.1 μM [77]. The model compound **70** (a fragment of the calixarene **28**) in concentrations 10 nM – 0.4 mM didn't practically influence the enzymatic activities as Na^{+}, K^{+}-ATPase and basal Mg^{2+}-ATPase. Hence the influence of calixarene **28** on the Na^{+}, K^{+}-ATPase activity is caused by cooperative action of two fragments **70** on the calixarene scaffold. Calixarenes **28** and **35** increase affinity of the Na^{+}, K^{+}-ATPase to the cardiac glycoside ouabaine [78].

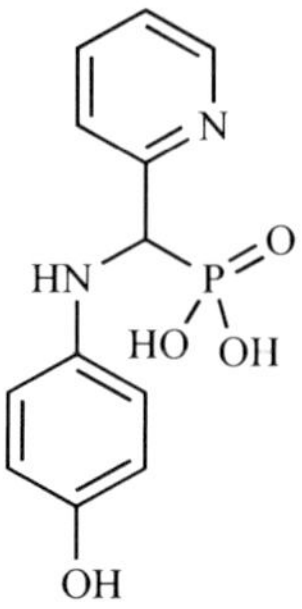

70

Table 9. Inhibition of activity of Na^+, K^+-ATPase of myometrium cell plasma membranes by the calixarenes 28, 35, 47 and ouabaine.

Inhibitors	Full inhibition concentration, μM	Pseudo constant of inhibition, $I_{0,5}$, nM	Hill coefficient, n_H
28	~ 10	54+6	0.40+0.08
35	~ 10	33+4	0.38+0.06
47	1000	98+8	0.12+0.03
ouabaine	1000	21000+5000	0.54+0.07

Further investigation of calixarene phosphonic derivatives **51** and **52** which are close structural analog of **47** showed significant loss of their activity after alkylation of lower rim. Probably, strong intramolecular hydrogen bonds in **47** favor to retain pinched cone conformation and so to possessing high inhibition of Na^+, K^+-ATPase of plasma membrane myometrium cell [79].

3.5. Chloride Ions Conductivity

The anion carriers and channels can change pH regulation inside the cell by transporting anions across the lipid bilayer of plasma membrane [80]. This may result in acidification of cytoplasm that triggers the apoptosis of the cell since disruption of normal ion and pH gradient have been considered its early event [81]. The action of series calix[4]arenes on solvent-containing planar bilayer membranes made of cholesterol and egg phosphatidylcholine (egg PC) or synthetic 18 carbons tail phospholipid DOPC has been investigated in a voltage-clamp mode. A steady-state voltage-dependent transmembrane current was achieved after addition of calix[4]arene **47** from the side of membrane the positive potential was applied to. This current exhibited anion selectivity passing more current at negative potentials applied from the side of membrane to which calix[4]arene **47** was introduced. The kinetics and temperature-dependence determined for calix[4]arene **47**-mediated chloride-ionic transport suggest a carrier mode of facilitated diffusion [82].

3.6. Chloride Channel Blockers

Chloride channels are ubiquitously distributed, biophysically varied and functionally diverse. Despite the known contribution of chloride channels to the cell physiology and the pathology of several diseases, high affinity ligands are not available to regulate activity of these channels. It was shown that tetrasulfonato-calix[4]arenes **27** and **71** are high affinity blockers of the outwardly rectifying chloride channels (ORCC) [83].

The K_i of calixarene **27** was 800 nM and CCT-plot analysis yielded block time constant (τ_b) of 260 ms. Substitution of four hydroxyl groups of **27** with four methoxy groups to obtain tetrasulfonato-tetramethoxycalix[4]arene **71** caused change conformation to 1,3-alternate. The most important was that conformational switch yielded a molecule where sulfonates were separated by near 10.5 Å. This subtle change in distance led to a 100-fold improvement in the K_i to a value 600 pM and prolonged τ_b to a value of 175 s.

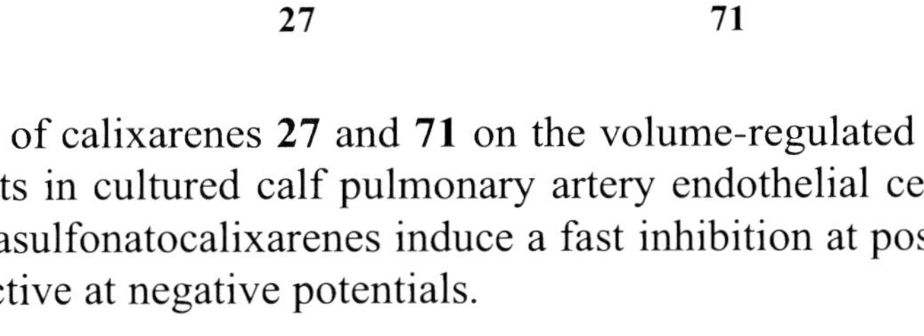

27 **71**

Next, the effect of calixarenes **27** and **71** on the volume-regulated anion channels (VRAC) currents in cultured calf pulmonary artery endothelial cells was studied [84]. These tetrasulfonatocalixarenes induce a fast inhibition at positive potentials but were ineffective at negative potentials.

Obtained results show that calix[4]arenes induce a voltage-dependent reversible inhibition of VRAC, probably by an occlusion of the pore region at moderately positive potentials. The pH dependence of the calixarene effect suggests a role for positively charged residues, presumably histidine, in the channel pore [82].

Calixarene **71** at 3 nM also inhibits a novel chloride channel activity associated with expression of p64-protein in HeLa cells *in vitro* [85]. This inhibition also has reversible behavior. Neither inhibitor affected the K^+ channel activity observed in these membranes.

Homologues of calix[4]arene **27** calix[8]arene **72** and calix[6]arene **73** as well as **67** were tested for chloride channel blockade too [86].

The activity of **27, 71-73** was investigated on a single chloride channels incorporated into planar lipid bilayers. Plasma membrane vesicles derived from rat colonic enterocytes were used as a source of chloride channels. The *p*-sulfonato-calixarenes resulted in long-lived zero conductance closures (i.e., block) of the colonic chloride channel. Chloride channel blockade by the calixarenes was completely reversed upon exchange of the *cis* chamber with calixarene-free buffer. The duration of the block period caused by the calixarenes increased with the size of the *p*-sulfonatocalix[n]arene as follow: n = 4, 3-4 s; n = 6, 30-40 s; n = 8, >> 100 s. The results for calixarene derivatives **27, 71-73** and simple calixarenes tested on chloride channels are summarized in Table **10** [84].

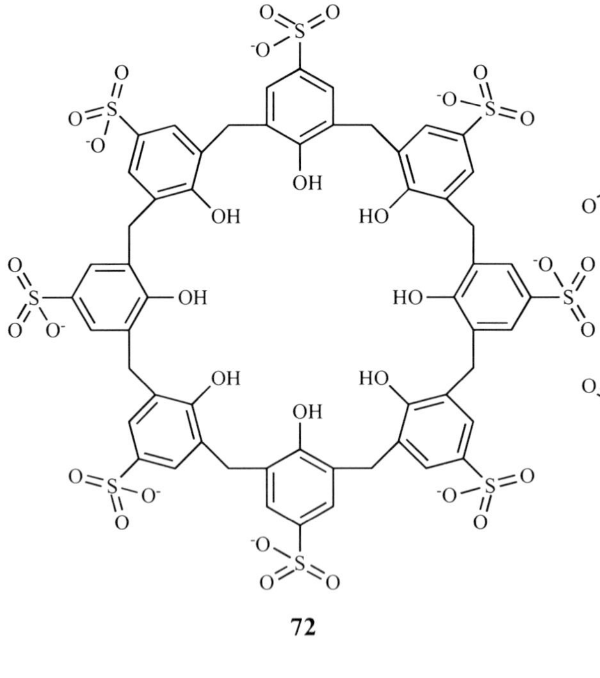

72

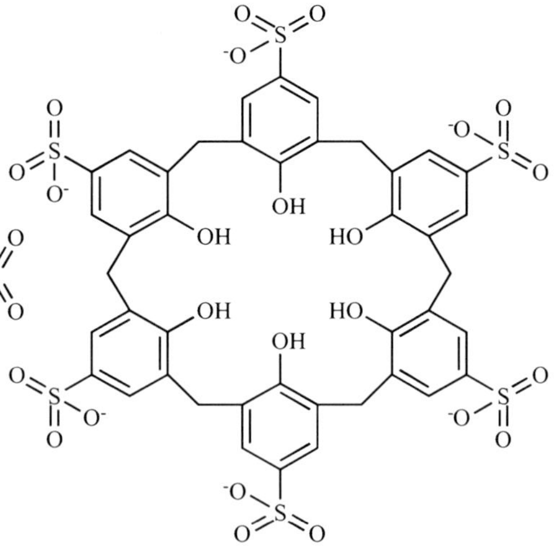

73

Table 10. Influence of calixarene-sulfonates onto chloride channels from rat enterocytes.

Compound	Maximum block period, s	K_i, nM
27	3-4	800-900
Tetrahydroxycalix[4]arene	No block to 0.1 mM	–
Tetramethoxycalix[4]arene	No block to 0.1 mM	–
71	>> 100	< 3
72	30-40	400-500
73	>> 100	< 50

4. BIOACTIVE PROPERTIES OF CALIXARENE DERIVATIVES

4.1. Antibacterial Activity

The development of biologically active molecules based on molecular recognition is an attractive and challenging research topic in medicinal and supramolecular chemistry. An interesting class of biologically active molecules, which explicate their action through a relatively simple molecular recognition process, is the Vancomycin group antibiotics. The molecular mechanism of the antibiotics action has been elucidated during recent years. It is now clear that the antibiotic binds to the cell wall mucopeptide precursors terminating in sequence D-alanyl-D-alanine thus inhibiting the growing of the cell wall and causing its lysis [87]. Biologically active Vancomycin mimics **74a,b** have been synthesized by linking two opposite aromatic nuclei of calix[4]arene platform with a polyfunctional bridge containing D- or L-alanine units and diethylentriamine segment [88].The calixarenes **74a,b** show an anti Gram-positive activity from moderate to good although slightly inferior to Vancomycin (Table **11**). Compounds **74a,b**, like Vancomycin, are inactive against fungi and yeast. Likewise, Vancomycin and **74a,b** are inactive (MIC > 128 mg/l) against a cell wall lacking bacteria, *Acholeplasma laidlawii,* suggesting that the antibacterial target should be involved in the cell wall biosynthesis.

Cytotoxicity assays suggest, that calixarenes **74,b,** similar to Vancomycin are not acting as a general toxin but are acting specifically as bactericidal compounds.

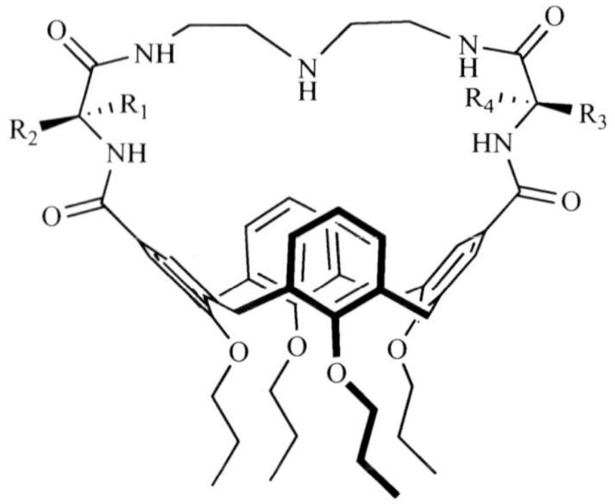

74 a,b

a: $R_1 = R_3 = H$, $R_2 = R_4 = CH_3$

b: $R_1 = R_3 = CH_3$, $R_2 = R_4 = H$

Subsequent detailed investigation by NMR diffusion measurements suggests that significant antibiotic activity of **74a,b** may originate from the ability of these compounds interact with Ala-Ala residue. As compounds **74** contains several functional groups of various types, the nature of their interactions with terminal D-alanyl-D-alanine sequence of cell wall mucopeptide precursors may be complex. The interaction may involve electrostatic, π-hydrogen, hydrophobic, and dipole-dipole interactions as well as hydrogen bonds. The log K_a, estimated from NMR diffusion measurements in $CDCl_3$ with 3% DMSO-d6 for **74b**·N-acetyl-L-ala-L-alaine complex was 3.4 [89].

Table 11. ***In vitro*** **activities of peptide calixarenes 74a,b and Vancomycin against selected Gram-positives.**

Organism	Minimum inhibitory concentration, MIC, mg/l.		
	Vancomycin	74a	74b
S. aureus 663	2	8	8
S. aureus 853	2	16	8
S. aureus 1131	2	4	4
S. epidermidis	2	4	8
B. cereus	2	16	NT

S. aureus 663 – penicillin sensitive, *S. aureus 853* – penicillin resistant, *S. aureus 1131* – methicillin resistant strain, NT = not tested.

Binding studies of silver complexes of a cation binding calixresorcin[4]arenes bis-crown with propyl, nonyl, decyl and undecyl alkyl chains showed that both 1 : 1

and 1 : 2 (host–Ag^+) complexes are present in solution. Langmuir–Blodgett films of the amphiphilic **75**–Ag complex were transferred onto a substrate and shown to possess antibacterial activity against *E. coli.* [90]. It is logical to assume that silver is released from the bis-crown complexes in the presence of the thiol groups on the bacterial surface, which leads to bacterial death and inhibits the overall bacterial growth as observed in these experiments. the inhibitory effect observed for **75**–Ag(I) is relevantly higher with a bacterial growth of only 7 ± 2% while the same LB film in the absence of silver shows a growth of 88 ± 7% and silver nitrate (equimolar quantity) shows growth as high as 72 ± 4%.

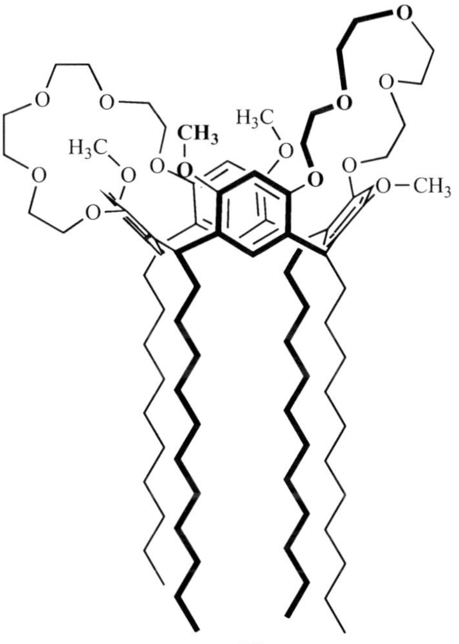

75

Calixarenes, modified with well-known bactericidal groups also considered as antibacterial agents. A new water-soluble calixarene **76** incorporating, *via* a labile bound, a quinolone antibiotic subunit was synthesized and investigated [91]. This calixarene was designed as a nalidixic acid prodrug and displays this behavior, that is, the release of quinolone in biological medium, as assessed by analytical studies. *In vitro* antibacterial evaluation (Table **12**) showed efficacy on *E. coli* and two *S. aureus* reference strains, with a clear gain of activity when compared to its two hydrolysis by-products. So, calixarene really serves as prodrug with function to delivery active component to the bacteria cell and didn't amplify itself the antibacterial potential of connected quinolone unit.

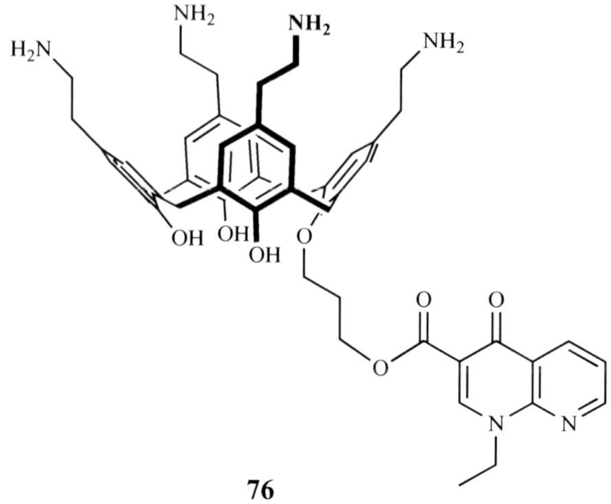

76

Table 12. MIC values of calixarene 76 and nalidixic acid.

Organism	Calix[4]arene 76		Nalidixic acid	
	mM/L	µg/mL	mM/L	µg/mL
E. coli ATCC 25922	25	35	100	23
S. aureus ATCC 25923	50	70	200	46
S. aureus ATCC 29213	50	70	200	46
E. faecalis ATCC 29212	100	140	200	46
P. aeruginosa ATCC 27853	100	140	200	46

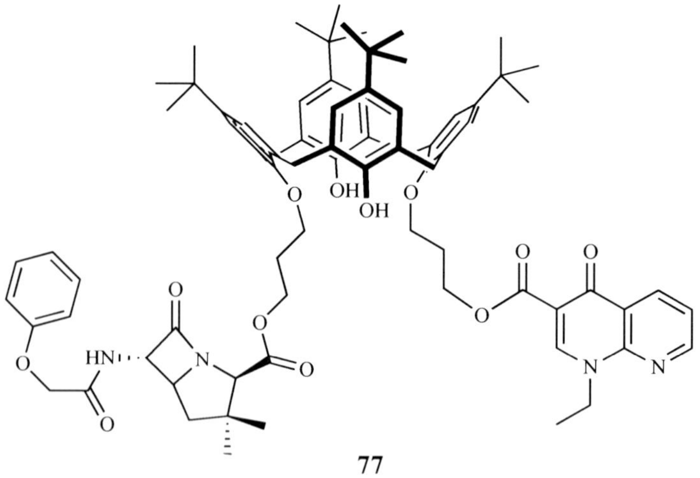

77

A series of calixarenes modified by two nalidixic acid groups or penicilline fragments was synthesized, but didn't show any antibacterial action [92-94]. Nevertheless, "mixed" calixarene **77** with one penicilline and one nalidixic group demonstrated antibacterial action on *S. aureus* ATCC 25923 strain [95].

Among the series of water-soluble anionic calixarenes bearing in alternate positions at the lower rim two 2,2'-bipyridine or 2,2'-bithiazole subunits the calixarene species **78** integrating two 2,2'-bithiazole-5,5'-dicarboxylate subunits exhibits activity against the *S. aureus* ATCC 25923 strain MIC 128 μg/ml. Other bacteria (Gram-negative *Escherichia coli* ATCC 25922 and *Pseudomonas aeruginosa* ATCC 27853 and Gram-positive *Staphylococcus aureus* ATCC 29213 and *Enterococcus faecalis* ATCC 29212) were insensitive to the presence of **78** [96].

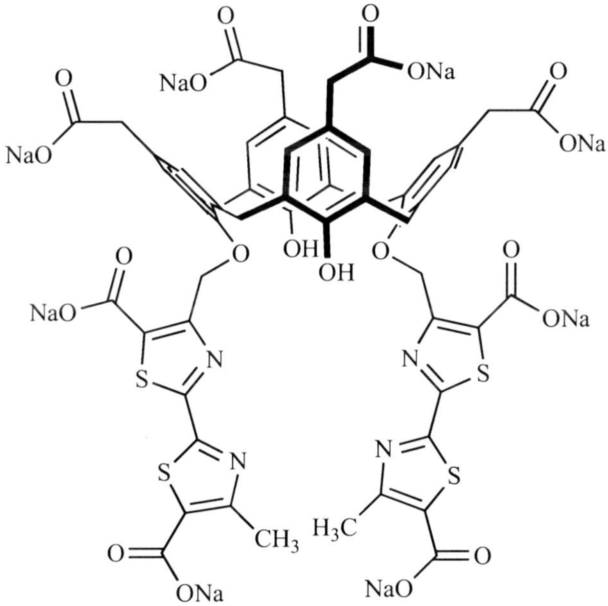

78

Antiseptic and antibacterial drugs such Aethonium and Decamethoxinum are bis-cationic quaternized compounds which effectively binds to bacteria cell membrane and disrupts its electric potential. One of the first cationic calixarene studied as antibacterial agent was tetraguanidinocalix[4]arene tetrakis-trifluoroacetate **79** [97-99].

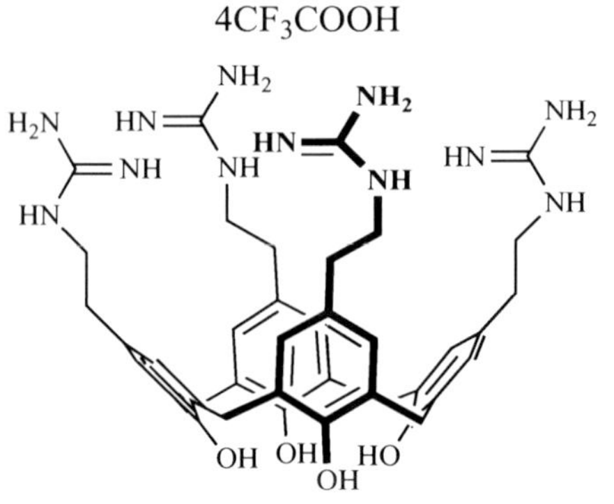

79

Calix[4]arene **79** was investigated *versus* hexamidine an antiseptic agent and synthaline A (bis-guanidine which used for medication of Diabetes Mellitus II). Calixarene **79** and hexamidine showed a broad antibacterial spectrum, both on Gram-positive and on Gram-negative bacteria (MIC 4 µg/mL against *E. coli* and 8 µg/mL against *S. aureus* for *para*-guanidinoethylcalix[4]arene), to a lesser degree against *E. faecalis* and *P. aeruginosa* (MIC 32 µg/mL). These compounds were bacteriostatic (MBC 256 µg/mL); and their MICs and MBCs obtained for clinical isolates were similar to those obtained with reference strains (Table **13**).

Table 13. MIC and MBC (µg/mL) for calixarene 79, hexamidine and synthaline A.

Organism	Calixarene 79		Hexamidine		Synthaline A	
	MIC	MBC	MIC	MBC	MIC	MBC
Reference strains						
E. coli **ATCC 25922**	4	256	8	256	64	256
S. aureus **ATCC 25923**	8	256	4	128	64	>512
S. aureus **ATCC 29213**	8	256	<1	128	16	>512
E. faecalis **ATCC 29212**	32	>256	2	128	256	>512
P. aeruginosa **ATCC 27853**	32	256	32	>512	128	>512
Isolated strains						
Penicillinase-producing *E. coli*	2	256	8	64	64	>512
Meticillin resistant *S. aureus*	8	256	2	64	64	64
Inactivating vancomicin *E. faecium*	8	128	2	64	256	>512
Inactivating vancomicin *E. faecalis*	64	256	4	256	>256	>512
Overexpression of efflux pumps *P. aeruginosa*	64	>256	64	>512	256	>512

Broad antibacterial spectrum of **79** is similar to hexamidine, whereas hexamidine and synthaline A have significant toxic effects that increased with time and concentration and in a range of 100–1000 times that for calixarene Another investigation demonstrated that calixarene **79** can present a good alternative not only hexamidine but also chlorhexidine, which are characterized by higher cytotoxicity. Also, it is suggested that the bacterial cell wall is a potential target of **79** [100].

Parallel investigation devoted to improving antibacterial effect showed that the introduction of two bipyridyl or bithiazolyl arms at the lower rim of calix[4]arene platform **79** does not greatly modify the antibacterial properties, maintaining the micromolar activity levels encountered for commercial products for both the reference Gram positive and negative strains. Obtained new calixarene compounds had a stronger impact on eukaryotic cell viability at 24 h, thus diminishing their selectivity indexes [101].

New calixarene derivative tetrakis-morpholinomethylcalix[4]arene **80** was synthesized and screened for bioactivity. The results show that a variety of microorganisms including Gram positive; *Staphylococcus albus* ATCC 10231, *Streptococcus viridans* ATCC 12392, Gram negative: *Bacillus procynous* ATCC 51189, *Enterobacter aerogenes* ATCC 13048, *Klebsiella aerogenous* ATCC 10031, *Escherichia coli* ATCC 8739, *Sallmonella* ATCC 6017 and Fungi: *Aspergillus niger* ATCC 16404, *Aspergillus fumagatus* ATCC 90906 are susceptible to compound **80**. The MIC of 4 μg/mL and 8 μg/mL was determined against most of the bacterial and fungal strains [102].

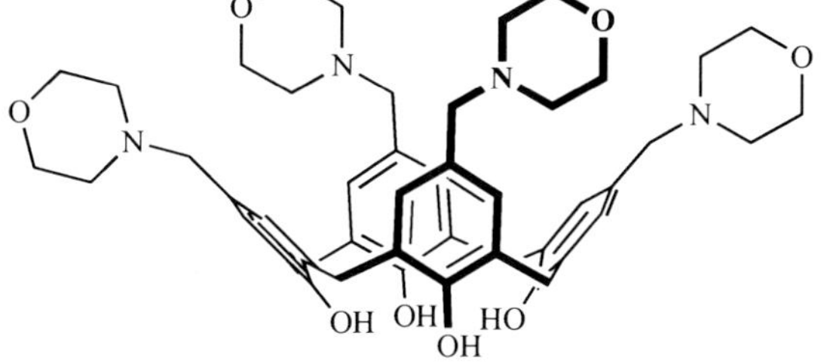

80

The first example of multivalent conjugate **81** in which four α-l-C-fucosyl units are clustered onto a calix[4]arene platform was designed as a new potential *Pseudomonas aeruginosa* biofilm inhibitor [103]. The anti-biofilm activity calixarene **81** against PAO1 strain was assayed and it was found to be dose-dependent. The data showed that ranging from 12.5 to 200 μM concentration, the percentage of biofilm inhibition increases from 35% to 73%. The presence of the fucose cluster improves the biofilm inhibitor efficiency as proven by the lower inhibitor activity of the analogous glycyl-calix[4]arene derivative **82** lacking in the fucose moieties.

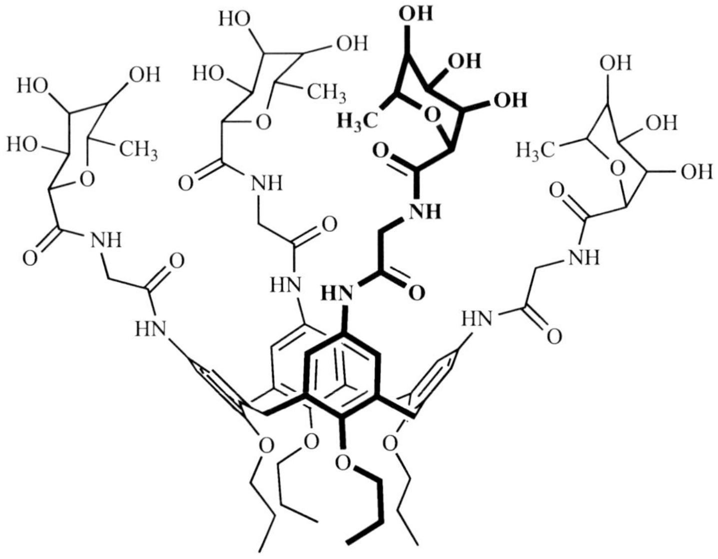

81

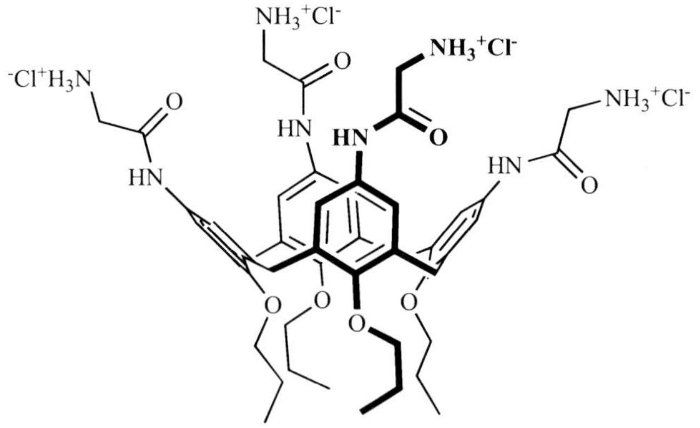

82

The antibacterial and antibiofilm activities of several amphiphillic tetracationic calixarenes functionalized with methylimidazolium **83a,b**, choline **84a,b** and ethylenediamine **85a,b** groups were tested and revealed that their MIC values are close to those of antibiotics and better or equal to those of QAS (Table **14**). Obtained

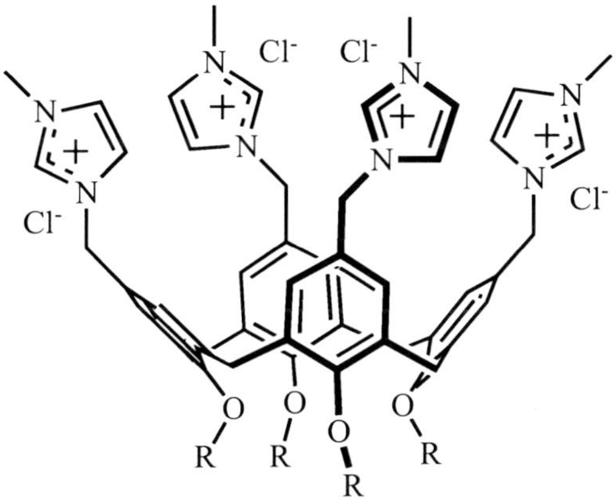

83 a R = n-C_3H_7,
b R = n-C_8H_{17}

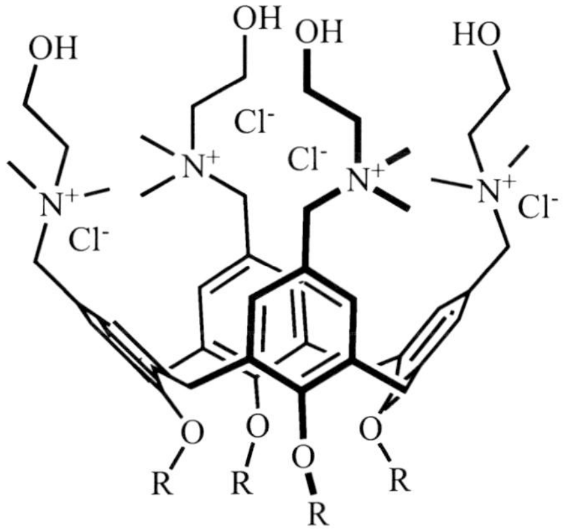

84 a R = n-C_3H_7,
b R = n-C_8H_{17}

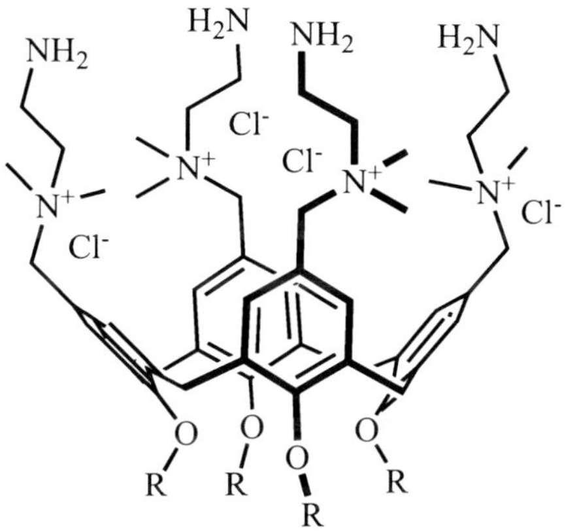

85 a R = n-C_3H_7,
b R = n-C_8H_{17}

Table 14. MIC values of calixarenes 83a-85a, µg/mL (µM).

Calixarene	*S. aureus* ATCC 6538	*P. aeruginosa* ATCC 9027	*K. pneumoniae* ATCC 10031	*E. coli* ATCC 25922
83a	≤10 (≤9)	50 (45)	≤10 (≤9)	≤10 (≤9)
84a	≤10 (≤9)	1000 (870)	100 (90)	100 (90)
85a	≤10 (≤5)	50 (26)	500 (260)	500 (260)

results demonstrated (i) the clear dependence of antibacterial activity from the alkyl chain length on the lower rim of calixarene. Tetrapropoxycalixarenes have 10-

50 times higher antibacterial activity than tetraoctyl calixarenes with same cationic residues on the upper rim. (ii) The nature of the cationic groups at their macrocycle upper rim, defines their mode of interaction with bacterial membrane structures. Calixarene **83a** grafted with more lipophillic and planar methylimidazolium groups showed best activity among all investigated calixarenes. At the same time, tetrapropoxycalixarene **85a** shows high inhibitory activity against *Staphylococcus aureus* biofilms [104]. Another deep investigation of calixarenes **84a,b** also confirmed previously described results [105].

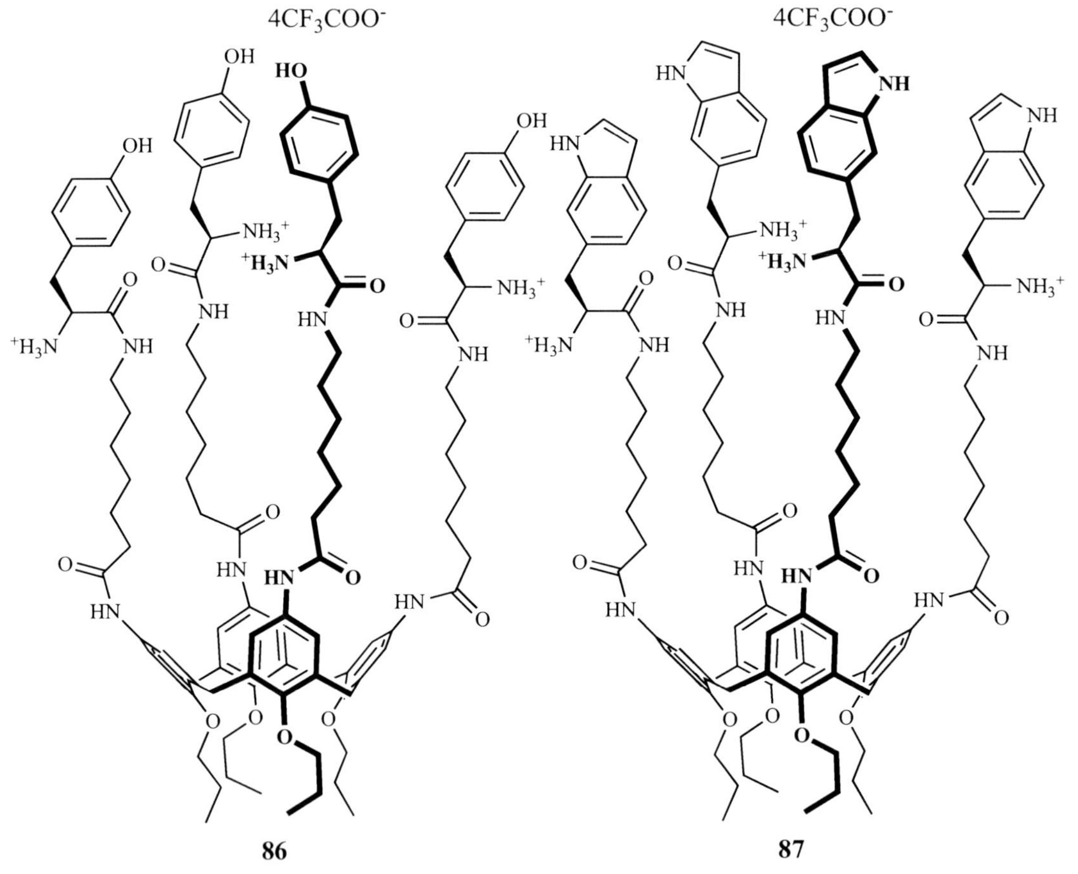

86 **87**

A probable general mechanism antibacterial activity of polycationic calixarenes against Gram-negative species was investigated on calixarenes **86** and **87** [106].

It was approved by NMR and UV titrations that calixarenes **86** and **87** can recognize and bind to lipopolysaccharides (which are the primary and principal constituent of the outer cellular membrane) of Gram-negative bacteria The investigated macrocycles possess facial amphiphilicity, a necessary requirement for disrupting bacterial membranes and generating active compounds against a broad spectrum of bacteria.

4.2. Antituberculosis Properties

It was found that attachment of hydrophilic polyoxyethylene chains to various hydrophobic alcohols and phenols gave water soluble products which, although inactive *in vitro*, influenced an experimental tuberculosis infection. Also was reported that attachment polyoxyethylene chains to calixarenes, gave compounds having striking and prolongated effect on experimental murine tuberculosis caused *Mycobacterium tuberculosis* after parenteral administration [107]. The macrocycles **88a** have atnituberculosis and **88b** have "protuberculosis" effects correspondingly [108, 109].

Investigations were carried out with resident peritoneal macrophages from female mice (albino Parkes strain) which were treated with **88a** (5.1 mg/mL) and **88b** (5.2 mg/mL) and then with *M. tuberculosis*.

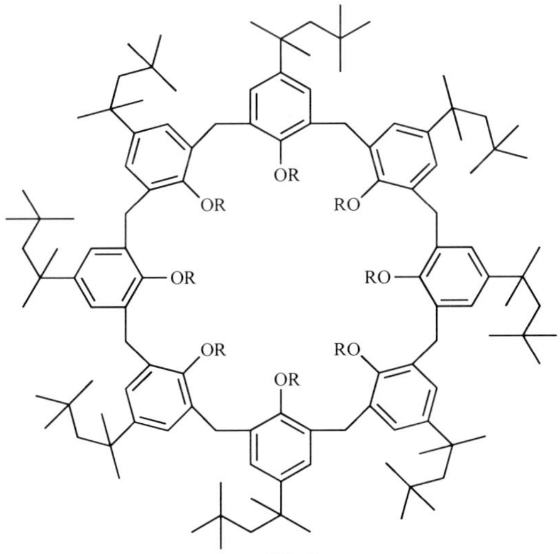

88 a,b

R = $(CH_2CH_2O)_nC_2H_5$
a: n = 12.5
b: n = 60.0

Growth of *M. tuberculosis* inside macrophages was inhibited by **88a** and stimulated by **88b**. Lipase activity was influenced by calixarenes similarly. This observation strengthens the evidence that lipids and lipid metabolism are involved in the interactions induced by these substances in experimental animals with *M. tuberculosis* infections. It indicates a unique mechanism, apparently different from that of all other agents against this disease, a mechanism in which the Host itself plays the essential part. Since resistance to conventional chemotherapeutic agents seems to be increasing, new approaches are needed.

Various polycharged calix[4]arenes were tested as anti-mycobacterial agents against *Mycobacterium tuberculosis* H37Rv strain and the isoniazid-resistant strain MYC5165 [110]. The sulfonate, carboxylate and phosphonate anionic species displayed no activity. Cationic derivatives integrating four aminoethyl groups at the upper rim and two 6,6'-dimethyl-2,2'-bipyridyl- or 4,4'-dimethyl-2,2'-bithiazolyl subunits at the lower rim were also found inactive. Cationic derivatives **79**, **89-90** integrating guanidinoethyl groups at the upper rim exhibited MIC values against *M. tuberculosis* H37Rv of 0.8-1.6 μM (Table **15**), similar to those of current commercial anti-tuberculosis agents. These macrocycles were also evaluated against the isoniazid-resistant strain MYC5165, resulting in micromolar or sub-micromolar MIC and IC_{50}, ca. 4–125 times more active than isoniazid.

Table 15. MIC of calixarenes 79, 89, 90 and anti-mycobacterial agents against two strains of *M. tuberculosis*.

Calixarene	$H_{37}Rv$		MYC1565	
	μg/mL	μM	μg/mL	μM
79	1.0	0.8	1.0	0.8
89	2.69	1.6	0.17	0.1
90	1.51	0.8	0.75	0.4
Izoniazid	0.8	2.5	1.7	12.5
Cyprofloxacin	0.08	0.6	0.8	2.5

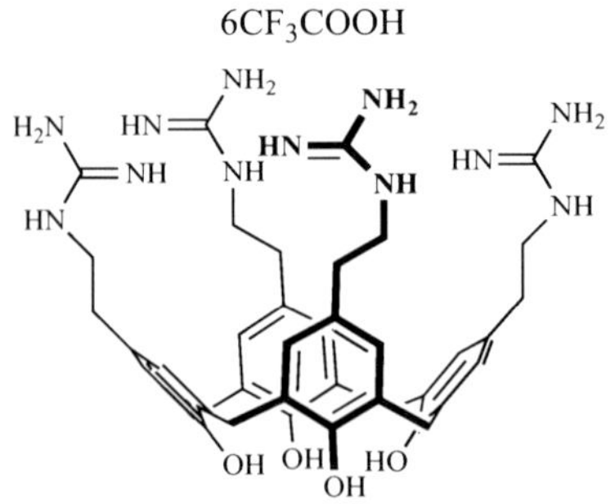

79

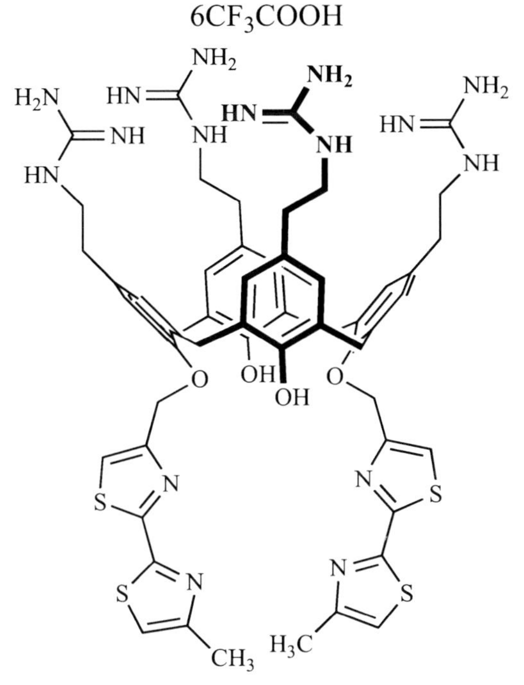

89

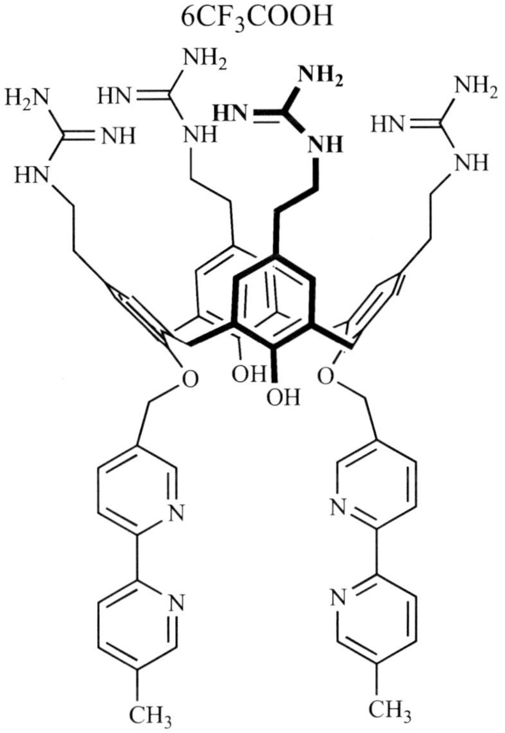

90

4.3. Antifungal Activity

Tetrasulfonatocalixarenes **27, 72-73**, **91** displayed range of fungicidal activity against *Fusarium olani f. sp. mori* (F.s.-26, Japan MAFF #840046), having an inhibition rate of approximately 60–70%. Also these calixarenes showed excellent and selective antimicrobial activity against fungal strains *Rosellinia necatrix* (Japan MAFF #840051), and *Colletotrichum dematium* (Japan MAFF #840066) [111]. At present, there are no pesticides against F.s-26, indicating that calixarenes could be a promising platform for new pesticide developments. Normally pesticides contain a large quantity of harmful halogens. The advantage of calixarenes is absent of halogens in high amounts.

Polycationic methyl- and adamantyl-imidazolium calixarene derivatives **92a,b 93** are characterized by tangible fungicidal effect against yeasts fungus *Candida tenius* but lesser against mold fungus *Aspergillus niger* [112]. Minimal fungistatic concentration (MFsC) as well as minimal fungicidal concentration (MFC) were determined (Table **16**), but mechanism of action wasn't proposed.

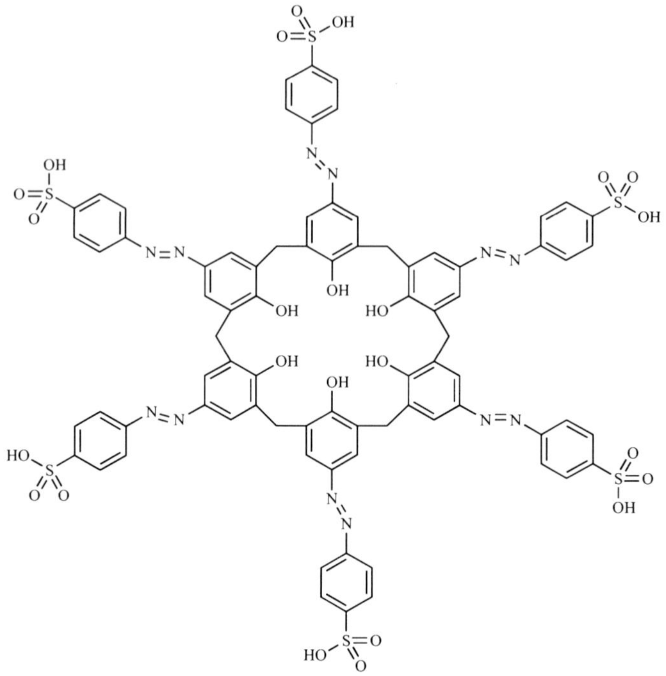

91

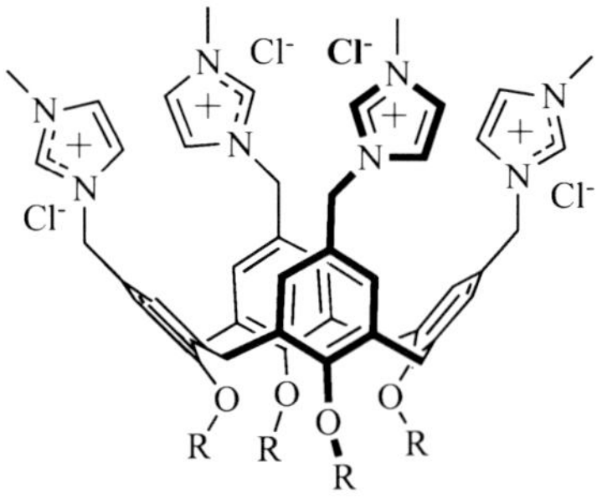

92 a R = n-C_6H_{13},
b R = n-$C_{12}H_{25}$

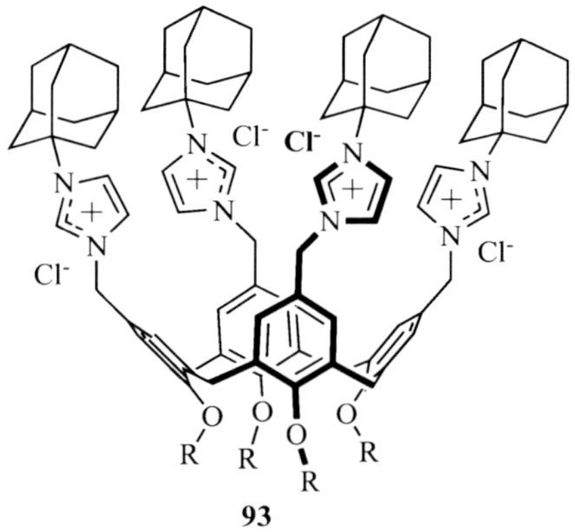

93

Table 16. MFsC and MFC of calixarenes 88a,b 89 obtained by limiting dilution method.

Calixarene	*Candida tenius*		*Aspergillus niger*	
	MFsC, μg/mL	MFC, μg/mL	MFsC, μg/mL	MFC, μg/mL
92a	15.6	62.5	250.0	-
92b	62.5	125.0	-	-
93	31.2	125.0	250.0	-

New conjugates bearing a calix[4]arene **94a,b** scaffold covalently linked to four amphotericin B molecules were synthesized [113]. These macromolecules adopt a cone conformation that mimics the structure of a transmembrane pore. The antifungal activity of the conjugates **94a,b** was superior or similar to that of native amphotericin B (AmB), with MIC values of 0.10 and 0.25 μM, respectively (Table **17**).

Table 17. MIC and EH_{50} values for AmB and calixarenes 94a, b.

Compound	MIC BY4741, μM	EH_{50}, μM
AmB	0.30	4.0
94a	0.10	50
94b	0.25	40

MIC: minimal, inhibitory concentration; BY4741 – *S. cerevisiae* cell line; EH_{50}: concentration, causing 50% hemoglobin release in human erythrocytes.

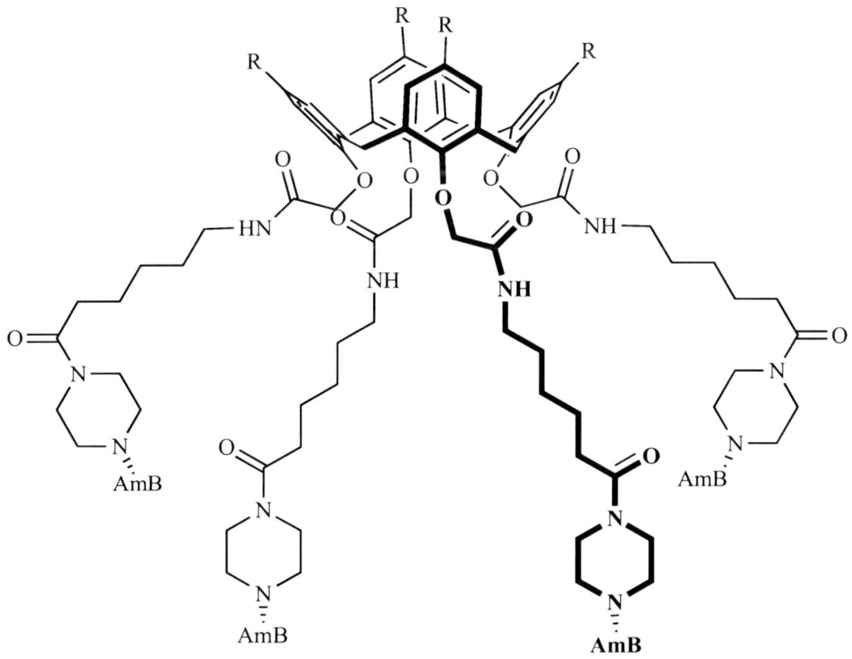

94 a: R = H
b: R = t-Bu

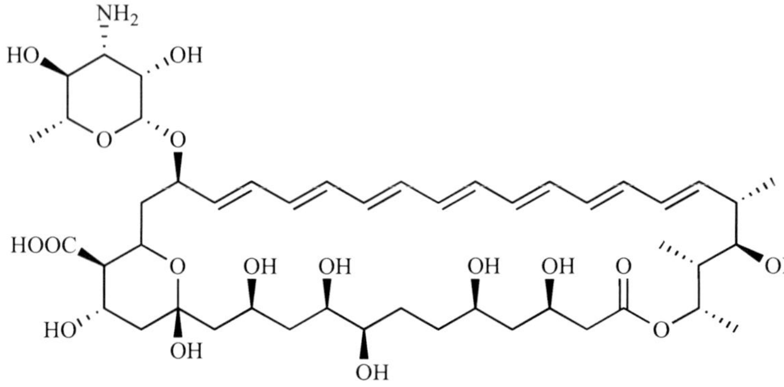

AmB

Conjugates **94a** and **b** are not only active as native AmB against *S. cerevisiae* but also retain the ability to induce K^+ leakage from vesicles, suggesting an efficient channel formation in the membrane.

4.4. Antiviral Activity

Currently, the most widely used antiviral agents are nucleoside analogs. This class acts by disrupting viral replication, either by inhibiting enzymes required for nucleic acid processing, or by producing defective viral genome, such as by premature termination of replication. Another drug therapy of viral infections includes attempts to block viral entry into cells, alter protein synthesis at the Host ribosomes, complexation of viral DNA/RNA and immunomodulation. These methods have undesirable side effects.

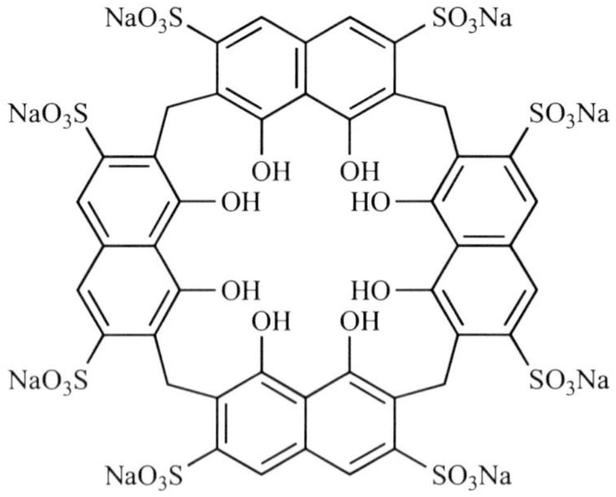

95

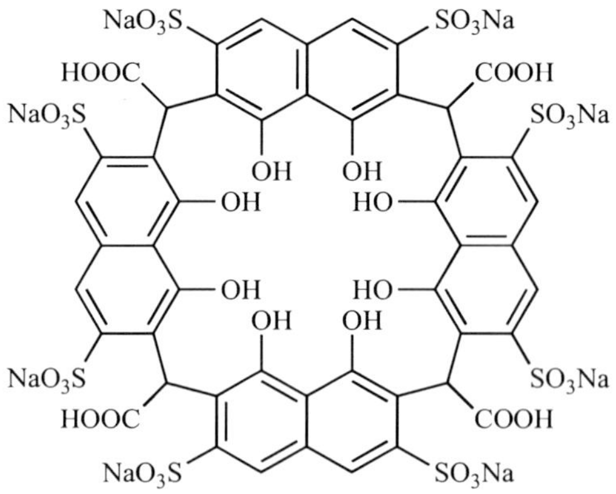

96

Calix[4]arene derivatives **95**, **96** and others, also known as chromotropic acid derivatives are patented as good antiviral agents [114]. The compounds are effective inhibitors of cell infection caused by enveloped viruses. Binding studies carried out in particular with respect to herpes simplex virus (HSV) viruses indicate that the anti-viral activity of the compounds is dependent on binding to virus envelope components, which inhibits virus attachment to infectable cells. The apparent inability of the compounds **95**, **96** to inhibit infection of non-enveloped viruses is consistent with this mechanism.

The compounds **95**, **96** were tested *in vitro* against herpes viruses: herpes simplex virus-1 (HSV-1), herpes simplex virus-2 (HSV-2), human immunodeficiency virus (HIV), influenza A and B, respiratory syncitial viruses (RSV), adenovirus and rhinovirus. VERO cells and human HEp_2, $HeLaS_3$ and H_9 cell lines were used for investigation. It was shown that compounds **95**, **96** effectively inhibit cytopathic effect caused by viral infection against all envelope viruses. The ED_{50} against HSV-1 was 0.7-2.7 μg/mL, against HSV-2 was 1-2.7 μg/mL. Moreover, these results are better than results of well-known drugs: acyclovir, ganciclovir, phosphonoformate and phosphomethoxyethyladenine. Against influenza A virus the value of ED_{50} is higher, but still not significant 5-7.9 μg/mL, against HIV ($HTLV\text{-}III_B$ and RF-II strains) ED_{50} equal 0.07-0.50 μg/mL [114].

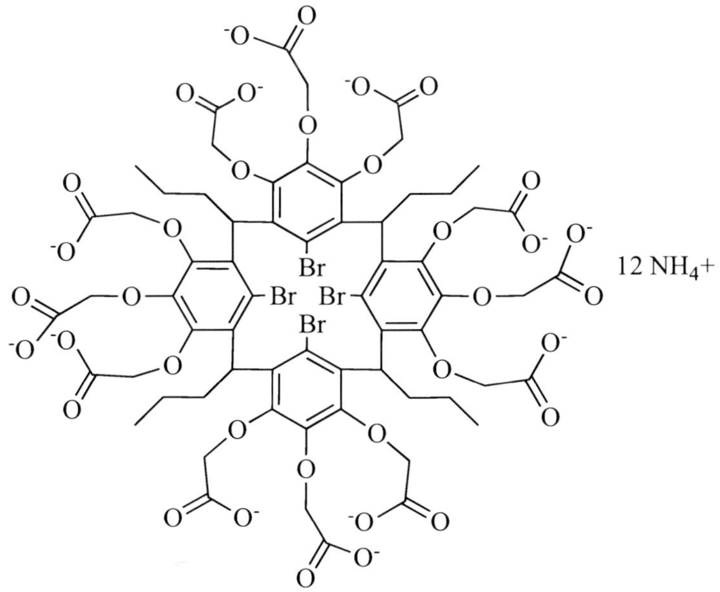

97

Calix[4]pyrogallolarene carboxylic acids **97** and **98** also have antiviral activity against HIV-1 and HIV-2. Calixarene **97** reduce cytopathic effect by 50% at concentration 10 and 0.03 μM respectively [115], compound **98** at 0.5 and 1 μM [116, 117].

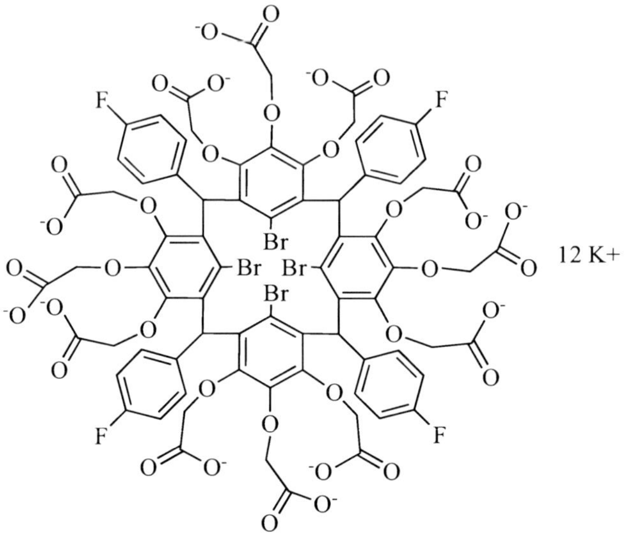

98

Mechanism of antiviral activity was described in general. Calixarenes inhibit the fusion of virus particles to host cells and integrases of viruses.

A potential anti-HIV and anti-Hepatitis C virus (HCV) drug candidate is highly desirable as coinfection has become a worldwide public health challenge. Calixarenes **99-101** functionalized with isophthalate groups demonstrate remarkable anti-HIV and anti-HCV activities. Structural activity relationship studies reveal that maintaining the cone conformation of the scaffold is important for antiviral activity. In addition, aromatic isophthalate spacers at the upper rim are essential for anti-HIV activities and the diacid groups are also necessary for the observed anti-HCV effects [118].

Furthermore, a **99a** was identified as potent compound that possesses dual inhibition for both HIV and HCV *in vitro*. Moreover, it retains potency against different HIV strains (HIV-IIIB, LaiM184 V, NL-43) in different cell lines (T-lymphocytes MT-2 and CEMx174 and HeLa cells) while maintaining low cytotoxicity (Table **18**).

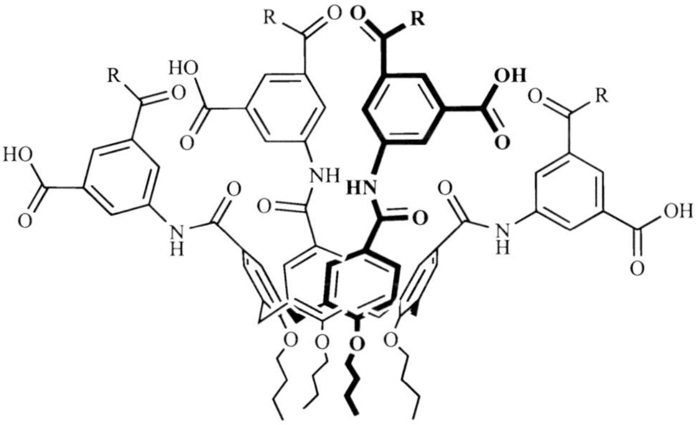

99 a R = H
b R = CH_2-Ph
c R = NH-cyclohexyl

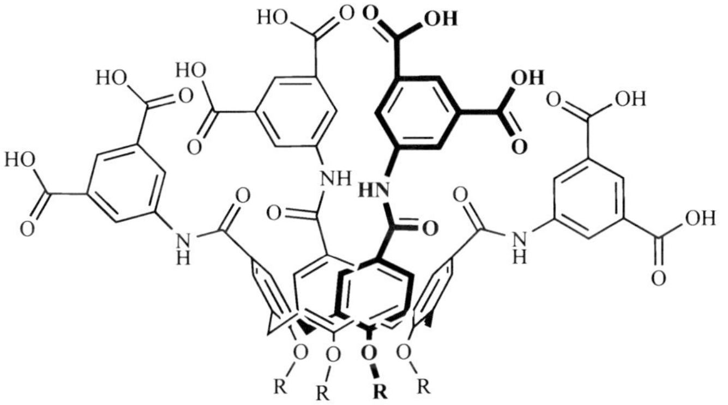

100 a R = H
b R = CH_2-Ph

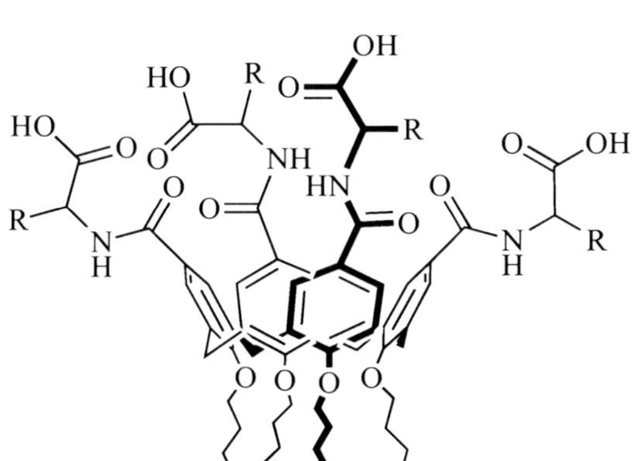

101 a R = COOH
b R = CH_2-COOH

Table 18. Antiviral activity and cytotoxicity of calixarenes 99-101.

	Inhibition of HIV, IC_{50}, μM	Inhibition of HCV, IC_{50}, μM	Cytotoxicity MT-2, μM	Cytotoxicity CEMx174, μM
99a	0.36 ± 0.12	1.8 ± 0.3	>50	50
99b	>50	25.6	>50	12.5
99c	>50	4.1	>50	>50
100a	6.7 ± 2.9	>50	>50	ND
100b	0.30	1.4	18	0.46
101a	3.0 ± 2.8	1.5 ± 0.3	>50	82
101b	2.4 ± 1.4	4.1 ± 1.8	>50	100

ND – not detected.

Amino-adamantylcalix[4]arenes **102a,b** with free and alkylated narrow rim were synthesized and evaluated for activity against HSV-2 at VERO-B cell culture [119].

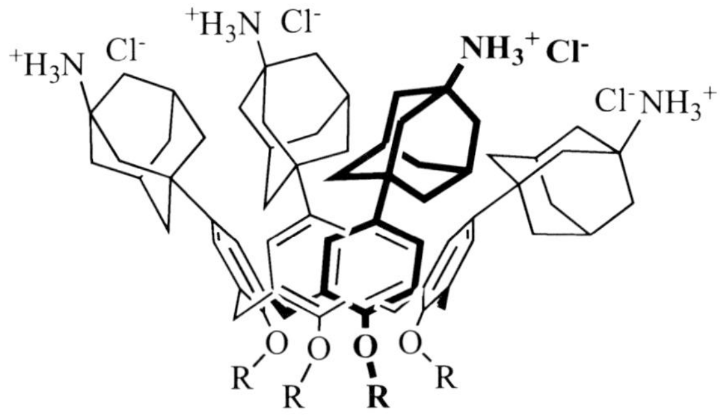

102 a R = H
b R = n-Bu

Tetrabutoxy-calix[4]arene **102b** is inactive while **102a** show reducing of virus cytopathic action at 10-125 μg/mL. But its activity lesser than acyclovir, MIC for **102a** equal 10 μg/mL, for acyclovir 0.25 μg/mL, chemotherapeutic indexes 12.5 and 2000 correspondingly.

A series anionic water-soluble calix[4]arene species, incorporating sulfonate, carboxylate or phosphonate groups and two 2,2'-bithiazole subunits in alternate position at the lower rim, were synthesized and evaluated as anti-HIV agents on various HIV strains and cells of the lymphocytic lineage (HIV-1 III B/MT4, HIV-1 LAI/CEM-SS, HIV-1 Bal/PBMC), using azidothymidine (AZT) as reference compound [120]. Highest antiviral activity showed tetrasulfonate **103**, tetraphosphonate **104**, octacarboxylate **78** and tetrarboxylate **105** calixarene derivatives (Table **19**). The parent unsubstituted derivatives showed a modest antiviral activity, so the introduction of bisthiazole units gave in result increasing of activity.

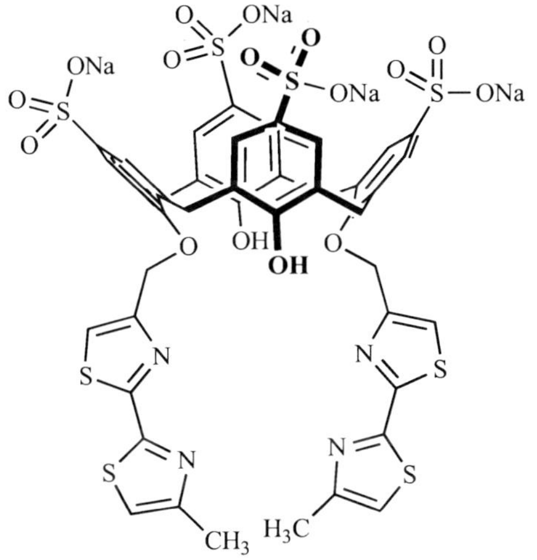

103

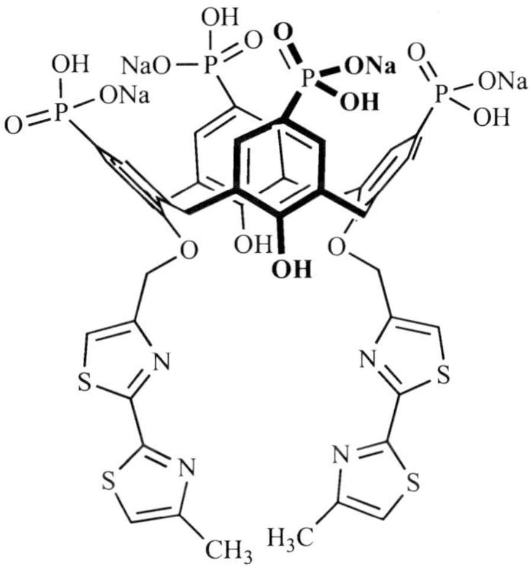

104

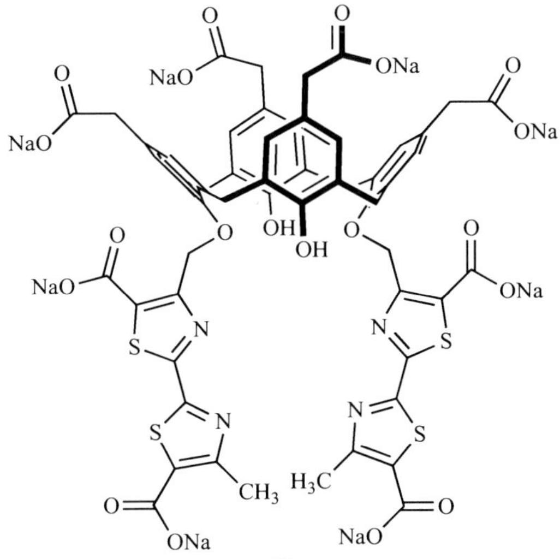

78

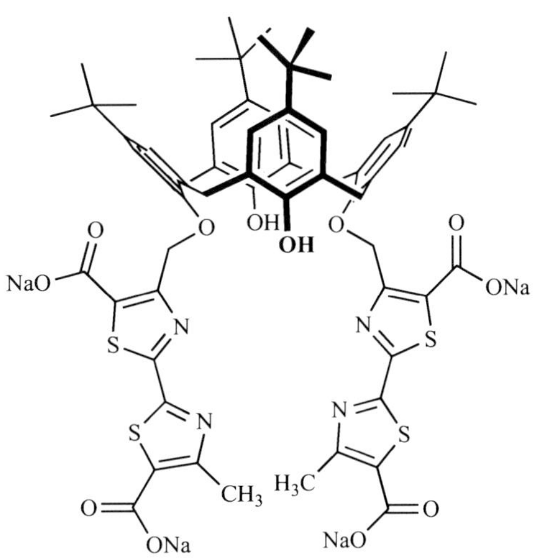

105

Table 19. Antiviral activity and cytotoxicity of anionic calixarenes 78, 103-105 and AZT.

Compound	MT4 (HIV-1 IIIB)		CTM-SS (HIV-1 LaI)		PBMC (HIV-1 Bal)	
	IC_{50}, μM	Cytotoxicity, μM	IC_{50}, μM	Cytotoxicity, μM	IC_{50}, μM	Cytotoxicity, μM
78	14	>100	5.3	>100	>37	37
103	1.6	>100	2.2	>100	1.5	>100
104	14	>100	15	>100	57	92
105	15	>100	19.5	>100	14.5	>100
AZT	0.01	>100	0.0051	>100	0.0077	75

These compounds do not display a significant cellular toxicity at 100 μM (for most of them, the highest concentration tested), except on PBMC in case of the carboxylate **78** and phosphonate **104** derivatives.

The major gains were obtained in the sulfonated family for compound **103**, with micromolar activities and apparent selectivity indexes superior to 45. But its effectiveness is lesser than AZT which was used at early 2000s years.

4.5. Anticoagulant and Antithrombotic Properties

Inappropriate coagulation of blood (thrombosis) may occur within the circulatory system in pathological states such as atherosclerosis or in response to a variety of insults, including surgery and implantation of medical devices. Prevention of thrombosis is considered a crucial part of the treatment regimen for patients at risk for the thrombi development.

In general, agents affecting blood hemostasis are divided into three categories: agents who interfere with portions of the coagulation cascade (anticoagulants), agents, which interfere with platelet activation and aggregation (antiplatelet drugs) and agents, which promote disintegration of blood clots (thrombolytics). Currently available anticoagulant drugs are limited to the heparin-like compounds, which are active only when given intravenously, and to the oral coumarin drugs. Heparin is an endogenous polysulfonated glycosaminoglycan which serves as catalyst for the reaction between antithrombin and various activated factors in the coagulation cascade. But, both heparin and coumarin drugs are limited in their usefulness by their hemorrhagic potential, because effective antithrombotic doses of these drugs also concomitantly produce excessive systemic anticoagulation.

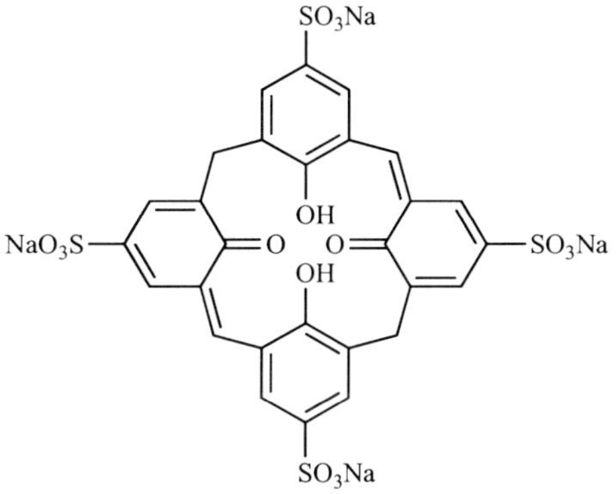

106

Calixarenes **27, 95, 106** functionalized with several sulfonate groups were patented as effective anticoagulant and antitrombothic agents [121].

Table 20. Effect of *in vivo* administration of calixarenes 27, 95 and 106 on PT, APTT, and Fibrinogen assay.

Calixarene	Dose, mg/kg	PT, %	APTT, %	Fibrinogen, Δ mg/l
27	5	112	167	70
	10	128	167	N
95	5	120	>600	-210
	10	188	>600	-590
106	5	125	193	N
	10	130	>600	-260

N – no effect. PT- prothrombin time assay, APTT – activated partial thromboplastin time.

Results of *in vivo* treatment of these compounds which were given intravenously to rats are shown in Table **20**. Samples of blood were taken after 5 hours after administration. The compounds can prolong overall clotting time.

The simultaneous prolongation of PT and APTT suggests that **27** may be directed against the common pathway of blood coagulation. Potential targets include Factor X, Factor V, prothrombin (Factor II) or phospholipids. Additionally, **27** activate heparin cofactor-II activity and inhibite collagen-induced platelet aggregation. Also, this compound is notable for its ability to inhibit thrombus formation at an oral dose which increases bleeding less than two fold over control.

Both **27** and **95** were shown to be active *in vivo* when administered either parenterally or orally. Peak effects after oral administrations were dose dependent and were observed between 0.5 to 4 hours after oral administration of 450 mg/kg of **27**. The elimination half-time, following intravenous administration of 2.5 and 5 mg/kg is about 3-4 hours.

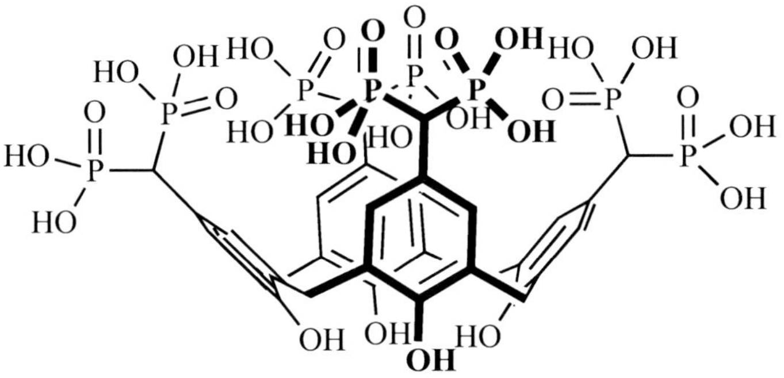

107

The effect on fibrin polymerization of calixarene tetrakis-methylene-bisphosphonic acid **107** was studied in two kinds of assay [122].

Turbidity analysis showed that **107** decreased the maximum rate of fibrin polymerization in the thrombin–fibrinogen reaction by 50% at a molar ratio of compound to starting fibrinogen of 1.7 : 1 (Fig. **11**). The lag-time was also increased strongly in the presence of **107**, indicating either a decrease of the rate of protofibril formation or, conceivably, an increase of protofibril critical length (Fig. **11B**). Similar results were obtained when calixarene **107** inhibited the re-association of dispersed desAB fibrin (Fig. **12**); in this case, $IC_{50} = 1.26 \cdot 10^{-6}$ M. Electron microscopy confirmed that **107** inhibits the first stage of fibrin polymerization (i.e. the formation of protofibrils).

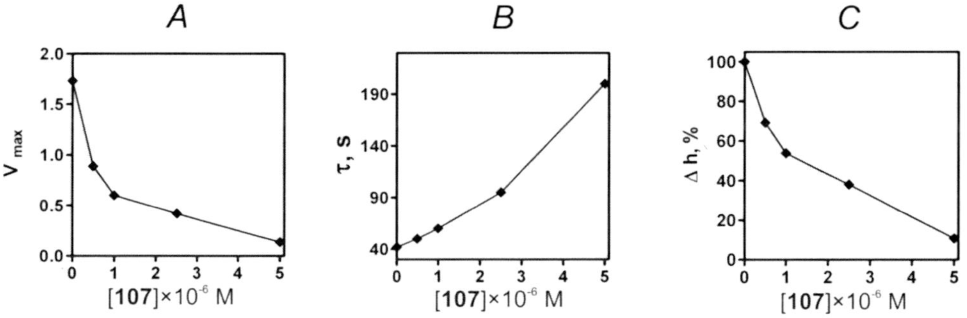

Figure 11: Turbidity analysis of the influence of **107** on fibrin polymerization in the fibrinogen + thrombin reaction. The dependence on calixarene **107** concentration is shown for (**A**) the maximum rate of fibrin polymerization V_{max}, (**B**) the lag-time τ, and (**C**) the final turbidity of fibrin clots Δh.

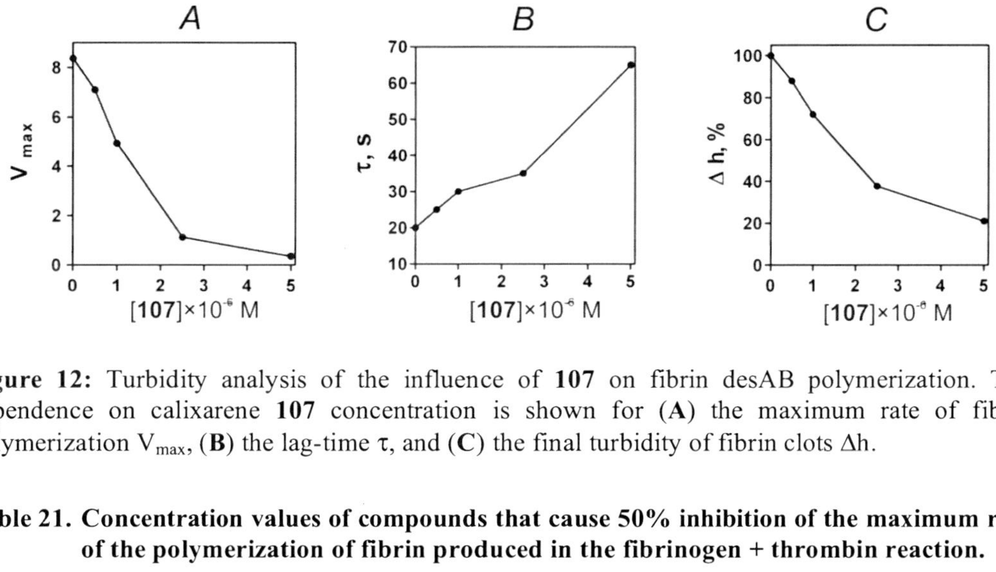

Figure 12: Turbidity analysis of the influence of **107** on fibrin desAB polymerization. The dependence on calixarene **107** concentration is shown for (**A**) the maximum rate of fibrin polymerization V_{max}, (**B**) the lag-time τ, and (**C**) the final turbidity of fibrin clots Δh.

Table 21. Concentration values of compounds that cause 50% inhibition of the maximum rate of the polymerization of fibrin produced in the fibrinogen + thrombin reaction.

Compound	IC_{50}
107	1.26×10^{-6} M
36	1.31×10^{-4} M
37	0.72×10^{-4} M
38	$>1.00 \times 10^{-4}$ M

It is of interest that the inhibitory activity of **36**, which has the two methylene bis-phosphonic acid substituents, is much less (Table **21**) (IC_{50}=$1.31\cdot10^{-4}$ M, indicating that the calixarene scaffold and the number of phosphoryl groups in the molecule play a crucial role in the inhibitory effect.

Probably, this inhibition is a result of the blocking of site 'A' (Aa17-19, GlyProArg) by the calixarene in a 'knob-hole' manner. HPLC study [123] of the complex formation between **107** and the synthetic peptide Gly-Pro-Arg-Pro, a synthetic analogue of the A knob was carried out. The association constants of calixarene **107** complexes with amino acids Gly, Pro, Arg and tetrapeptide Gly-Pro-Arg-Pro in a methanol–water mobile phase (50 : 50, v/v) presented in Table **22**. The highest association constant 3395 M^{-1}, is observed for tetrapeptide Gly-Pro-Arg-Pro, analyze of retention time indicates an 1:1 ratio of calixarene to Gly-Pro-Arg-Pro in the complex.

Table 22. The association constants of calixarene 107 with amino acids and tetrapeptide.

Guest	K_A, M^{-1}, (RSD, %)
Gly	280 (10)
Pro	814 (26)
Arg	2576 (21)
Gly-Pro-Arg-Pro	3395 (19)

Furthermore, calixarene **107** doubles both the PT and APTT in normal human blood plasma at concentrations of $7.13 \cdot 10^{-5}$ M and $1.10 \cdot 10^{-5}$ M, respectively.

4.6. Anticancer Agents

The cells, especially tumor cells, show nonspecific affinities to hydrophobic molecules. This is the basic of photodynamic therapy of tumors by the use of porphyrin and related sensitizers. Oligosaccharides play essential roles in various cellular activities as antigens, growth signals, targets of bacterial/viral infection, and glues in cell adhesion and cancer transfer. The usage of highly saccharide-functionalized calixresorcinarenes **108a,b** for the capture and molecular delivery to hepatocytes, which are well-known to have receptors for the terminal galactose residues, was proposed [124]. Their interactions with rat hepatoma cells (RLC-16) were investigated by fluorescence microscopy. Obtained data show that galactose cluster **108a** is captured by the cells, while glucose cluster **108b** is not. Moreover, these calixresorcinarenes form complexes with variety of hydrophobic dyes. Thus, the included Guest-dye molecules are thereby either delivered to the target cells (complex with **108a**) or protected in solution away from the cells (complex with **108b**).

108a

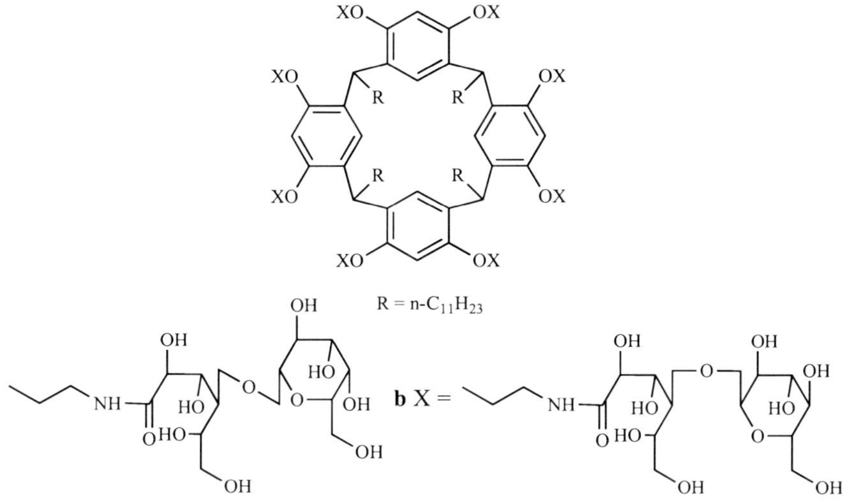

The possibility of a potential anti-cancer drug to interconnect a layer of cancer cells is a new way to cancer therapy. Formation of such a "cancer-net" around a solid tumor might, conceivably, retards growth and/or inhibits metastasis. To be effective, of course a cancer-net drug must selectively adhere to pathological cells in preference to healthy cells. Thus, it is possible, that molecules with two galacto-calixarene macrocycles, covalently joined by a long spacer, might "glycotarget" a group of cancer cells and thereby interconnect them. Compounds **109** sprung from this idea [125].

Cell experiments were carried out on epithelial-like rat liver cells (Riken, RLC-16, Japan). After compounds **109a-c** were added (0.05 mM) subsequent inspection of the cell suspensions by light microscope and light scattering gave no convincing evidence for aggregation. Despite of this setback, the concept of a cancer-net drug is the concept clearing the way for the future work in the area.

Platelet-derived growth factor (PDGF) and its receptor PDGFR are required for tumor growth and angiogenesis, so disruption of the PDGF–PDGFR interaction should lead to starvation of tumors and reduction of tumor growth. Potent PDGF antagonists have been discovered through the synthesis of a series of calix[4]arene-based compounds that are designed to bind to the three-loop region of PDGF. The effect of lower-rim alkylation, linker and number of interacting head groups on the calix[4]arene scaffold on PDGF affinity and cellular activity has been investigated using calixarenes **30**, **110-117** [126].

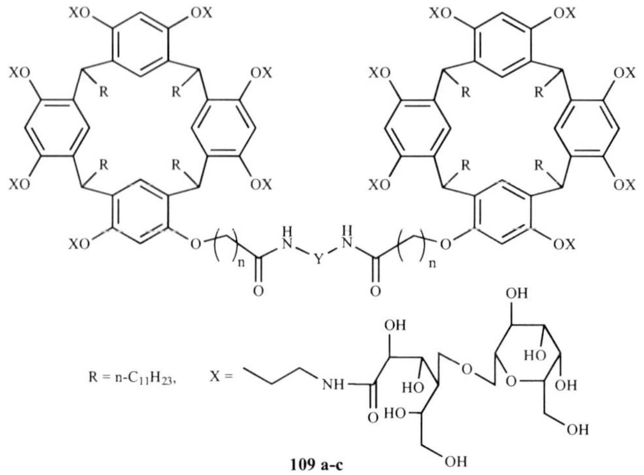

109 a-c

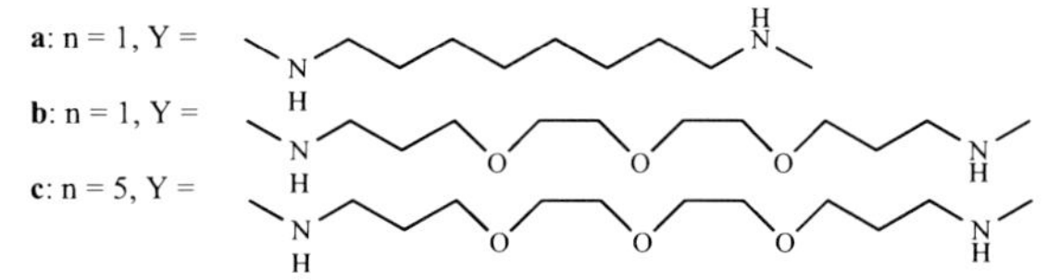

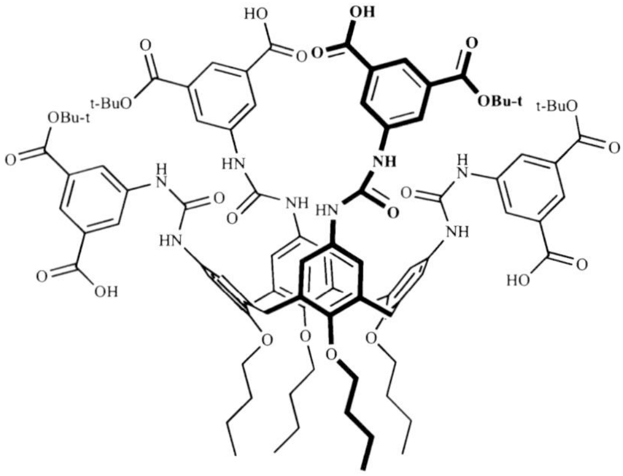

110

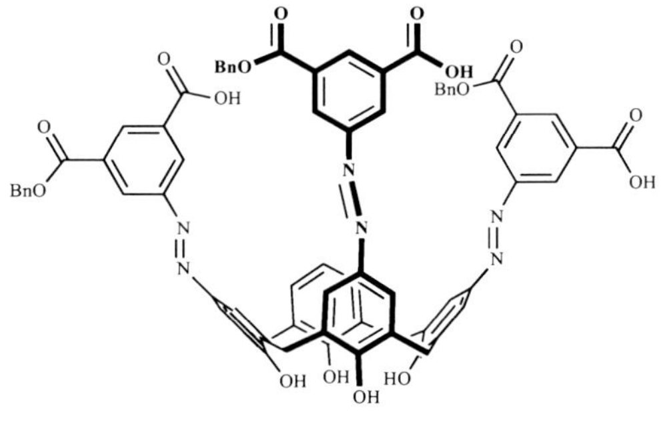

111

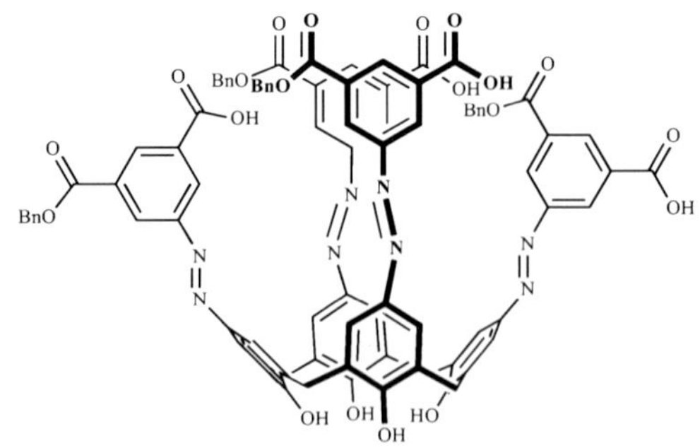

112

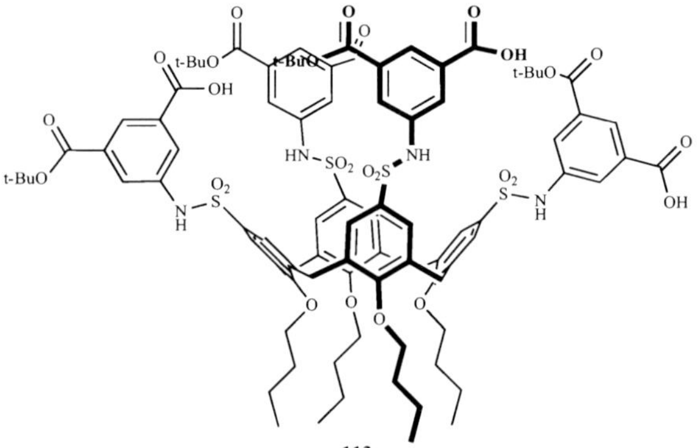

113

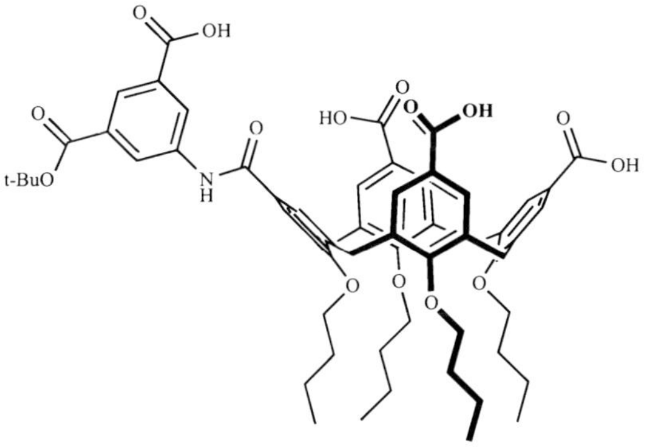

114

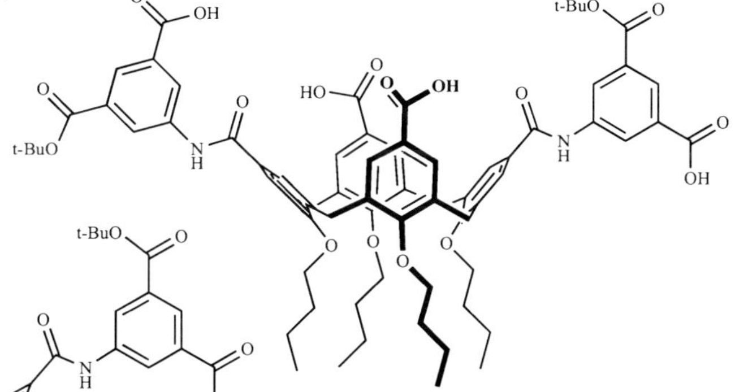

115

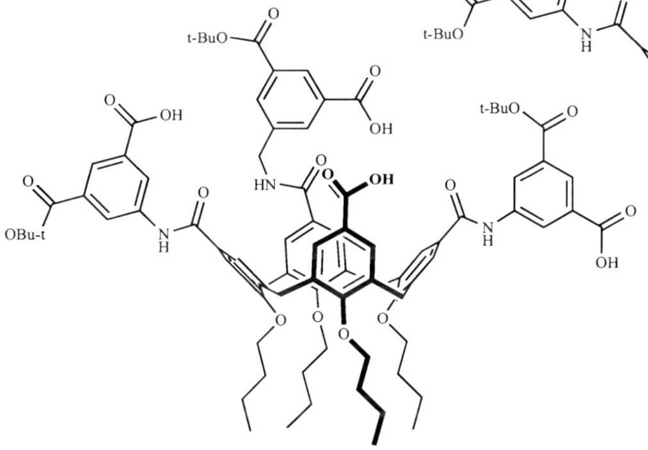

116

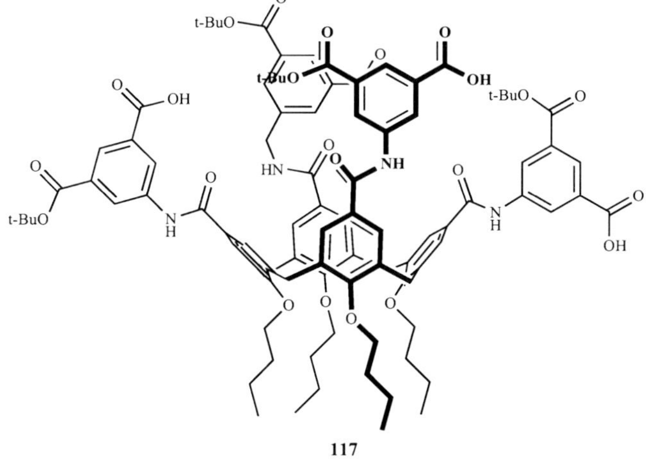

117

The dissociation constants towards PDGF and an ability to inhibit PDGF-stimulated PDGFR tyrosine phosphorylation were investigated with respect to the library of compounds that address different aspects of the structural variation of the parent **30.** The IC_{50} values on NIH 3T3 cells also provided a direct evaluation of the cellular activity of presented PDGF antagonists (Table **23**).

Table 23. The dissociation constants K_d of PDGF-BB subunit with antagonists and the corresponding IC_{50} values for the inhibition of PDGFR autophosphorylation.

Calixarene	K_d, nM^{-1}	IC_{50}, μM
30	17.3	0.19 ± 0.06
110	–	>10
111	23.8	0.48 ± 0.1
112	22.6	0.11 ± 0.04
113	13.7	>10
114	137	>10
115	39.4	>10
116	21.1	0.43 ± 0.35
117	9.26	0.35 ± 0.31

It was concluded that three isophthalate arms turned out to be sufficient to obtain good binding affinity and also to give competitive cellular activity (pairs **111, 112** and **116, 117** show close properties). The linker between the isophthalate and calix[4]arene scaffolds had an insignificant effect on the affinity, but notably influenced the cellular efficacy. Compounds **30** and **112** (with diazo and amide linkers) all have satisfactory activity in cell culture. In contrast, compounds **110** and **113** with sulfonamide and urea linkers gave very low activity. The alkylation at the lower rim hydroxyls was found unnecessary because the four OH groups form an array of hydrogen bonds to stabilize the calix[4]arene scaffold into a cone conformation well pre-organized to interact with the loop regions of PDGF-BB.

The inhibition of angiogenesis is a promising avenue for cancer treatment. Although some angiostatic compounds are in the process of development and testing, these often prove ineffective *in vivo* or have unwanted side effects. The design, synthesis, and evaluation of a small library of nonpeptidic, calixarene-based protein surface topomimetics were carried out. These calixarenes have chemical substituents (hydrophobic and positively charged residues) to approximate the molecular dimensions and amphipathic features of the antiangiogenic peptide anginex, which, like many antiangiogenic proteins, consists primarily of an antiparallel β-sheet structure as the functional unit.

Two of the calixarenes **118** and **119** were potent angiogenesis inhibitors *in vitro*, as determined by endothelial cell proliferation, migration, and chick embryo chorioallantoic membrane assays [127]. Moreover, both compounds were highly effective at inhibiting tumor angiogenesis and growth in two mouse models (MA148 human ovarian carcinoma and B16 murine melanoma).

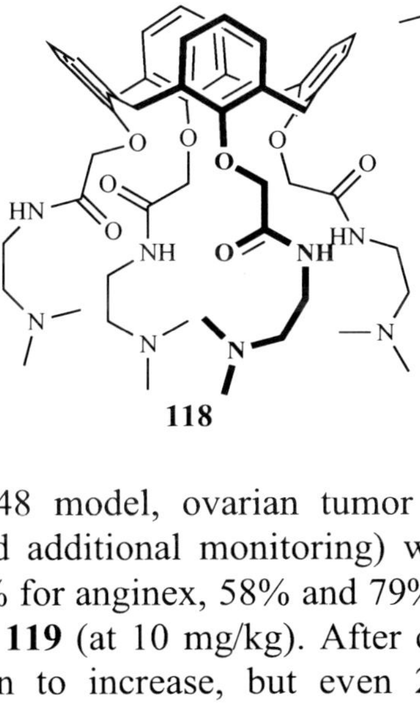

118

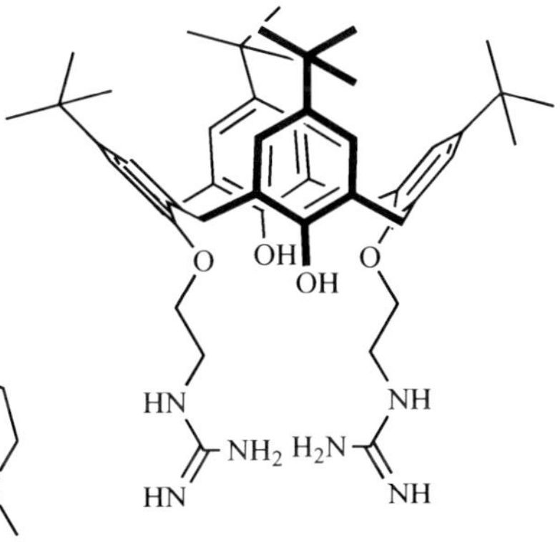

119

In the MA148 model, ovarian tumor growth at the end of the study (after treatment and additional monitoring) was reduced compared to that in control mice: by 62% for anginex, 58% and 79% for **118** (2.4 and 10 mg/kg respectively) and 59% for **119** (at 10 mg/kg). After cessation of treatments, the rate of tumor growth began to increase, but even 2 weeks later, tumor growth inhibition remained pronounced. Against the more aggressive B16 melanoma, tumor growth curves for all treated groups were also different from those of tumors in control mice. At the end of the study, **118** had inhibited growth on average by 77% relative to control treatment when administered *via* minipumps, whereas **119** inhibited growth by 70%, when administered by injection. In comparison, anginex inhibited tumor growth by 57-66% dependable from administering method.

Although anginex and some other angiogenesis inhibitors including endostatin, angiostatin, and Avastin are proteins, nonpeptidic compounds generally make for superior pharmaceutical drugs, primarily because they often exhibit greater *in vivo* exposure, can be administered orally, lack an immune response, and/or are less expensive to produce. It is for these reasons that protein surface topomimetics may stand as the new generation of antiangiogenic agents for use in the clinic against cancer.

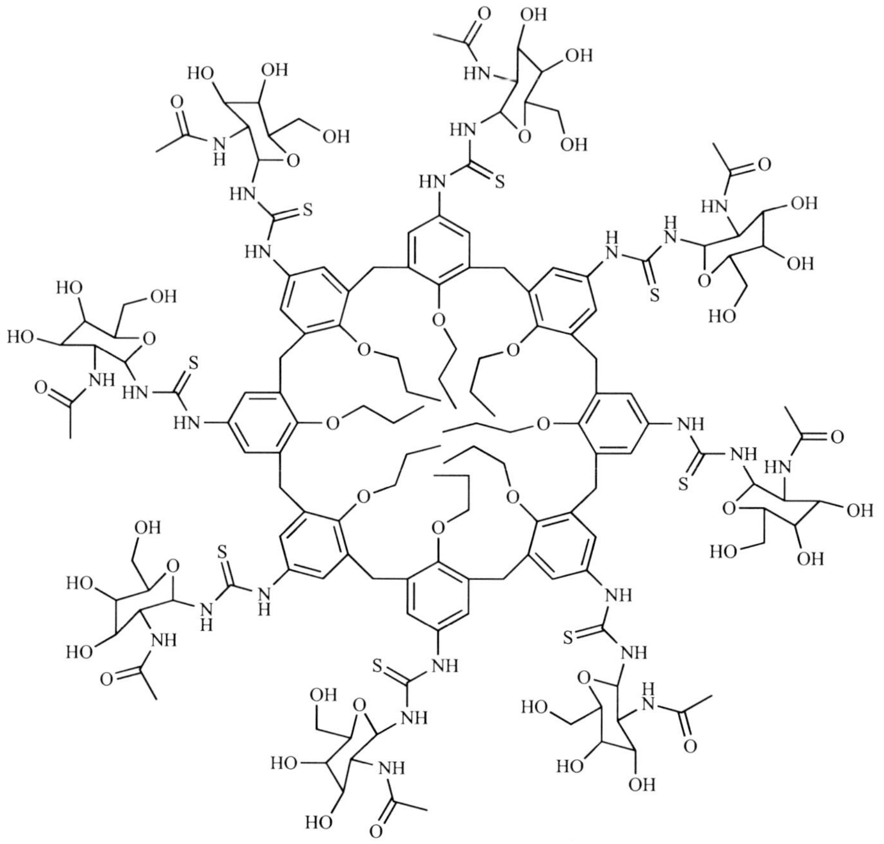

120

It was found that β-1,4-GalTase V is highly expressed in glioma and promotes invasion, growth, and survival of glioma cells [128, 129]. A glycocalix[8]arene exposing N-acetylglucosamine (GlcNAc) residues **120** inhibited rat C6 glioma cell migration as assessed in a scratch wound model [130]. This effect was related to inhibition of focal adhesion kinase phosphorylation, and specifically observed in the area bordering the scratch wound. Compound **120** inhibited also C6 cell proliferation, an effect unrelated to its ability to interact with Gal-Tase as it was mimicked by different calix[8]arene derivatives, all characterized by multivalency and ureido groups. Compound **120** did not induce apoptotic death, but caused a different distribution of C6 cells within the cell cycle. It is noted, that calix[8]arene skeleton appears to be a very good platform for the arrangement of multiple GlcNAc moieties in a spatial organization suitable to elicit GalTase recognition, because GlcNAc alone was effective only at very high concentrations and at later time points.

5. CALIXARENES IN DNA TRANSFECTION

The efficient introduction of genetic material into cells is an area of considerable interest both to provide tools for biological research and for the development of gene-based therapeutics. According to *in vitro* and *in vivo* studies, naked DNA and RNA exhibit very poor capacity to transfect cells, because of a number of barriers and their decomposition by nucleases in biological media [131, 132]. Several scaffolds have been used for the synthesis of cationic lipids (cytofectines) which are useful in promoting gene transfection. Recently, macrocyclic molecules, such as fullerenes, cyclodextrins and calix[4]arenes were included in these scaffolds because they have a well-defined 3D topology due to their persistent shape and they bear multiple groups for interacting with nucleic acids [133-136].

The group of Aoyama synthesized a number of calix[4]resocinarenes **121a-c**, modified with eight disaccharide residues: maltose, lactose and cellobiose, which are well known for binding with proteoglycans at the cellular surface [137].

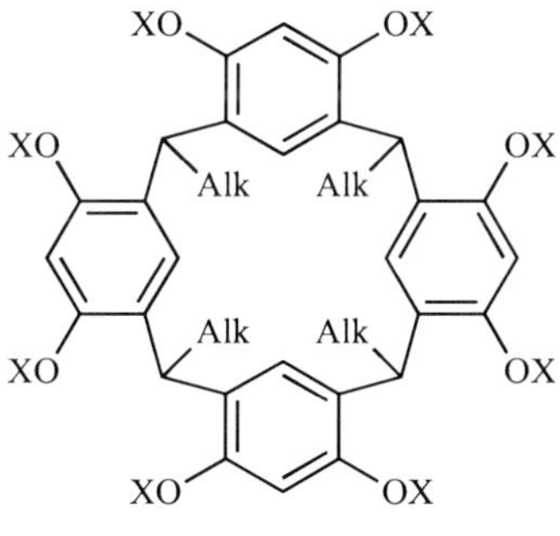

Alk = n-$C_{11}H_{23}$

121

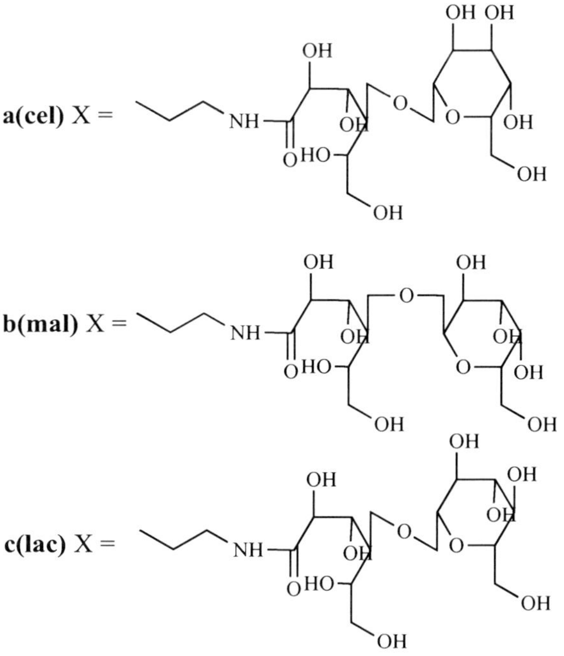

In water solutions, compounds **121** were reported to irreversibly self-assemble into micellar nanoparticles of 4.3-4.8 nm diameter [138]. The high affinity between saccharide-covered micelles and phosphate ions provided rationale for studies of their interactions with nucleic acids. Macrocycle **121a** with 8 cellobiose fragments in the presence of plasmid DNA can form nanoparticles of 54 nm diameter, which the authors called "artificial glycoviruses" (Fig. **13**). In contrast, maltose (**121b**) and lactose (**121c**) derivatives formed much larger particles (200-300 nm). Transfection experiments on HeLa and HepG2 cells showed that the transfection efficiency of small-sized cellobiose **121a** aggregates was up to 20- and 1000-fold larger than those of lactose **121b** and maltose **121c** based particles [139]. Cellobiose nanoparticles were more efficient than lipofectamine and only 2-3-fold less efficient than polyethyleneimine (jetPEI) [140]. Cytotoxicity data for cellobiose micellar glycocluster nanoparticles showed 85% of cell viability in standard transfection conditions [141].

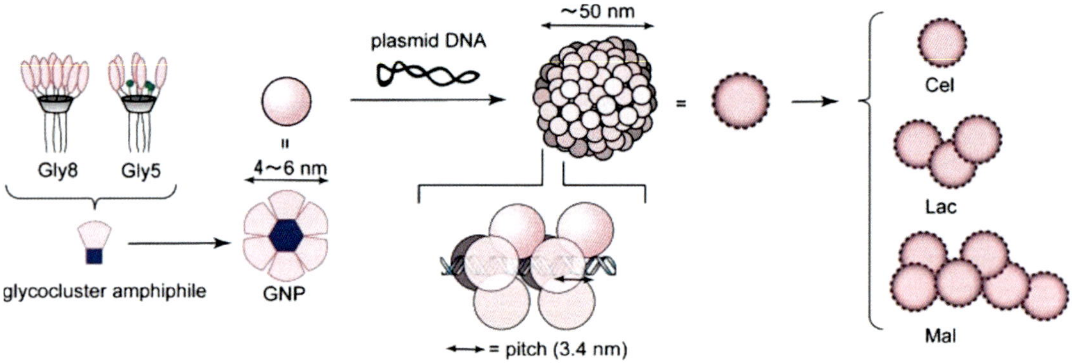

Figure 13: Hierarchical growth of glycocluster amphiphile through nanoparticle to "artificial glycoviruse" and its aggregates. (Reprinted with permission from [140]. Copyright 2015 American Chemical Society).

Amphiphilic tetraaminotetradodecyloxycalixarene **122** [142], form solid lipid nanoparticles (SLN) of 155 nm diameter. The binding of calixarenes to plasmid DNA was studied by gel electrophoresis and DLS [143]. The obtained particles were not able to promote transfection, so SLNs were covered sequentially with cationic polysaccharide chitosan and plasmid DNA for obtaining successful transfection. In comparing to Lipofectamine SLNs coated with multiple chitosan-DNA layers were 25-35 times more effective in ransfection MDCK cell line, if chitosan was the last layer. The authors noted the low toxicity of the studied particles, though no dedicated experiment to determine the cytotoxicity was performed [142].

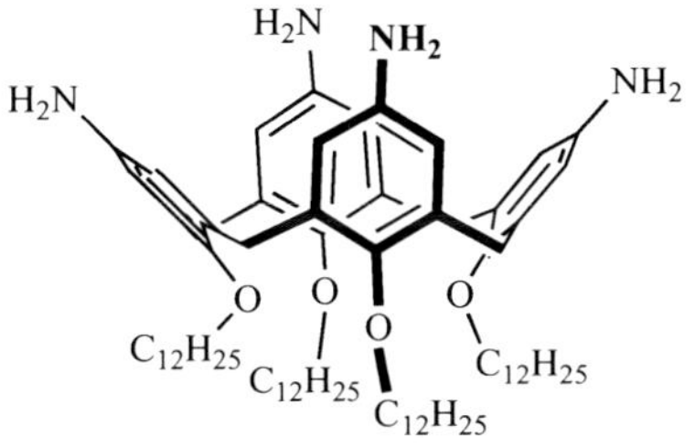

122

Polycationic multicalixarenes **123** and **124a,b** were synthesized and tested as for DNA-binding-ability as well as for ability for plasmid DNA transfection [144].

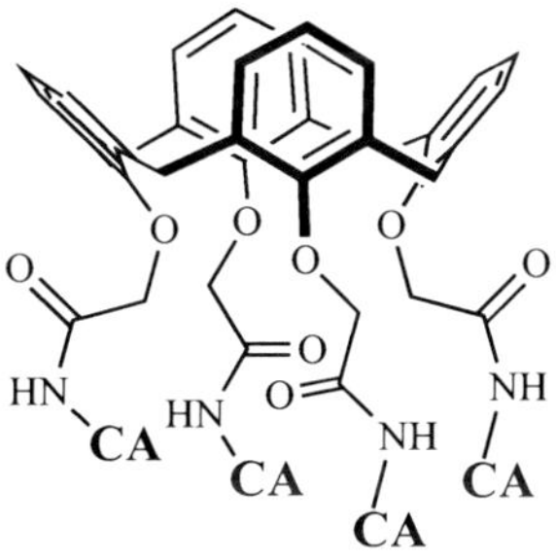

123

CA = 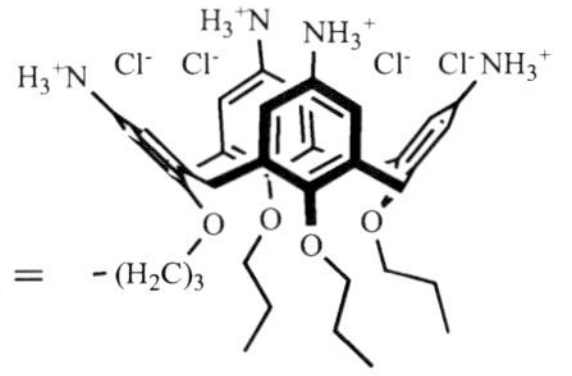

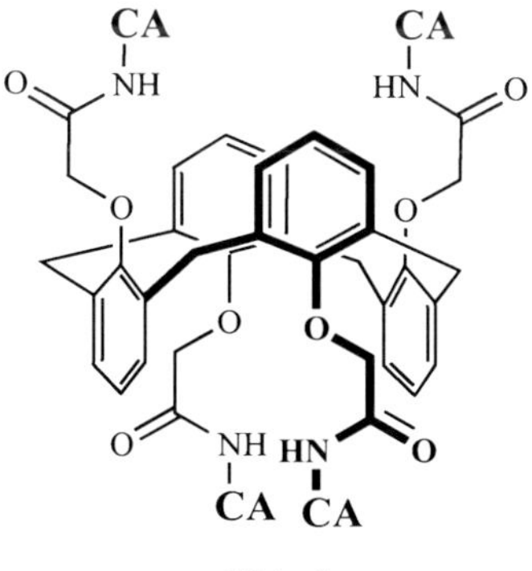

124a,b

a: CA = 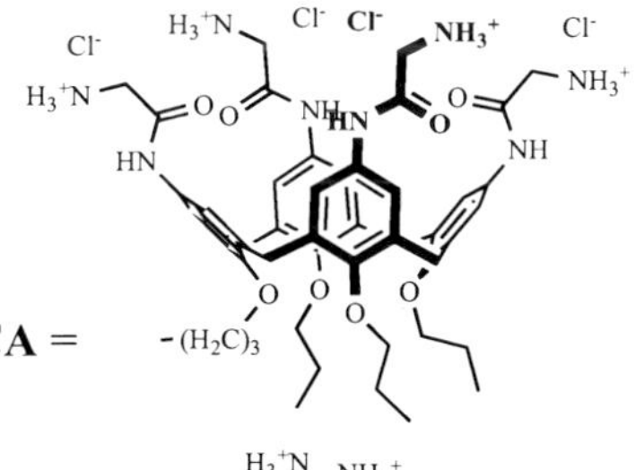

b: CA = 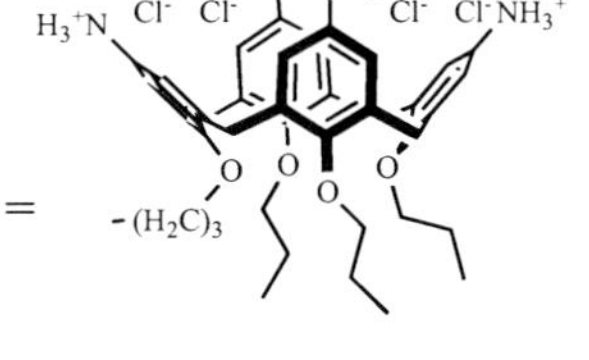

The DNA-binding properties of calixarenes **123** and **124a,b** were evaluated through gel electrophoresis studies using pGEM-EYFP or pEGFP-N1 plasmid. Screening at a high ligand concentration (2mM) indicated that **123** and **124a,b** were able to bind DNA through charge neutralisation. Observation at low concentration shows that binding ability is decreased in order **124a>123>124b,** and reaches the minimum of concentration for **124a** close to 0.2 mM. Investigation of macrocycle's **123, 124a,b** ability to transfect DNA plasmid revealed, that only calixarene **124a** (which didn't contain arylammonium groups) shows significant transfection. The transfection efficiency was 10 % for calixarene **124a**, less than 1% for **124b** and nearly zero for **123**, whereas the efficiency of the commercial agent FuGene was ~50% under these conditions. Interestingly, that a building block of compound **124a** the tetracationic amphiphilic calixarene **82**, also showed good affinity to DNA, but its gene transfection activity was negligible.

Calixarenes bearing guanidinium groups at the lower rim of macrocycle **125a-c** show both DNA-binding properties and cell transfection ability as it's described in [145].

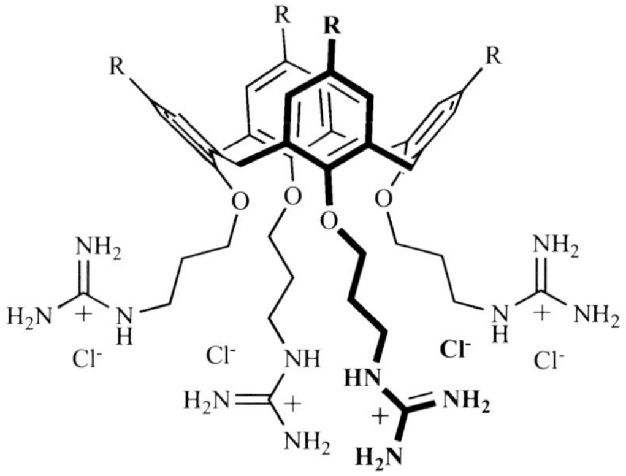

125a-c

a: R = *t*-Bu
b: R = H
c: R = *n*-Hex

The macrocycles **125** are good water soluble in the range of concentrations from 1mM for **125c** to 20mM for **125b**. The gel electrophoresis experiments show higher plasmid-binding ability of **125** compared to compounds **123** and **124**. Calixarene **125b** completely bind DNA plasmid at 50μM, whereas **125a** and **125c** at 100-200μM. Atomic force microscopy experiments show that **125a** and **125c** bearing alkyl groups at the upper rim form highly condensed nanomeric aggregates of single DNA plectonemes at concentration 0.6 and 2.5 μM respectively. However, the alkyl depleted **125b** at 5 μM and higher concentrations didn't give any condensates because of high solubility and full charge compensations at high concentration.

Transfection experiments were performed using plasmid DNA pEGFP-C1 and RD-4 human rhabdomyosarcoma cells. No transfection occurs when **125a-c** are used alone, but formulation of calixarene with dioleoylphosphatidylethanolamine (DOPE, biologically available "helper lipid") is quite effective. The amount of transfected cells is more than 48% for **125b** and only 30% for commercially available lipofectamine [146]. More lipophillic **125a** and **125c** show low activity, (3-7%). Later studies showed that in the presence of DOPE, the DNA complexes of **125b** became well-defined compact structures and these formulations were less toxic than those with calixarenes **125a** and **125c** [147]. The superior transfection properties of **125b** were attributed to the appropriate flexibility of the chains linking the guanidinium groups and the optimal lipophilic-hydrophilic balance.

Rather different transfection properties were found for upper rim functionalized polyguanidinocalixarenes **126-131** [148, 149]. Methoxycalixarenes **126, 130** and **131**, which do not adopt fixed conformation, and tetrapropocycalix[4]arene **128** which is in the 1,3-alternate conformation showed no transfection activity alone and with addition of DOPE, probably, due to their inability to form compact particles with DNA. In contrast, the cone-shaped amphiphilic calixarenes **127** and **129** in combination with DOPE revealed some transfection activity (<5 % of the transfected cells). It should be noted that calixarene **129** alone at high concentrations (40 μM) was able to promote the transfection, while **127** did not work without helper lipid DOPE.

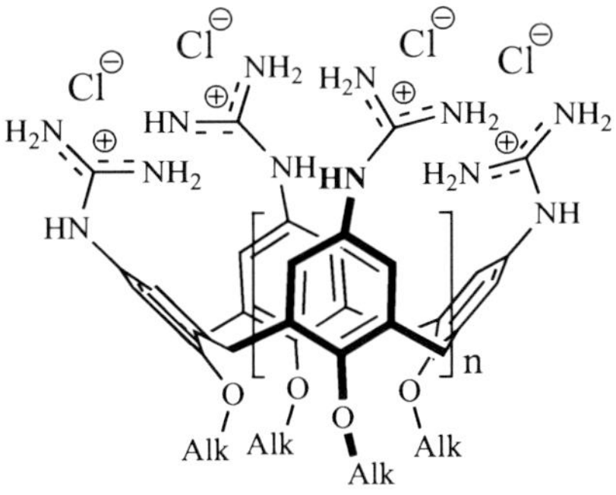

126 n = 1, Alk = CH_3,
127 n = 1, *cone*, Alk = C_3H_7,
128 n = 1, *1,3-alt*, Alk = C_3H_7,
129 n = 1, *cone*, Alk = C_8H_{17}
130 n = 3, Alk = CH_3
131 n = 5, Alk = CH_3

A major drawback of these calixarenes **126-131** is their relatively high toxicity at concentrations above 25 μM (for instance, calixarene **129** at 40 μM concentration showed close to 100% cell death). Notably, calixarenes **127** and **129**, which showed appreciable transfection activity, were more toxic than inactive calixarenes **126**, **130** and **131**, and 1,3-alternate **128**.

Recently, calixarene bearing four arginine residues at the upper rim together with four hexyls groups at the lower rim **132a** or just four arginine residues at the lower rim **133** were synthesized [150]. According to AFM data, amphiphile **132a** condensed DNA into 50–60 nm nanoparticles, whereas non-amphiphillic calixarene **133** caused the formation of large aggregates. Remarkably, calixarene **132a** showed excellent transfection efficiency, which was even better than that of Lipofectamine and polyethyleneimine in a large variety of cell lines (COS-7, HEK, ISHIKAWA, EADSCs, C2C12, HeLa, N2a, RD-4, VERO). In contrast, calixarene **133** exhibited almost no transfection activity. Calixarene **132b** (close analog of **132a**) also showed poor transfection activity, though it is significantly improved in the presence helper lipid DOPE.

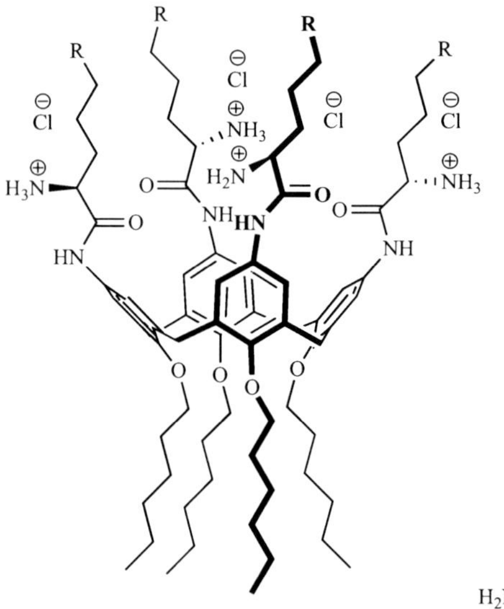

132 a R = $NH(C{=}NH_2^+)NH_2, \bar{C}l$
b R = $CH_2NH_3^+ Cl$

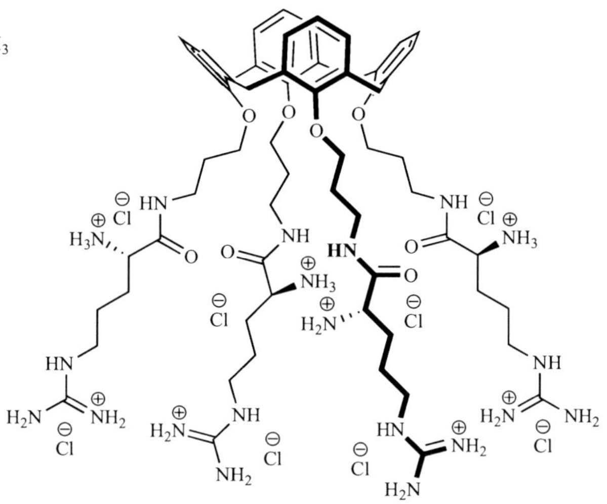

133

This drastic difference between **132a** and **133b** indicates that clusters of arginine residues within an amphiphilic calixarene are crucial to achieve highly efficient transfection, probably by improving capacity of the DNA complexes to cross the membrane barriers of the cell. Finally, the most efficient calixarene **132a** also showed relatively low cytotoxicity, which was comparable to that of lipofectamine [148].

Tetraalkylamonium calixarenes derivatives **134** and **136** were observed to self-assemble into micelle-like structures of 3-4 nm diameter in aqueous solutions [151]. Further studies of calix[4]arenes **134-136** in cone conformation and 1,3-alternate ones as well as conformationally mobile calix[6]arenes **137** and calix[8]arenes **138** revealed their ability to host-guest complexes with anionic nucleotides, such as AMP, GMP, CMP, UMP, TMP, ADP and ATP. Also, calixarenes **134-138** were found to bind with high affinity to model DNA sequences and natural calf thymus DNA (CT-DNA) [152]. But any experiments on cells with these complexes and aggregates were not carried.

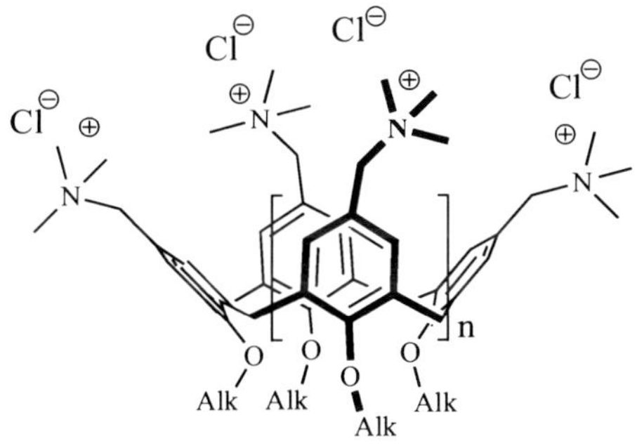

134, n = 1, *cone* Alk = C_3H_7
135, n = 1, *1,3-alt*, Alk = C_3H_7
136, n = 1, *cone* Alk = C_6H_{13}
137, n = 3, Alk = CH_3
138, n = 5, Alk = CH_3

A series of amphiphilic calixarenes **83b, 84a-d, 85b** bearing cationic groups at the upper rim and the alkyl chains at the lower rim were synthesized [153-155]. The molecular design of these calixarenes was directed to increase their affinity for DNA (compared to **134-138**) due to the formation of additional hydrogen bonds between the hydroxyl group of choline residues and the carbohydrate and phosphate groups of DNA in the case of **84** or additional electrostatic interaction between protonated aminogroups of **85b** and phosphate groups of DNA or to improve π-stacking interaction with the nucleic bases of DNA in the case of **83b**.

The effect of the length of aliphatic chains and the head group structure on their self-assembly, interaction with DNA, properties of their DNA complexes, the transfection efficiency and cytotoxicity was studied. Gene transfection experiments were performed in HeLa and COS-7 cells. It was found that longer alkyl chains favor formation of smaller particles with plasmid DNA of remarkably small sizes ~40 nm. These nano-aggregates are characterized by higher transfection efficiency and lower cytotoxicity.

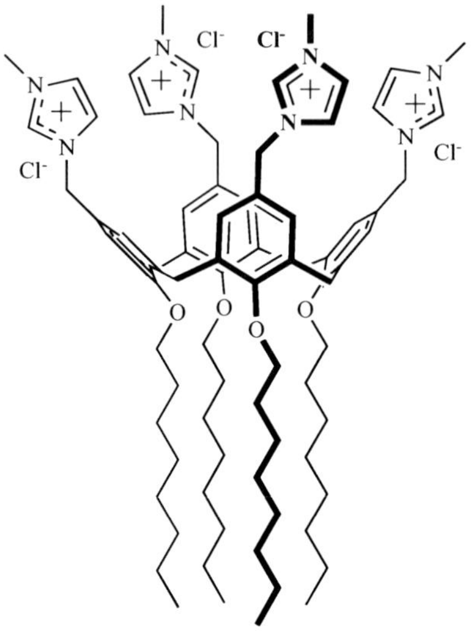

83b

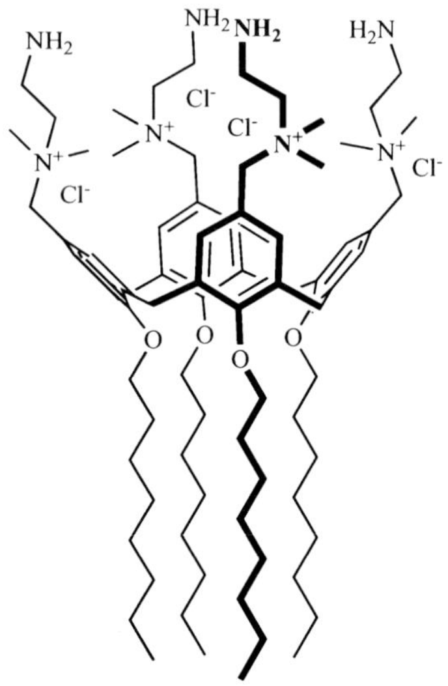

85b

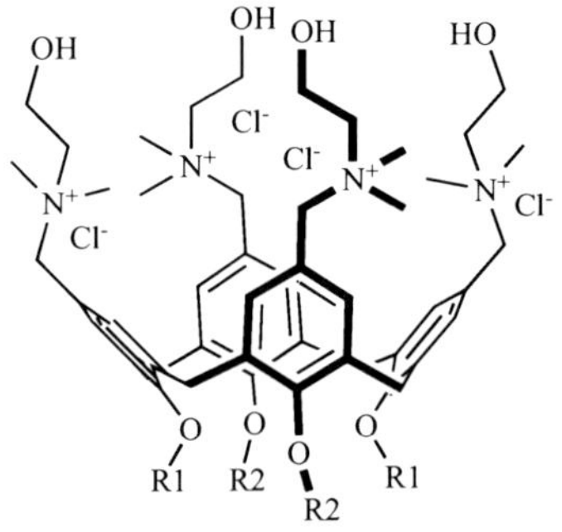

84 a R1 = R2 = n-C_3H_7,
b R1 = R2 = n-C_8H_{17}
c R1 = R2 = n-C_6H_{13}
d R1 = R2 = n-$C_{12}H_{25}$
e R1 = n-C_3H_7, R2 = n-$C_{16}H_{33}$

The most promising results are observed for **84d**-DOPE complex at N/P 2 ratio, which shows much higher rate of transfected cells (~80%) compared to firstly investigated **84b** (~36%) and even to jetPEI (~70%), which is one of the most efficient transfection agents. The rate of transfected cells grows in the following order: **84a<85b<84c<83b<84e~84b<84d**. In the presence of serum only **84d** molecule showed detectable transfection among the studied calixarenes. Moreover, **84d** was the only calixarene that showed detectable transfection without DOPE. Thus, long alkyl chains and the largest hydrophobic domain of **84d** improve significantly the transfection efficiency. On the other hand, presence of four additional amino groups, which could be protonated at pH<7 affected only stoichiometry of the complexes with DNA without affecting their transfection efficiency or cytotoxicity. The obtained results provide new insights for designing new artificial viruses based on macrocyclic molecules.

6. CALIXARENES IN MAGNETIC RESONANCE IMAGING

The complexation of lanthanide ions by calix[4]arene ligands **139** has been studied with the purpose of new luminescent probes development [156]. Very little is known on the use of calixarene lanthanide complexes as contrast agents for magnetic resonance imaging, although this possibility has been investigated. A new calixarene-based ligand is able to form complexes with Gd(III) in water, these complexes are characterized by high values of relaxivity.

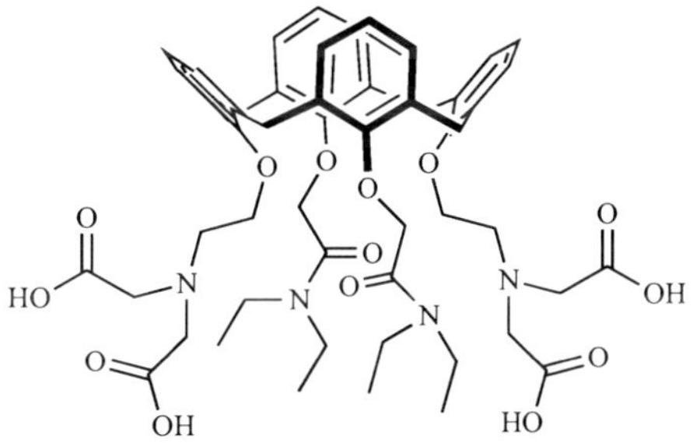

139

From relaxometric competition experiments a lower limit of the stability constant was measured as $1{\cdot}10^{13}$ M^{-1} for Gd(III)·**139**, 1:1 complex in water at pH 6.8, a value sufficiently high to ensure the integrity of the complex in the presence of protein. From the analysis of the data of the proton relaxation enhancement an affinity constant with human serum albumin (HSA) was calculated as $2.4{\cdot}10^{4}$ M^{-1}.

The high binding affinity of Gd(III)·**139** to HSA is the result of a cooperative effect of the hydrophobic interaction brought about by the aromatic macrocyclic moiety and of the coordination of the protein donor groups (probably Asp or Glu).

Calix[8]arene conjugates **140a-c**, **141** useful for imaging in MRI and computed tomography (CT) are described [157]. Imaging moieties useful for CT imaging include those comprising iodine atoms **(140a)**. Imaging moieties useful for MRI include organic moieties comprising fluorine atoms (**141**); nitroxyl spin labeled moieties (**140b**); and Gd^{3+}-chelating groups (**140c**).

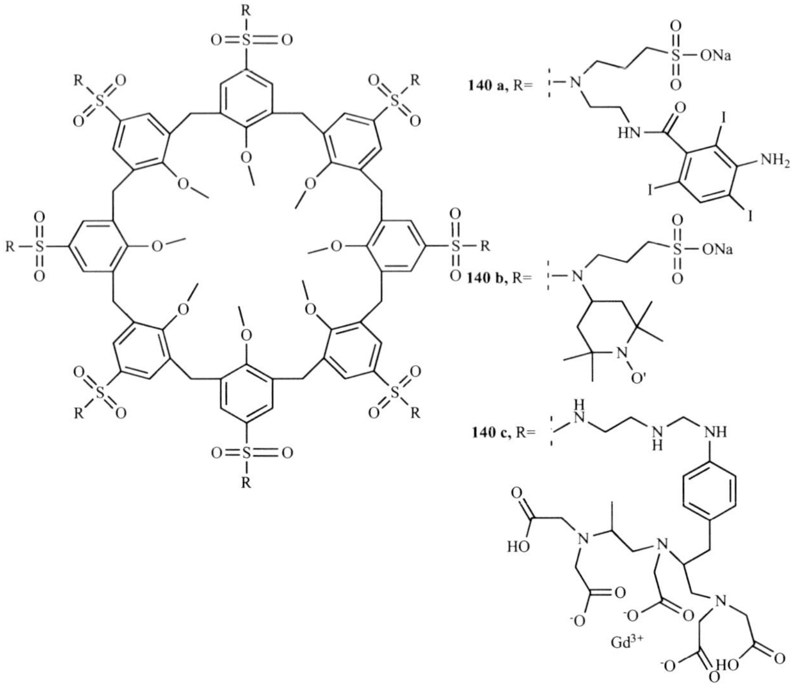

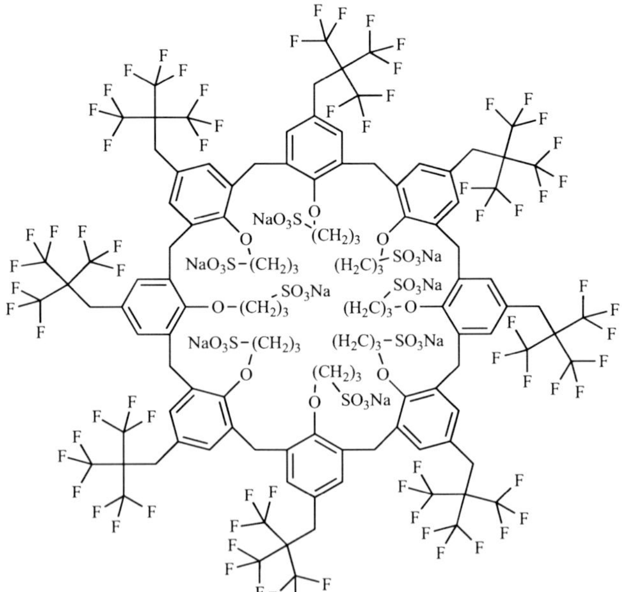

141

7. TOXICITY OF CALIXARENES

A brief report in a review by *Gutsche* in 1985 stated that the mother calixarenes had shown no activity in the Ames test [158]. Studies of calixarene sulfonates using inhibition of cell growth (ICG) on human fibroblast cells showed that whilst the calix[6] and [8] derivatives (**73** and **72** respectively) have the same level of toxicity as glucose, there is some slight toxicity associated with calix[4]arene sulfonate **27** [158]. Investigators noted that certain calix[4]arene phosphoric acid derivative **142** showed no effects on the cell growth of human fibroblasts [159].

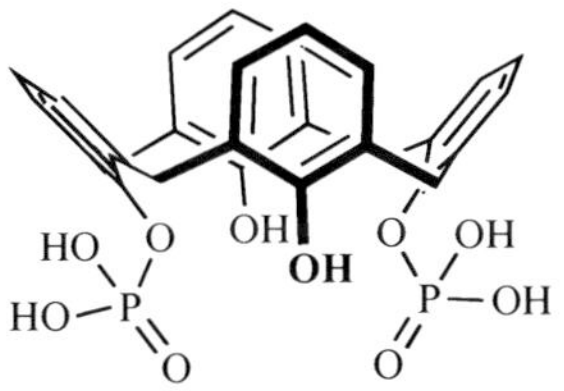

142

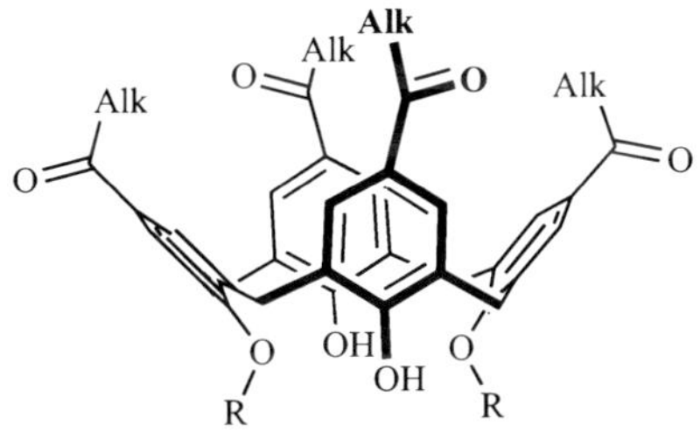

143a-d

	Alk	R
a	C_7H_{15}	H
b	C_9H_{19}	H
c	$C_{11}H_{23}$	H
d	$C_{11}H_{23}$	$PO(OH)_2$

More recently, it was reported that solid lipid nanoparticles based on amphipihilic calix[4]arene derivatives **143a-d** show zero haemolytic effects at concentrations ranging up to 300 mg/l [160], this is in contrast to the behaviour of SLNs based on amphiphilic cyclodextins which show significant haemolytic effects [161]. Similarly, in a recent publication it was shown that for a series *para*-sulphonato-calix[4, 6 and 8]arenes **27, 73** and **72** and their monosubstituted derivatives, haemolytic effects are observed in the order *para*-sulphonato-calix[8]arenes > *para*-sulphonato-calix[6]arenes > *para*-sulphonato-calix[4]arenes, with 5% haemolysis occurring at concentrations of 20, 50 and 200 mM, respectively [162]. It was shown that *in vivo*, para-^{35}S-sulfonato-calix[4]arene **27** shows no acute toxicity for single injected doses equivalent to 2–5 g in humans, and that the molecule is rapidly cleared from mice *via* elimination in urine. In view of the lack of accumulation in the liver, it is unlikely that para-^{35}S-sulfonato-calix[4]arene is metabolized [163]. Results on cytotoxicity showed that **75** is effective in inhibiting several Gram-positive and Gram-negative bacteria, without increasing toxicity on HaCaT and MRC-5 cell lines [99]. The hemotoxicity of the conjugates **94a,b** was considerably lower (at least 10 times) than the hemotoxicity of AmB [113]. Also, treatment with calixarenes **118, 119** on the mouse did not cause observable toxicity as assessed by behavior, body weight change, and hematocrit level [127]. DNA-tranfection calixarenes **123, 124a,b** show lack of cell toxicity (in Chinese Hamster Ovary and Human Embryotic Kidney cell lines) over the full pharmaceutically relevant concentration range [144], also calixarene **125b** exhibits low toxicity showing 75-80% RD-4 cell viability at 48h from transfection, similar results were obtained for **125a** [145]. The cytotoxicity of amphilillic polycationic calixarene **83-85** decreases with increase in the length of aliphatic chains for all concentrations studied, i.e. the following tendency was observed: **84a>84c>84b~85b~83b>84d>84e**, where survival rate was 32% for **84e** at 10 μM. However, when the incubation was done in the presence of serum the cytotoxicity of all calixarenes significantly decreased, survival rate for **84d** and **84e** near 80% at 40 μM [155].

CONCLUSION

Cup-shaped nano-sized calixarenes due to unique ability to selective complexation with bio-polymers, such as enzymes, proteins and to formation of membrane channels show broad spectrum of bio-activity. High antiviral, antibacterial, anticancer, antitrombotic activities, gene transfection were patented and documented. At the same time, the calixarenes (as well their nanoparticles) have demonstrated neither toxicity nor immune responses, thus increasing the interest to them in the field of biopharmaceutical applications. Undoubtedly, the utilization of calixarenes will show a large growth in the biotechnology, pharmaceutical investigations, diagnostics, nanomedicine etc.

CONFLICT OF INTEREST

The authors confirm that this chapter contents have no conflict of interest.

ACKNOWLEDGEMENTS

The authors express their thanks to Prof. S.Kosterin and Prof. A.Vovk for helpful discussion. The work was supported by the National Academy of Sciences of Ukraine.

DISCLOSURE

This is an updated Chapter of the earlier review article [by R. V. Rodik, V. I. Boyko, V. I. Kalchenko Calixarenes in bio-medical researches Curr. Med. Chem., 2009, 16: 1630-1655], with additional text and references.

ABBREVIATIONS

Å	=	Angstrom
Ach	=	acetylcholine
AmB	=	amphotericin B
ApA	=	adenylic diribonucleotide
ApG	=	adenyl-guanylic diribonucleotide
APTT	=	activated partial thromboplastin time
ATP	=	adenosine-5'-triphosphate
BIAP	=	bovine intestinal alkaline phosphatase
CCT	=	card calibration table
CpC	=	cytidylic diribonucleotide
cyt c	=	cytochrome c
DMSO	=	dimethyl sulfoxide
CT-DNA	=	calf thymus deoxyribonucleic acid
DNA	=	deoxyribonucleic acid
DOPC	=	dioctanoyl-L-α-phosphatidylcholine
DOPE	=	dioleoylphosphatidylethanolamine
ED_{50}	=	effective dose with a 50% survival rate

EPNP	=	ethyl-*p*-nitrophenyl phosphate
FAB	=	fragment antigen binding
GpA	=	guanyl-adenylic diribonucleotide
GpG	=	guanyl-cytidylic diribonucleotide
GpG	=	guanylic diribonucleotide
HeLa	=	cell line taken from cervical cancer of Henrietta Lacks
HEPES	=	4-(2-hydroxyethyl)-1-piperazineethanesulfonic acid
HIV	=	human immunodeficiency virus
HPLC	=	high performance liquid chromatography
HPNP	=	2-nydroxypropyl-*p*-nitrophenyl phosphate
HSA	=	human serum albumin
HSV	=	herpes simplex virus
IC_{50}	=	the concentration is needed to inhibit a given biological process by half
ICG	=	inhibition of cell growth
LMWH	=	low molecular weight heparin
LUV	=	large unilamellar vesicles
MIC	=	minimum inhibitory concentration
MRI	=	magnetic resonance imaging
NADH	=	reduced form of nicotinamide adenine dinucleotide
ORCC	=	outwardly rectifying chloride channels
PC	=	phosphoryl-choline

PDGF	=	platelet-derived growth factor
PDGFR	=	receptor of platelet-derived growth factor
pEGFP	=	plasmid of enhanced green fluorescent protein
PKAP	=	porcine kidney alkaline phosphatase
PNPCC	=	choline *p*-nitrophenylcarbonate
PT	=	prothrombin time assay
RNA	=	ribonucleic acid
RLC-16	=	line of rat hepatoma (liver cancer) cells
RSV	=	respiratory syncytial viruse
SLN	=	solid lipid nanoparticles
Tris	=	trishydroxymethylaminomethane
UFH	=	unfractionated heparin
UpU	=	uridylic diribonucleotide
VRAC	=	volume-regulated anion channels

REFERENCES

[1] Moriuchi, T.; Hirao, T. Highly ordered structures of peptides by using molecular scaffolds. *Chem. Soc. Rev.*, **2004**, *33*, 294-301.

[2] Koyack, M.J.; Cheng, R.P. Design and synthesis of β-peptides with biological activity. In: *Protein Design. Methods and Applications*; Guerois, R.; López de la Paz M. Eds.; Humana Press: New York, *Methods in Molecular Biology*, **2008**; Vol. *340*, pp. 95-109.

[3] Kilcoyne, M.; Joshi, L. Carbohydrates in therapeutics. *Cardiovasc. Hematol. Agents Med. Chem.*, **2007**, *5*, 186-197.

[4] Wrodnigg, T.M.; Sprenger, F.K. Bioactive carbohydrates and recently discovered analogues as chemotherapeutics. *Mini Rev. Med. Chem.*, **2004**, *4*, 437-459.

[5] Cal, K.; Centkowska, K. Use of cyclodextrins in topical formulations: practical aspects *Eur. J. Pharm. Biopharm.*, **2008**, *68*, 467-478.

[6] Duchêne, D.; Bochot, A.; Yu, S.-C.; Pépin, C.; Seiller, M. Cyclodextrins and emulsions. *Int. J. Pharm.*, **2003**, *266*, 85-90.

[7] Sansone, F.; Segura, M.; Ungaro, R. Calixarenes in bioorganic and biomimetic chemistry. In: *Calixarenes 2001*; Asfari Z.; Boehmer V.; Harowfield J.; Vicens J. Eds.; Kluwer Academic Publishers: Dodrecht, **2001**; pp. 496-512.

[8] Bohmer, V. Calixarenes, macrocycles with (almost) unlimited possibilities. *Angew. Chem. Int. Ed. Engl.*, **1995**, *34*, 713-745.

[9] De Namor, A.F.D.; Pugliese, A.; Casal, A.R.; Llerena, M.B.; Aymonino, P.J.; Velarde, F.J.S. The various factors involved in the extraction of alkali metal picrates by calixarene ester derivatives in the mutually saturated water-dichloromethane solvent system. *Phys. Chem. Chem. Phys.*, **2000**, *2*, 4355-4360.

[10] Talanova, G.G.; Hwang, H.-S.; Talanov, S.; Bartsch, R.A. Calix[4]arcncs with a novel proton-ionizable group: synthesis and metal ion separations. *Chem. Commun.*, **1998**, (3), 419-420.

[11] Matthews, S.E.; Parzuchowski, P.; Garcia-Carrera, A.; Dozol J.-F.; Grüttner, C.; Böhmer, V. Extraction of lanthanides and actinides by a magnetically assisted chemical separation technique based on CMPO-calix[4]arenes. *Chem. Commun.*, **2001**, (5), 417-418.

[12] Pelizzi, N.; Casnati, A.; Friggeri, A.; Ungaro R. Synthesis and properties of new calixerene-based ditopic receptors for the simultaneous complexation of cations and carboxylate anions. *J. Chem. Soc., Perkin Trans. 2*, **1998**, 1307-1312.

[13] Budka, J.; Lhotak, P.; Michlova, V.; Stibor, I. Urea derivatives of calix[4]arene 1,3-alternate: an anion receptor with profound negative allosteric effect. *Tetrahedron Lett.*, **2001**, *42*, 1583-1586.

[14] Arena, G.; Contino, A.; Gulino, F.G.; Magri, A.; Sciotto, D.; Ungaro, R. Complexation of small neutral organic molecules by water soluble calix[4]arenes. *Tetrahedron Lett.*, **2000**, *41*, 9327-9330.

[15] Gutsche C. D. *Calixarenes Revisited*; The Royal Society of Chemistry: Cambridge, **1998**.

[16] *Calixarenes for Separations*, Lumetta G.J.; Rogers R.D.; Gopalan A.S. Eds.; American Chemical Society: Washington, **2000**.

[17] Nimse, S.B.; Kim, T. Biological applications of functionalized calixarenes. *Chem. Soc. Rev.*, **2013**, *42*, 366-386.

[18] Da Silva, E.; Lazar, A.N.; Coleman, A.W. Biopharmaceutical applications of calixarenes. *J. Drug. Sci. Tech.*, **2004**, *14*(1), 3-20.

[19] Ukhatskaya, E.V.; Kurkov, S.V.; Matthews, S.E.; Loftsson, T. Encapsulation of drug molecules into calix[n]arene nanobaskets. Role of aminocalix[n]arenes in biopharmaceutical field. *J. Pharm. Sci.*, **2013**, *102*, 3485–3512.

[20] Perret, F.; Coleman, A. W. Biochemistry of anionic calix[n]arenes. *Chem. Commun.*, **2011**, *47*, 7303–7319.

[21] Tauran, Y.; Kim, B.; Coleman, A.W. Bio-applications of calix[n]arene capped silver nanoparticles. *J. Nanosci. Nanotechnol.*, **2015**, *15*, 1-19.

[22] Rodik, R.V.; Klymchenko, A.S.; Mely, Y.; Kalchenko, V.I. Calixarenes and related macrocycles as gene delivery vehicles. *J. Incl. Phenom.*, **2014**, *80*, 189-200.

[23] Coleman, A.W.; Perret, F.; Moussa, A.; Dupin, M.; Guo, Y.; Perron, H. Calix[n]arenes as protein sensors *Top. Curr. Chem.*, **2007**, *277*, 31–88.

[24] Giuliani, M.; Morbioli, I.; Sansone, F.; Casnati. A. Moulding calixarenes for biomacromolecule targeting. *Chem. Commun.*, **2015**, *51*, 14140-14159.

[25] Perret, F.; Lazar, A.N.; Coleman, A.W. Biochemistry of *para*-sulfonato-calix[n]arenes. *Chem. Commun.*, **2006**, (23), 2425-2438.

[26] Casnati, A.; Sansone, F.; Ungaro, R. Peptido- and glycocalixarenes: playing with hydrogen bonds around hydrophobic cavities. *Acc. Chem. Res.*, **2003**, *36*, 246-254.

[27] Young, M.J.; Chin, J. Dinuclear Copper (II) complex that hydrolyzes RNA. *J. Am. Chem. Soc.*, **1995**, *117*, 10577-10582.

[28] Molenveld, P.; Engbersen, J.F.J.; Kooijman, H.; Spek, A.L.; Reinhoudt, D.N. Efficient catalytic phosphate diester cleavage by the synergetic action of two Cu(II) centers in a dinuclear *cis*-diaqua Cu(II) calix[4]arene enzyme model. *J. Am. Chem. Soc.*, **1998**, *120*, 6726-6737.

[29] Molenveld, P.; Kapsabelis, S.; Engbersen, J.F.J.; Reinhoudt, D.N. Highly efficient phosphate diester transesterification by a calix[4]arene-based dinuclear Zinc(II) catalyst. *J. Am. Chem. Soc.*, **1997**, *119*, 2948-2949.

[30] Molenveld, P.; Stikvoort, W.M.G.; Kooijman, H.; Spek, A.L.; Engbersen, J.F.J.; Reinhoudt, D.N. Dinuclear and trinuclear Zn(II) calix[4]arene complexes as models for hydrolytic metallo-enzymes. Synthesis and catalytic activity in phosphate diester transesterification. *J. Org. Chem.*, **1999**, *64*, 3896-3906.

[31] Molenveld, P.; Engbersen, J.F.J.; Reinhoudt, D.N. Specific RNA dinuclear cleavage by a synthetic calix[4]arene-based trinuclear metallo(II)-phosphodiesterase. *Angew. Chem. Int. Ed. Engl.*, **1999**, *38*, 3189-3192.

[32] Cacciapaglia, R.; Casnati, A.; Mandolini, L.; Reinhoudt, D.N.; Salvio, R.; Sartori, A.; Ungaro, R. Catalysis of diribonucleoside monophosphate cleavage by water soluble Copper(II) complexes of calix[4]arene based nitrogen ligands. *J. Am. Chem. Soc.*, **2006**, *128*, 12322-12330.

[33] Cacciapaglia, R.; Casnati, A.; Mandolini, L.; Peracchi, A.; Reinhoudt, D.N.; Salvio, R.; Sartori, A.; Ungaro, R. Efficient and selective cleavage of RNA oligonucleotides by calix[4]arene-based synthetic metallonucleases. *J. Am. Chem. Soc.*, **2007**, *129*, 12512-12520.

[34] Dospil, G.; Schatz, J. Synthesis and characterization of imidazole-substituted calix[4]arenes as simple enzyme-mimics with acyltransferase activity. *Tetrahedron Lett.*, **2001**, *42*, 7837-7840.

[35] Magrans, J.O.; Ortiz, A.R.; Molins, M.A.; Lebouille, P.H.P.; Sanchez-Quesada, J.; Prados, P.; Pons, M.; Gaco, F.; de Mendoza, J. Designed non-peptidic receptor that mimics the phosphocholine binding site of the McPC603 antibody. *Angew. Chem. Int. Ed. Engl.*, **1996**, *35*, 1712-1715.

[36] Cuevas, F.; Di Stefano, S.; Magrans, J.O.; Prados, P.; Mandolini, L.; de Mendoza, J. Toward an artificial acetylcholineesterase. *Chem. Eur. J.*, **2000**, *6*, 3228-3234.

[37] Pollack, S.J.; Jacobs, J.W.; Schultz, P.G. Selective chemical catalysis by an antibody. *Science*, **1986**, *234*, 1570-1573.

[38] Jackson, D.Y.; Prudent, J.R.; Baldwin, E.P.; Schultz, P.G. A mutagenesis study of a catalytic antibody. *Proc. Natl. Acad. Sci. USA*, **1991**, *88*, 58-62.

[39] Breslow, R.; Dong, S.D. Biomimetic reactions catalyzed by cyclodextrins and their cerivatives. *Chem. Rev.*, **1998**, *98*, 1997-2011.

[40] Murakami, Y. Functionalized cyclophanes as catalysts and enzyme models. *Top. Curr. Chem.*, **1983**, *115*, 107-155.

[41] Shinkai, S.; Mori, S.; Koreishi, H.; Tsubaki, T.; Manabe, O. Hexasulfonated calix[6]arene derivatives: a new class of catalysts, surfactants, and Host molecules. *J. Am. Chem. Soc.*, **1986**, *108*, 2409-2416.

[42] Gutsche, C.D.; Alam, I. Calixarenes. 23. The complexation and catalytic properties of water soluble calixarenes. *Tetrahedron*, **1988**, *44*, 4689-4694.

[43] Tian-Ming, Y.; Zhi-Feng, Y.; Li, W.; Jin-Ying, G.; Si-De, Y.; Xian-Fa, S. Supramolecular interaction between water-soluble calix[4]arene and ATP--the catalysis of calix[4]arene for hydrolysis of ATP. *Spectrochimica Acta Part A.*, **2002**, *58*, 3033-3038.

[44] Shkrabak, O.A.; Kalchenko, O.I.; Rodik, R.V.; Veklich, T.O.; Kalchenko, V.I.; Kosterin, S.O. Calixarene-dependent hydrolysis of ATP. I. Kinetics and complexation of the calixarene C-107 with adenosinetriphosphate. *UBJ[Ukrainian]*, **2008**, *80*(2), 90-100.

[45] Shkrabak, O.A.; Veklich, T.O.; Rodik, R.V.; Boyko, V.I.; Kosterin, S.O. Calixarene-dependent hydrolysis of ATP. II. The catalytic properties of reaction stimulated by calixarene C-107. *UBJ[Ukrainian]*, **2008**, *80*(3), 55-64.

[46] Sayin, S.; Yilmaz, E.; Yilmaz, M. Improvement of catalytic properties of *Candida Rugosa* lipase by sol–gel encapsulation in the presence of magnetic calix[4]arene nanoparticles *Org. Biomol. Chem.*, **2011**, *9*, 4021–4024.

[47] Hamuro, Y.; Calama, M.C.; Park, H.S.; Hamilton, A.D. A calixarene with four peptide loops: an antibody mimic for recognition of protein surfaces. *Angew. Chem. Int. Ed. Engl.*, **1997**, *36*, 2680-2683.

[48] Wei, Y.; McLendon, G.L.; Hamilton, A.D.; Case, M.A.; Purring, C.B.; Lin, Q.; Park, H.S.; Lee, C.-S.; Yu, T. Disruption of protein-protein interactions: design of a synthetic receptor that blocks the binding of cytochrome c to cytochrome c peroxidase. *Chem. Commun.*, **2001**, (17), 1580-1581.

[49] Mecca, T.; Consoli, G.M.L.; Geraci, C.; La Spina, R.; Cunsolo, F. Polycationic calix[8]arenes able to recognize and neutralize heparin. *Org. Biomol. Chem.*, **2006**, *4*, 3763–3768.

[50] Schneider, H.-J.; Schneider, U. Inhibition of choline acetate hydrolysis in the presence of a macrocyclic polyphenolate. *J. Org. Chem.*, **1987**, *52*, 1613-1615.

[51] Schneider, U.; Schneider, H.-J. Synthese und eigenschaften von makrocyclen aus resorcinen sowie von entsprechenden derivaten und wirt-gast-komplexen. *Chem. Ber.*, **1994**, *127*, 2455-2469.

[52] Bevza, A.A.; Labintseva, R.D.; Rodik, R.V.; Cherenok, S.O.; Kosterin, S.O.; Kalchenko, V.I. Effect of calix[4]arenes on the activity of actomyosin ATPase and actomyosin subfragment-1 ATPase from the myometrium. *UBJ[Ukrainian]*, **2009**, *81*(6), 49-58.

[53] Vovk, A.I.; Kalchenko, V.I.; Cherenok, S.A.; Kukhar, V.P.; Muzychka, O.V.; Lozynsky, M.O. Calix[4]arene methylenebisphosphonic acids as calf intestine alkaline phosphatase inhibitors. *Org. Biomol. Chem.*, **2004**, *2*, 3162-3166.

[54] Cherenok, S.; Vovk, A.; Muravyova, I.; Shivanyuk, A.; Kukhar, V.; Lipkowski, J.; Kalchenko, V. Calix[4]arene α-aminophosphonic acids: asymmetric synthesis and enantioselective inhibition of an alkaline phosphatase. *Org. Lett.*, **2006**, *8*, 549-552.

[55] Vovk, A.I.; Kononets, L.A.; Tanchuk, V.Yu.; Cherenok, S.O.; Drapailo, A.B.; Kalchenko, V.I.; Kukhar, V.P. Inhibition of Yersinia protein tyrsine phosphatase by calixarene phosphonous and methylenebisphosphonous acids. *Bioorg. Med. Chem. Lett.*, **2010**, *20*, 483-487.

[56] Koren, S.; Fantus, I.G. Inhibition of the protein tyrosine phosphatase PTP1B: potential therapy for obesity, insulin resistance and type-2 diabetes mellitus. *Best Pract. Res. Clin. Endocrinol. Metab.*, **2007**, *21*, 621-640.

[57] Trush, V.V.; Cherenok, S.O.; Tanchuk, V.Yu.; Kukhar, V.P.; Kalchenko, V.I.; Vovk, A.I. Calix[4]arene methylenebisphosphonic and hydroxymethylenebisphosphonic acids as inhibitors of protein tyrosine phosphatase 1B. *Bioorg. Med. Chem. Lett.*, **2013**, *23,* 5619-5623.

[58] Trush, V.V.; Tanchuk, V.Y.; Cherenok, S.O.; Kalchenko, V.I.; Vovk, A.I. Calix[4]arene α-hydroxymethylphosphonic acids as potential inhibitors of protein tyrosine phosphatases. *Zh. Org. Farm. Khim.*, **2014**, *12*(1), 39-42.

[59] Trush, V.; Cherenok, S.; Tanchuk, V.; Kalchenko, V.; Vovk, A. Evaluation of inhibition of protein tyrosine phosphatase 1B by calixarene-based α-ketophosphonic acids. *Chem. Biol. Lett.*, **2015**, *2*(1), 1-5.

[60] Vovk, A.; Trush, V.; Kharchenko, S.; Tanchuk, V.; Kalchenko, V. Phosphonate monoesters on thiacalix[4]arene framework as potential inhibitors of protein tyrosine phosphatase 1B. *Org. Biomol. Chem.*, **2015**, *13,* 8803-8806.

[61] Cherenok, S.O.; Yushchenko, O.A.; Tanchuk, V.Yu.; Mischenko, I.M.; Samus, N.V.; Ruban, O.V.; Matvieiev, Yu.I.; Karpenko, Yu.A.; Kukhar, V.P.; Vovk, A.I.; Kalchenko, V.I. Calix[4]arene-α-hydroxyphosphonic acids. Synthesis, stereochemistry and inhibition of glutathione S-transferase. *ARCIVOC*, **2012**, *iv*, 278-288.

[62] Vovk, A.I.; Mischenko, I.M.; Cherenok, S.O.; Tanchuk, V.Yu.; Kalchenko, V.I.; Kukhar, V.P. Phosphorylated calix[4]arenes as inhibitors of glutathione S-transferase. *Phosphorus, Sulfur Silicon Relat. Elem.*, **2011**, *186*, 961-963.

[63] Park, H.S.; Lin, Q.; Hamilton, A.D. Protein surface recognition by synthetic receptors: a route to novel submicromolar inhibitors for α-chymotripsin. *J. Am. Chem. Soc.*, **1999**, *121*, 8-13.

[64] Park, H.S.; Lin, Q.; Hamilton, A.D. Modulation of protein-protein interactions by synthetic receptors: Design of molecules that disrupt serine protease-proteinaceous inhibitor interaction. *Proc. Natl. Acad. Sci. USA*, **2002**, *99*, 5105-5109.

[65] Mecca, T.; Consoli, G.M.L.; Geraci, C.; Cunsolo, F. Designed calix[8]arene-based ligands for selective tryptase surface recognition. *Bioorg. Med. Chem.*, **2004**, *12*, 5057-5062.

[66] Murillo, O.; Watanabe, S.; Nakano, A.; Gokel, G.W. Synthetic models for transmembrane channels: structural variations that altercation flux. *J. Am. Chem. Soc.*, **1995**, *117*, 7665-7669.

[67] Jin, T.; Kinjo, M.; Koyama, Y.; Kobayashi, Y.; Hirata, H. Selective Na^+ transport through phospholipid bilayer membrane by a synthetic calix[4]arene carrier. *Langmuir*, **1996**, *12*, 2684-2689.

[68] Jin, T.; Kinjo, M.; Kobayashi, Y.; Hirata, H. Ion transport activity of calix[n]arene (n = 4, 5, 6, 7, 8) esters toward alkali-metal cations in a phospholipid bilayer membrane. *J. Chem. Soc. Faraday Trans.*, **1998**, *94*, 3135-3140.

[69] de Mendoza, J.; Cuevas, F.; Prados, P.; Meadows, E.S.; Gokel, G.W. Synthetic cation-transporting calix[4]arene derivative active in phospholipid bilayers. *Angew. Chem. Int. Ed. Engl.*, **1998**, *37*, 1534-1537.

[70] Tanaka, Y.; Kobuke, Y.; Sokabe, M. A non-peptidic ion channel with K^+ selectivity. *Angew. Chem. Int. Ed. Engl.*, **1995**, *34*, 693-694.

[71] Kobuke, Y.; Nagatani, T. A supramolecular ion channel based on amphiphilic cholic acid derivatives. *Chem. Lett.*, **2000**, *29,* 298-299.

[72] Yoshino, N.; Satake, A.; Kobuke, Y. An artificial ion channel formed by a macrocyclic resorcin[4]arene with amphiphilic cholic acid ether groups. *Angew. Chem. Int. Ed. Engl.*, **2001**, *40*, 457-459.

[73] Klyachina, M.A.; Boyko, V.I.; Yakovenko, A.V.; Babich, L.G.; Shlykov, S.G.; Kosterin, S.O.; Khilya, V.P.; Kalchenko, V.I. Calix[4]arene N-chalconeamides: synthesis and influence on Mg^{2+},ATP-dependent Ca^{2+} accumulation in the smooth muscle subcellular structures. *J. Incl. Phenom.*, **2008**, *60*, 131-137.

[74] Rodik, R.V.; Boyko, V.I.; Danylyuk, O.B.; Suwinska, K.; Tsymbal, I.F.; Slinchenko, N.V.; Babich, L.G.; Shlykov, S.O.; Kosterin, S.O.; Lipkowski, J.; Kalchenko, V.I. Calix[4]arenesulfonylamidines. Synthesis, structure and influence on Mg^{2+}, ATP-dependent calcium pumps. *Tetrahedron Lett.*, **2005**, *46*, 7459-7462.

[75] Labyntseva, R.D.; Slinshenко, N.M.; Veкlich, T .O.; Rodik, R.V.; Cherenoк, S.O.; Boyko, V.I.; Kalchenko, V.I.; Koѕтerin, S.O. Comparative investigation of calixarenes influence on Mg^{2+}-dependent ATP-hydrolase enzymatic systems from smooth muscle cells of the uterus. *UBJ [Ukrainian]*, **2007**, *79*(3), 44-54.

[76] Veklich, T.A.; Shkrabak, A.A.; Slinchenko, N.N.; Mazur, I.I.; Rodik, R.V.; Boyko, V.I.; Kalchenko, V.I.; Kosterin, S.A. Calix[4]arene C-90 selectively inhibits Ca^{2+},Mg^{2+}-ATPase of myometrium cell plasma membrane. *BIOCHEMISTRY(Moscow)*, **2014**, *79*(5), 417-424.

[77] Veklich, T.O.; Kosterin, S.O.; Rodik, R.V.; Cherenok, S.O.; Boyko, V.I.; Kalchenko, V.I. Effect of calixarene-phosphonic acid on Na^{+}, K^{+}-ATPase activity in plasma membranes of the smooth-muscle cells. *UBJ[Ukrainian]*, **2006**, *78*(1), 70-86.

[78] Veklich, T.O.; Shkrabak, O.A.; Kosterin, S.O.; Rodik, R.V.; Cherenok, S.O.; Boyko, V.I.; Kalchenko, V.I. The calixarenes C-97 and C-107 stimulate influence of ouabain on the Na^{+},K^{+}-ATPase activity in plasmatic membrane of smooth muscle cells. *UBJ[Ukrainian]*, **2006**, *78*(6), 53-63.

[79] Veklich, T.O.; Shkrabak, A.A.; Cherenok, S.O.; Kalchenko, V.I.; Kosterin, S.O. Comparative investigation of the effect of calix[4]arene C-99 and its analogs on Na^{+},K^{+}-ATPase activity of uterus myocite plasma membrane. *UBJ*, **2012**, *84*(6), 49-57.

[80] Busschaert, N.; Gale, P.A. Small-molecule lipid-bilayer anion transporters for biological applications. *Angew. Chem., Int. Ed.*, **2013**, *52*, 1374-1382.

[81] Gale, P.A.; Pérez-Tomás, R.; Quesada, R. Anion transporters and biological systems. *Acc. Chem. Res.*, **2013**, *46*, 2801-2813.

[82] Shatursky, O.Y., Kasatkina, L.A., Rodik, R.V., Cherenok, S.O., Shkrabak, A.A., Veklich, T.O., Borisova, T A., Kosterin, S.O., Kalchenko, V.I. Anion carrier formation by calix[4]arene-bis-hydroxymethylphosphonic acid in bilayer membranes. *Org. Biomol. Chem.*, **2014**, *12*, 9811-9821.

[83] Singh, A.K.; Venglarik, C.J.; Bridges, R.J. Development of chloride channel modulators. *Kidney Int.*, **1995**, *48*, 985-993.

[84] Droogmans, G.; Prenen, J.; Eggermont, J.; Voets, T.; Nilius, V. Voltage-dependent block of endotelial volume-regulated anion channels by calix[4]arenes. *Am. J. Physiol.*, **1998**, *275*, c646-c652.

[85] Edwards, J.C.; Tulk, B.; Schlesinger, P.H. Functional expression of p64, an intracellular chloride channel protein. *J. Membrane Biol.*, **1998**, *163*, 119-127.

[86] Atwood, J.L.; Bridges, R.J.; Juneja, R.K.; Singh, A.K. Calixarene chloride-channel blockers. U. S. Patent, February 6, 5,489,612, 1996.

[87] Cho, Y.R.; Maguire, A.J.; Try, A.C.; Westwell, M.S.; Groves, P.; Williams, D.H. Cooperativity in ligand binding expressed at a model cell membrane by the vancomycin group antibiotics. *Chemistry & Biology*, **1996**, *3*, 207-215.

[88] Casnati, A.; Fabbi, M.; Pelizzi, N.; Pochini, A.; Francgesco, S.; Ungaro, R.; di Modugno, E.; Tarzia, G. Synthesis, antimicrobial activity and binding properties of calix[4]arene based Vancomycin mimics. *Bioorg. Med. Chem. Lett.*, **1996**, *6*, 2699-2704.

[89] Frish, L.; Sansone, F.; Casnati, A.; Ungaro, R.; Cohen, Y. Complexation of peptidocalix[4]arene, a Vancomycin mimic, with Alanine-containing Guests by NMR diffusion measurements. *J. Org. Chem.*, **2000**, *65*, 5026-5030.

[90] Helttunen, K.; Moridi, N.; Shahgaldian, P.; Nissinen, M. Resorcinarene bis-crown silver complexes and their application as antibacterial Langmuir–Blodgett films. *Org. Biomol. Chem.*, **2012**, *10*, 2019-2025.

[91] Dibama, H.M.; Clarot, I.; Fontanay, S.; Salem, A.B.; Mourer, M.; Finance, C.; Duval, R.E.; Regnouf-de-Vains, J-B. Towards calixarene-based prodrugs: Drug release and antibacterial behaviour of a water-soluble nalidixic acid/calix[4]arene ester adduct. *Bioorg. Med. Chem. Lett.*, **2009**, *19*, 2679-2682.

[92] Ben Salem, A.; Regnouf-de-Vains, J.-B. Synthesis and characterisation of a new podand based on a calixarene and a β-lactam. *Tetrahedron Lett.* **2001**, *42*, 7033-7036.

[93] Ben Salem, A.; Regnouf-de-Vains, J.-B. Towards a new family of calixarene-based podands incorporating quinolone arms. An example using nalidixic acid. *Tetrahedron Lett.*, **2003**, *44*, 6769-6771.

[94] Korchowiec, B., Ben Salem, A., Corvis, Y., Regnouf-de-Vains, J.-B., Korchowiec, J., Rogalska, E. Calixarenes in a membrane environment: a monolayer study on the miscibility of three p-tert-butylcalix[4]arene β-lactam derivatives with 1,2-dimyristoyl-sn-glycero-3-phosphoethanolamine. *J. Phys. Chem. B,* **2007**, *111*, 13231-13242.

[95] Ben Salem, A.; Sautrey, G.; Fontanay, S.; Duval, R.E.; Regnouf-de-Vains, J.-B. Molecular drug-organiser: synthesis, characterization and biological evaluation of penicillin V and/or nalidixic acid calixarene-based podands. *Bioorg. Med. Chem.*, **2011**, *19*, 7534-7540.

[96] Mourer, M.; Fontanay, S.; Duval, R.E.; Regnouf-de-Vains, J.-B. Synthesis, characterization, and biological evaluation as antibacterial agents of water-soluble calix[4]arenes and phenol derivatives incorporating carboxylate groups. *Helv. Chim. Acta.* **2012**, *95*, 1373-1386.

[97] Grare, M.; Mourer, M.; Fontanay, S.; Regnouf-de-Vains, J-B.; Finance C.; Duval, R.E. In vitro activity of para-guanidinoethylcalix[4]arene agains susceptible and antibiotic-resistant Gram-negative and Gram-positive bacteria. *J. Antimicrob. Chemother.*, **2007**, *60*, 575-581.

[98] Sautrey, G.; Orlof, M.; Korchowiec, B.; Regnouf-de-Vains, J.-B.; Rogalska, E. Membrane activity of tetra-p-guanidinoethylcalix[4]arene as a possible reason for its antibacterial properties *J. Phys. Chem. B*, **2011**, *115*, 15002-15012.

[99] Formosa, C.; Grare, M.; Jauvert, E.; Coutable, A.; Regnouf-de-Vains, J.-B.; Mourer, M.; Duval, R.; Dague, E. Nanoscale analysis of the effects of antibiotics and CX1 on a Pseudomonas aeruginosa multidrug-resistant strain. *Sci. Rep.*, **2012**, *2*, 575.

[100] Grare, M.; Dibama, H.M.; Lafosse, S.; Ribon, A.; Mourer, M.; Finance, C.; Regnouf-de-Vains, J.-B.; Duval, R.E. Cationic compounds with activity against multidrug-resistant bacteria: interest of a new compound compared with two older antiseptics, hexamidine and chlorhexidine. *Clin. Microbiol. Infect.*, **2010**, *16*, 432-438.

[101] Mourer, M.; Dibama, H.M.; Fontanay, S.; Grare, M.; Duval, R.E.; Finance, C.; Regnouf-de-Vains, J.-B. p-Guanidinoethyl calixarene and parent phenol derivatives exhibiting antibacterial activities. Synthesis and biological evaluation. *Bioorg. Med. Chem.*, **2009**, *17*, 5496-5509.

[102] Soomro, A.M.; Oad, R.K.; Memon, S.; Qureshi, I. Bioactivity assessment of water soluble calix[4]arene derivative. *Pak. J. Anal. Environ. Chem.* **2012**, *13*(1), 36-39.

[103] Consoli, G.M.L.; Granata, G.; Cafiso, V.; Stefani, S.; Geraci, C. Multivalent calixarene-based C-fucosyl derivative: a new *Pseudomonas aeruginosa* biofilm inhibitor. *Tetrahedron Lett.*, **2011**, *52*, 5831-5834.

[104] Melezhyk, I.O.; Rodik, R.V.; Iavorska, N.V.; Klymchenko, A.S.; Mely, Y.; Shepelevych, V.V.; Skivka, L.M.; Kalchenko, V.I. Antibacterial properties of tetraalkylammonium and imidazolium tetraalkoxycalix[4]arene derivatives. *Anti-Infective Agents.* **2015**, *13*, 87-94.

[105] Ukhatskaya, E.; Kurkov, S. V.; Hjálmarsdóttir, M.A.; Karginov, V.A.; Matthews, S.E.; Rodik, R.V.; Kalchenko, V.I.; Loftsson, T. Cationic quaternized aminocalix[4]arenes: cytotoxicity, haemolytic and antibacterial activities. *Int. J. Pharm.*, **2013**, *458*, 25-30.

[106] Mecca, T.; Cunsolo, F. Calix[4]arene-based ligands as endotoxin receptors. *Tetrahedron,* **2007**, *63*, 10764-10767.

[107] Jain, M.K.; Jahagirdar, D.V. Effect of antituberculous calixarenes on phospholipase A2 susceptibility and on fusion of phospholipid bilayers. *Biochem. J.(Great Britain)*, **1985**, *227*, 789-794.

[108] Hart, P.D.; Armstrong, J.A.; Brodaty, E. Calixarenes with Host-mediated potency in experimental tuberculosis: further evidence that macrophage lipids are involved in their mechanism of action. *Infect. and Immun.*, **1996**, *64*, 1491-1493.

[109] Colston, M.J.; Hailes, H.C.; Stavropoulos, E.; Herve, A.C.; Herve, G.; Goodworth, K.J.; Hill, A.M.; Jenner, P.; Hart, P.D.; Tascon, R. E. Antimycobacterial calixarenes enhance innate defense

mechanisms in murine macrophages and induce control of *Mycobacterium tuberculosis* infection in mice. *Infect. and Immun.*, **2004**, *72*, 6318-6323.

[110] Mourer, M., Dibama, H.M.; Constant, P.; Daffé, M.; Regnouf-de-Vains, J.-B. Anti-mycobacterial activities of some cationic and anionic calix[4]arene derivatives. *Bioorg. Med. Chem.*, **2012**, *20*, 2035-2041.

[111] Lamartine, R.; Tsukada, M.; Wilson, D.; Shirata, A. Antimicrobial activity of calixarenes *C.R. Chimie,* **2002**, *5*, 163-169.

[112] Saberov, V.S.; Marichev, K.O.; Korotkikh, M.I.; Shvaika, O.P.; Rodik, R.V.; Drapailo, A.B.; Pekhtereva, T.M.; Komarovska-Porokhnyavets, O.Z.; Lubenets, V.I.; Novikov, V.P. Synthesis and the antimicrobial activity of precarbene and metalcarbene compounds of the imidazole series. *Zh. Org. Farm. Khim.[Ukrainian],* **2014**, *12*(2), 36-43.

[113] Paquet, V.; Zumbuehl, A.; Carreira, E. M. Biologically active Amphotericin B-calix[4]arene conjugates. *Bioconjugate Chem.*, **2006**, *17*, 1460-1463.

[114] Hwang, K.M.; Qi, Y.M.; Liu, S.-Y. Aryl macrocyclic compositions for treating viral infections. U. S. Patent 5,312,837, May 17, 1994.

[115] Takeshi, Y.; Kazuhiro, F.; Makiko O. Antimicrobial. Japan Patent 10,203,906,A2, August 4, 1998.

[116] Coveney D.; Costello B. Anti-viral compounds. U. S. Patent 7,338,977, March 4, 2008.

[117] Harris, S. J. Anti-viral compounds WO 0,244,121, June 6, 2002.

[118] Tsou, L.K.; Dutschman, G.E; Gullen, E.A.; Telpoukhovskai, M.; Cheng, Y.C.; Hamilton, A.D. Discovery of a synthetic dual inhibitor of HIV and HCV infection based on a tetrabutoxy-calix[4]arene scaffold. *Bioorg. Med. Chem. Lett.*, **2010**, *20*, 2137-2139.

[119] Motornaya, A.E.; Alimbarova, L.M.; Shokova, E.A.; Kovalev, V.V. Synthesis and antiherpetic activity of N-(3-amino-1-adamantyl)calix[4]arenes. *Pharm. Chem. J.,* **2006**, *40*, 68-72.

[120] Mourer, M.; Psychogios, N.; Laumond, G.; Aubertin, A.-M.; Regnouf-de-Vains, J.-B. Synthesis and anti-HIV evaluation of water-soluble calixarene-based bithiazolyl podands. *Bioorg. Med. Chem.*, **2010**, *18*, 36-45.

[121] Hwang, K.M.; Qi, Y.M.; Liu, S.-Y.; Thomas, B.; Lee, C.; Bruno, S.; Choy W.; Chen, J. Antitrombothic treatment with calix[n]arene compounds. U. S. Patent 5,409,959, April 25, 1995.

[122] Lugovskoy, E.V.; Gritsenko, P.G.; Koshel, T.A.; Cherenok, S.O.; Kalchenko, V.I.; Komisarenko, S.V. Calix[4]arene methylenebisphosphonic acids as inhibitors of fibrin polymerization. *FEBS Journal*, **2011**, *278*, 1244-1251.

[123] Kalchenko, O.I.; Lipkowski, J.; Kalchenko, V.I.; Vysotsky, M.A.; Markovsky, L.N. Effect of octakis(diethoxyphosphoryloxy)-tert-butyl-calix[8]arene in mobile phase on the reversed-phase retention behavior of aromatic compounds: host-guest complex formation and stability constants determination. *J. Chrom. Sci.*, **1998**, *36*, 269-273.

[124] Fujimoto, K.; Miyata, T.; Aoyama, Y. Saccharide-directed cell recognition and molecular delivery using macrocyclic saccharide clusters: masking of hydrophobicity to enhance the saccharide specificity. *J. Am. Chem. Soc.*, **2000**, *122*, 3558-3559.

[125] Menger, F.M.; Bian, J.; Sizova, E.; Martinson, D.E.; Seredyuk, V.A. Bolaforms with fourteen galactose units: a proposed site-directed cohesion of cancer cells. *Org. Lett.*, **2004**, *6,* 261-264.

[126] Zhou, H.; Wang, D.; Baldini, L.; Ennis, E.; Jain, R.; Carie, A.; Sebti, S.M.; Hamilton, A.D. Structure–activity studies on a library of potent calix[4]arene-based PDGF antagonists that inhibit PDGF-stimulated PDGFR tyrosine phosphorylation. *Org. Biomol. Chem.*, **2006**, *4*, 2376-2386.

[127] Dings, R.P.M.; Chen, X.; Hellebrekers, D.M.E.I.; van Eijk, L.I.; Zhang, Y.; Hoye, T.R.; Griffioen, A.W.; Mayo, K.H. Design of nonpeptidic topomimetics of antiangiogenic proteins with antitumor activities. *J. Natl. Cancer Inst.*, **2006**, *98*, 932-936.

[128] Passaniti, A.; Hurt, G. W. Metastasis-associated murine melanoma cell surface galactosyltransferase; Characterization of enzyme activity and identification of the major surface substrates. *Cancer Res.*, **1990**, *50*, 7261-7271.

[129] Penno, M.B.; Passaniti, A.; Fridman, R.; Hart, G.W.; Jordan, C.; Kumar, S.; Scott, A.F. *In vitro* galactosylation of a 110-kDa glycoprotein by endogenous cell surface galactosyltransferase correlates with the invasiveness of adrenal carcinoma cells. *Proc. Natl. Acad. Sci. USA*, **1989**, *86*, 6057-6061.

[130] Viola, S.; Consoli, G.M.L.; Merlo, S.; Drago, F.; Sortino, M.A.; Geraci, C. Inhibition of rat glioma cell migration and proliferation by a calix[8]arene scaffold exposing multiple GlcNAc and ureido functionalities. *J. Neurochem.*, **2008**, *107*, 1047-1055.

[131] Dominska, M.; Dykxhoorn, D.M. Breaking down the barriers: siRNA delivery and endosome escape. *J. Cell Sci.*, **2010**, *123*, 1183-1189.

[132] Nishikawa, M.; Huang, L. Nonviral vectors in the new millennium: Delivery barriers in gene transfer. *Hum. Gene Ther.*, **2001**, *12*, 861-870.

[133] Ortiz Mellet, C.; Benito, J.M.; Fernandez, J.M.G. Preorganized, macromolecular, gene-delivery systems. *Chem. Eur. J.*, **2010**, *16*, 6728-6742.

[134] Srinivasachari, S.; Fichter, K. M.; Reineke, T.M. Polycationic β -cyclodextrin "click clusters": monodisperse and versatile scaffolds for nucleic acid delivery. *J. Am. Chem. Soc.*, **2008**, *130*, 4618-4627.

[135] Díaz-Moscoso, A.; Gourriérec, L.L.; Gómez-García, M.; Benito, J.M.; Balbuena, P.; Ortega-Caballero, F.; Guilloteau, N.; Giorgio, C.D.; Vierling, P.; Defaye, J.; Mellet, C.O.; Fernández, J.M.G. Polycationic amphiphilic cyclodextrins for gene delivery: synthesis and effect of structural modifications on plasmid DNA complex stability, cytotoxicity, and gene expression. *Chem. Eur. J.*, **2009**, *15*, 12871-12888.

[136] Nierengarten, I.; Nothisen, M.; Sigwalt, D.; Biellmann, T.; Holler, M.; Remy, J.S.; Nierengarten, J.F. Polycationic pillar[5]arene derivatives: interaction with DNA and biological applications. *Chem. Eur. J.*, **2013**, *19*, 17552–17558.

[137] Fujimoto, T.; Shimizu, C.; Hayashida, O.; Aoyama, Y. Solution-to-surface molecular-delivery system using a macrocyclic sugar cluster. Sugar-directed adsorption of guests in water on polar solid surfaces. *J. Am. Chem. Soc.*, **1997**, *119*, 6676-6677.

[138] Hayashida, O.; Mizuki, K.; Akagi, K.; Matsuo, A.; Kanamori, T.; Nakai, T.; Sando, S.; Aoyama, Y. Macrocyclic glycoclusters. Self-aggregation and phosphate-induced agglutination behaviors of calix[4]resorcarene-based quadruple-chain amphiphiles with a huge oligosaccharide pool. *J. Am. Chem. Soc.*, **2003**, *125*, 594-601.

[139] Aoyama, Y.; Kanamori, T.; Nakai, T.; Sasaki, T.; Horiuchi, S.; Sando, S.; Niidome, T. Artificial viruses and their application to gene delivery. Size-controlled gene coating with glycocluster nanoparticles. *J. Am. Chem. Soc.*, **2003**, *125*, 3455-3457.

[140] Nakai, T.; Kanamori, T.; Sando, S.; Aoyama, Y. Remarkably size-regulated cell invasion by artificial viruses. Saccharide-dependent self-aggregation of glycoviruses and its consequences in glycoviral gene delivery. *J. Am. Chem. Soc.*, **2003**, *125*, 8465-8475.

[141] Aoyama, Y. Macrocyclic Glycoclusters: From Amphiphiles through Nanoparticles to Glycoviruses. *Chem. Eur. J.*, **2004**, *10*, 588-593.

[142] Nault, L.; Cumbo, A.; Pretôt, R.F.; Sciotti, M.A.; Shahgaldian, P. Cell transfection using layer-by-layer (LbL) coated calixarene-based solid lipid nanoparticles (SLNs). *Chem. Commun.*, **2010**, *46*, 5581-5583.

[143] Shahgaldian, P.; Sciotti, M.A.; Pieles, U. Amino-substituted amphiphilic calixarenes: Self-assembly and interactions with DNA. *Langmuir*, **2008**, *24*, 8522-8526.

[144] Lalor, R.; DiGesso, J.L.; Mueller, A.; Matthews, S.E. Efficient gene transfection with functionalized multicalixarenes. *Chem. Commun.*, **2007**, (46), 4907-4909.

[145] Bagnacani, V.; Sansone, F.; Donofrio, G.; Baldini, L.; Casnati, A.; Ungaro, R. Macrocyclic nonviral vectors: High cell transfection efficiency and low toxicity in a lower rim guanidinium calix[4]arene. *Org. Lett.*, **2008**, *10*, 3953-3956.

[146] ThermoFischer Scientific, Lipofectamine™ LTX Reagent. http://tools.thermofisher.com/downloads/F-069843-LpofctLTX_FHR.pdf [Accessed November 2015].

[147] Bagnacani, V.; Franceschi, V.; Fantuzzi, L.; Casnati, A.; Donofrio, G.; Sansone, F.; Ungaro, R. Lower rim guanidinocalix[4]arenes: Macrocyclic nonviral vectors for cell transfection. *Bioconjug. Chem.*, **2012**, *23*, 993-1002.

[148] Sansone, F.; Dudič, M.; Donofrio, G.; Rivetti, C.; Baldini, L.; Casnati, A.; Cellai, S.; Ungaro, R. DNA condensation and cell transfection properties of guanidinium calixarenes: Dependence on macrocycle lipophilicity, size, and conformation. *J. Am. Chem. Soc.*, **2006**, *128*, 14528-14536.

[149] Dudic, M.; Colombo, A.; Sansone, F.; Casnati, A.; Donofrio, G.; Ungaro, R. A general synthesis of water soluble upper rim calix[n]arene guanidinium derivatives which bind to plasmid DNA. *Tetrahedron,* **2004**, *60*, 11613-11618.

[150] Bagnacani, V.; Franceschi, V.; Bassi, M.; Lomazzi, M.; Donofrio, G.; Sansone, F.; Casnati, A.; Ungaro, R. Arginine clustering on calix[4]arene macrocycles for improved cell penetration and DNA delivery. *Nat. Commun.*, **2013**, 4.

[151] Arimori, S.; Nagasaki, T.; Shinkai, S. Self-assembly of tetracationic amphiphiles bearing a calix[4]arene core. Correlation between the core structure and the aggregation properties. *J. Chem. Soc., Perkin Trans. 2*, **1995**, 679-683.

[152] Shi, Y.; Schneider, H. J. Interactions between amino calixarenes and nucleotides or nucleic acids. *J. Chem. Soc., Perkin Trans. 2*, **1999**, 1797-1803.

[153] Mchedlov-Petrossyan, N.O.; Vilkova, L.N.; Vodolazkaya, N.A.; Yakubovskaya, A.G.; Rodik, R.V.; Boyko, V.I.; Kalchenko, V.I. The nature of aqueous solutions of a cationic calix[4]arene: A comparative study of dye-calixarene and dye-surfactant interactions. *Sensors*, **2006**, *6*, 962-977.

[154] Rodik, R.V.; Klymchenko, A.S.; Jain, N.; Miroshnichenko, S.I.; Richert, L.; Kalchenko, V.I.; Mély, Y. Virus-sized DNA nanoparticles for gene delivery based on micelles of cationic calixarenes. *Chem. Eur. J.*, **2011**, *17*, 5526-5538.

[155] Rodik, R.V.; Anthony, A.-S.; Kalchenko, V.I.; Mély, Y.; Klymchenko, A.S. Cationic amphiphilic calixarenes to compact DNA into small nanoparticles for gene delivery. *New J. Chem.*, **2015**, *39*, 1654-1664.

[156] Aime, S.; Barge, A.; Botta, M.; Casnati, A.; Fragai, M.; Luchinat, C.; Ungaro, R. Calix[4]arene Gd^{III} complex endowed with high stability, relaxivity, and binding affinity to serum albumin. *Angew. Chem. Int. Ed. Engl.*, **2001**, *40*, 4737-4739.

[157] Krishnan, A.M.; Lohrmann, R. Calixarene conjugate diagnostic agents for computerized tomography and method for using same. U. S. Patent 5,846,515, December 8, 1998.

[158] Gutsche, C.D. Calixarenes. In *Host Guest Complex Chemistry Macrocycles: Synthesis, Structures, Applications,* 2nd ed; Vögtle, F.; Weber, E. Eds.; Springer: Tokyo, **1985**, pp 120-145.

[159] Perret, F.; Mazzorana, M.; Shahgaldian, P.; Coleman, A.W. *PB-26. First step in the study of the cellular toxicity of the calixarene.* In XI International Symposium on Supramolecular Chemistry, ISSC XI: Book of abstracts, **2000**, July 30-August 4, Fukuoka, Japan; pp. 387.

[160] Shahgaldian, P.; Da Silva, E.; Coleman, A.W. A first approach to the study of calixarene solid lipid nanoparticle (SLN) toxicity. *J. Incl. Phenom.*, **2003**, *46*, 175-177.

[161] Memisoglu, E.; Bochot, A.; Oezalp, M.; Sen, M.; Duchene, D.; Hincal, A.A. Direct formation of nanospheres from amphiphilic β-cyclodextrin inclusion complexes. *Pharm. Res.*, **2003**, *20*, 117-125.

[162] Da Silva, E.; Shahgaldian, P.; Coleman, A.W. Haemolytic properties of some water-soluble para-sulphonato-calix-[n]-arenes. *Int. J. Pharm.*, **2004**, *273*, 57-62.

[163] Coleman, A.W.; Jebors, S.; Cecillon, S.; Perret, P.; Garin, D.; Marti-Battle. D.; Moulin, M. Toxicity and biodistribution of para-sulfonato-calix[4]arene in mice. *New. J. Chem.*, **2008**, *32*, 780-782.

Index

A

Atta-ur-Rahman, M. Iqbal Choudhary and Allen B. Reitz (Eds.)

B

C

D

E

F

G

H

I

N

O

P

Q

R

S

T

U

V

W

X